FIELD-EFFECT ELECTRONICS

FIELD-EFFECT ELECTRONICS

W. GOSLING
W. G. TOWNSEND
J. WATSON

Wiley Interscience
Division of John Wiley & Sons, Inc.
New York, Toronto

Sole Distributors for the United States and Canada,
Wiley-Interscience, Division of John Wiley & Sons Inc.,
New York, Toronto

Library of Congress Catalog Card Number 78-146057

ISBN 0 408 70123 4

Printed in England by Page Bros. (Norwich) Ltd., Norwich

CONTENTS

PREFACE

In 1964 one of the authors published a book on field-effect transistors. This was, in fact, the first book to be devoted wholly to this subject, but, like most pioneer efforts, was quickly eclipsed by rapid technical developments. Thus it did not seem appropriate to issue a second edition, but instead the decision was taken to write a completely new book of substantially greater size which would give an up-to-date and comprehensive coverage of the subject. The present volume is the outcome of that decision.

The book begins by reviewing the history of field effect devices—a far longer one than most suppose, dating back to 1925! A few years later, Julius Lillienfeld had proposed all the three types of transistor (junction FET, IGFET and bipolar transistor) which currently dominate electronics technology. The succeeding two chapters give an introduction to the theory of operation of junction and insulated gate FETs including that relevant to the more recent developments such as the use of III–V compound semiconductors, and both Schottky and silicon gate technologies.

The problem of noise in field effect devices seemed to us to merit a chapter to itself, in which we review and summarise the present state of knowledge. Chapter 5, on device characteristics, naturally leads on to another concerned with the elementary principles of amplifier design, followed by three concerned with special types of amplifier (those in which FETs are combined with bipolar transistors, AF amplifiers and direct coupled amplifiers). As an analogue switch or chopper the FET has the advantage of zero offset voltage, and has consequently been much used. Chapter 10 briefly reviews this important field, which is quite closely related to that of the following chapter: the use of the FET as a variable resistor.

Because the low levels of intermodulation distortion attainable make the FET particularly suitable as an RF amplifier or frequency converter, no book of this type could omit coverage of this topic (Chapter 12). Even more important are the applications of field effect devices in integrated circuits, considered in Chapters 13 and 14, where the particular problems and opportunities which arise with circuit integration, whether using the film or the monolithic approach, are fully treated for both digital and linear applications. Finally a chapter on the Photo-FET, most sensitive of all solid state light-sensitive devices, contributes to filling a notable gap in present published work.

A book of this kind necessarily depends on the co-operation of the research workers, principally in the semiconductor industry itself, who are so vigorously extending our understanding of field-effect devices and their applications. We acknowledge with gratitude the advice and assistance of our very many friends in the industry without which we could not have written, but hope that they will forgive us if we do not mention them by name here. A complete list would be very extensive, and if it were not complete injustice would be done. Detailed references will be found in the text, as will specific acknowledgements of our indebtedness to the industrial organisations which have so kindly allowed us to reproduce data and other written material. We would, however, like to express our thanks to Carol Evans, who prepared the manuscript and cheered us up in the periods of depression which are inseparable from authorship.

W.G.
W.G.T.
J.W.

1

INTRODUCTION

1.1. Origins of the field-effect transistor

The whole technology of electronics depends upon the availability of satisfactory amplifiers. An investigation of the equipment used in wireless telegraphy before the triode valve was available, for example, reveals great ingenuity on the part of individual experimenters, but at the same time a most severe limitation in the range of what was possible without devices capable of giving significant power gain.

Any amplifier which produces an electrical current as its output must depend upon the modification of the properties of an electrically conducting channel by some physical effect. Although a wide range of different effects could be employed, the simplest and possibly the most direct approach would be a device in which the conductance of a channel is modified directly by means of an electrostatic field. In this case the input to the amplifier can supply the controlling field, and because the output is also in electrical form the connection of a chain of such amplifiers in cascade is a simple matter. It seems reasonable to describe all such electrostatic-field-controlled amplifying devices as field-effect devices. Although strictly speaking this definition will include the thermionic valve, it seems nevertheless a not unreasonable one. A sub-group of this class is comprised of those devices which embody the field-effect principle in solid-state form. These are of course the field-effect transistors, with the technology and application of which this book is concerned.

Once the principle of modifying the flow of current in a conductor by means of an electrostatic field is accepted, certain design principles for device realisation become immediately apparent. One is that the field must penetrate throughout the conducting channel, and indeed it is clear that the dimensions of the channel must be such that there is significant field penetration throughout all, or almost all, of its width. If this were not so then there would be a substantial uncontrolled conducting path in shunt with the controlled path, and the efficiency of the amplifier would be poor. Unfortunately, electrostatic fields do not penetrate significantly into metallic conductors. A very simple calculation based on Poisson's equation and utilising the known density of charge carriers in metals shows that the penetration is very small indeed, with the field falling off rapidly as a function of distance into the interior of the metal. Thus if the field-effect device were to use a metal as the conducting channel it would have to be very narrow indeed. Much more promising are the semiconducting materials which have resistivities in the range 10 mΩ–1 MΩ-cm at room temperature, very much greater than those of the metals, which are typically 1μ Ω-cm at room

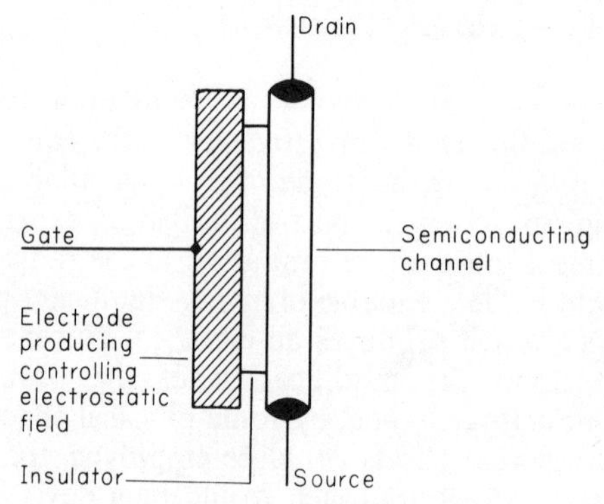

Fig. 1.1. A primitive field-effect transistor

temperature. Because the resistivities are so much higher it is reasonable to deduce that charge densities are correspondingly lower and this was indeed confirmed by early investigators of these materials. Thus, the penetration of an electrostatic field into a semiconductor is at least a thousand times greater than in the case of a metal.

The primitive notion of a field-effect transistor will therefore be, as in Fig. 1.1, a conducting channel which consists of a thin film of semiconductor to which ohmic contacts are made at each end, together with some means of producing an electrostatic field which can penetrate the channel and hence modulate its conductivity. Shockley proposed that the connections to the channel should be referred to as the source and the drain, indicating respectively the terminal which originates the flow of charge carriers in the channel and that which receives it. The separated electrode which produces the electrostatic controlling field was to be termed the gate. In the diagram the gate is shown as physically insulated from the channel, and indeed this constitutes one very important class of field-effect transistor, namely the insulated gate FET, sometimes referred to as the IGFET. However, because the conductive channel is a semiconductor there also exists another possible way in which the gate can be electrically isolated, hence permitting the application of an electrostatic potential to it and the production of a suitable controlling field. In order to discuss this alternative structure, which was in fact the earliest described, it is necessary to digress somewhat to consider semiconductor rectifiers.

Scientists had noticed the unusual electrical properties of semiconductors for the last hundred years, and Braun published his observations of the rectifying properties of lead sulphide and iron sulphide crystals in 1873. The first applications of semiconductors were in the various types of metal-semiconductor rectifiers. It was quickly realised that a most intriguing phenomenon was present in these devices. With the electric potential difference applied to its terminals in one direction the device conducted, and hence free or mobile carriers must be present in the whole of the device. However, with the polarity reversed no current flowed, and hence some region of the device must be non-conducting: that is, it must not contain mobile charge carriers. The reverse field must have swept away the carriers, which are of course electrons or holes. The region in which conduction could be controlled by a potential was located at the interface between the metal and the semiconductor, or in some cases between two semiconductors, and became known as the blocking layer. It was obvious to at least one early experimenter that this so-called blocking layer could serve as a means of isolating the gate from the channel in a type of field-effect transistor.

Thus the field-effect transistor consists of a conducting channel formed from a suitable semiconductor material, isolated from which (either by an insulating layer or the blocking layer of a rectifying junction) is a gate electrode which allows an electrostatic field to be applied to the channel. The operation of such a device can be grasped even in crudely unscientific terms. For example, if the semiconductor channel is *n*-type, a negative voltage applied to the gate can be seen as producing an electrostatic field which repels the electrons from the region of the gate and therefore tends to deplete that part of the semiconducting channel which is nearest to the gate. As the negative gate potential increases in magnitude, this depletion region will tend to spread deeper into the channel and hence to restrict the effective cross-sectional area of the channel in which current can flow. The effect can be seen as a kind of electrostatically controlled variable resistance. In practice, the operation of the field-effect device is considerably more complicated than this simple description suggests and a correct physical theory of the devices is developed in Chapters 2 and 3.

Even so, the very simple treatment given here is something which can be readily grasped, and stimulated early workers, who attempted to build devices of this kind. It often comes as a considerable surprise that the development of the transistor began in the mid 1920s but this was indeed the case, and there is a substantial patent history for the transistor even before the second world war. Both insulated-gate and junction-gate devices were proposed, and indeed the first worker in this field, Julius Edgar Lillienfeld, described not only the two basic types of field-effect transistor but also a bipolar transistor, in patents filed before 1930.

1.2. The patent history of early FET devices

A series of patents was taken out, and papers written, from about 1925 onwards, although the commercial field-effect transistor came only much later, after the point-contact and other types of bipolar transistor had been developed. This need not have been so, for the concept of the FET as a device with a conducting path of variable cross-section is not invalid and could easily be visualised in terms of a blocking layer being used to alter the conducting region of a device.

However, in 1925, Lillienfeld applied for a Canadian patent (and also in 1926 a similar patent in the U.S.A.) for protection of a device

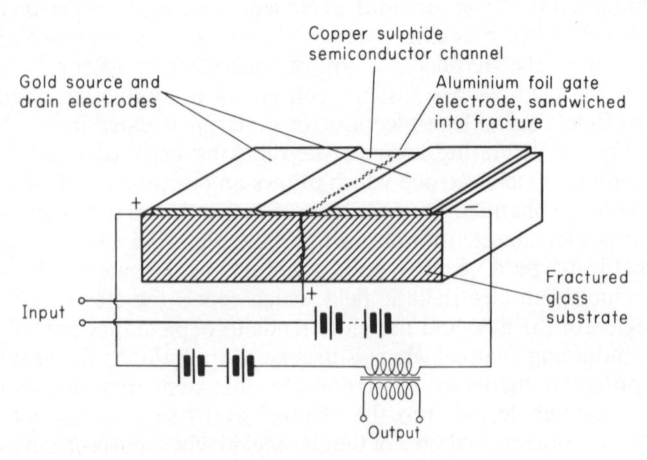

Fig. 1.2. J. E. Lillienfeld's original field-effect transistor

which he claimed could act as a solid-state amplifier (Fig. 1.2). His invention was a kind of FET and seems to be the first description of such a device. He stated that his device enabled the flow of current between two terminals to be controlled by the potential on a third electrode.

Lillienfeld indicates clearly that the aluminium electrode, the gate, should form a rectifier with the copper sulphide layer and draws a diagram of the potential in the layer. His theory of the process is approximately correct. The rectifier interferes with electron flow by creating a non-conducting region in the copper sulphide. No characteristic curves are given in the patent and it is not clear whether Lillienfeld had built such a device, although it would work, without doubt, if made with present day purities and tolerances. It was a great pity that the idea was not followed up by a well-equipped laboratory, but this was the day of the thermionic tube and interests were elsewhere.

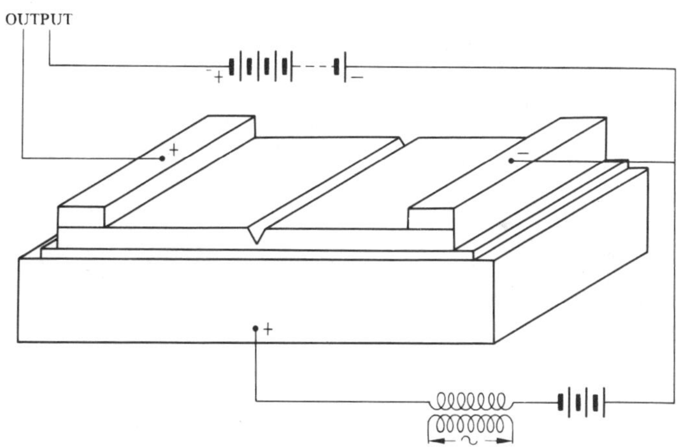

Fig. 1.3. Lillienfeld's insulated gate FET of 1928
(For details see Fig. 1.4)

In 1928, Lillienfeld again applied for a U.S. patent but this time the device was a metal-base bipolar transistor. It consisted of several layers of metals and semiconductors with rectifying properties between them. Lillienfeld was obviously now well aware of an empirical theory of operation and must be considered an outstanding inventor.

In 1928 Lillienfeld also patented the improved FET shown in Fig. 1.3. This seems to have been based on experimental discoveries and led to the formulation of the insulated gate electrode. He seems to have been interested in applying large voltages to his gate, which was insulated by a very thin layer of aluminium or magnesium oxide. His device in cross-section is shown in Fig. 1.4.

The groove in the oxide layer is introduced in order to produce a very thin cross-section both of oxide and of semiconductor adjacent to the insulating layer, so that the intense electric field could have most effect. There is no theoretical explanation but the device appears to be based on his detailed experimental work on surfaces. Again it is not clear whether he built a complete FET.

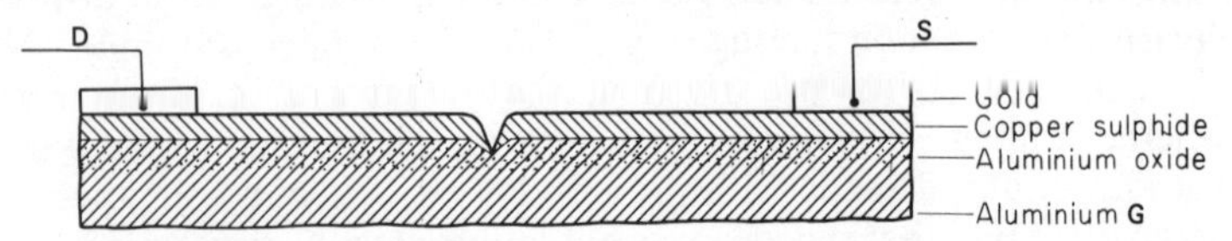

Fig. 1.4. Cross-section of the FET shown in Fig. 1.3. Note the groove in the semiconductor channel; it is in this locally thinner channel that the transistor action occurs

In 1930, H. C. Weber of the Industrial Development Corporation, Salem, applied for a U.S. patent for an electronic device for controlling electron streams in a solid. The device shown in Fig. 1.5 consisted of four layers, say Cu–CuS–CuO–CuS–Cu. Embedded in the copper oxide layer is an insulated spiral of fine wire. A potential applied to this wire relative to the copper end-electrodes was stated to control the current through the device frome one copper electrode to the other. No characteristics are given but its possible use as an amplifier and oscillator is indicated.

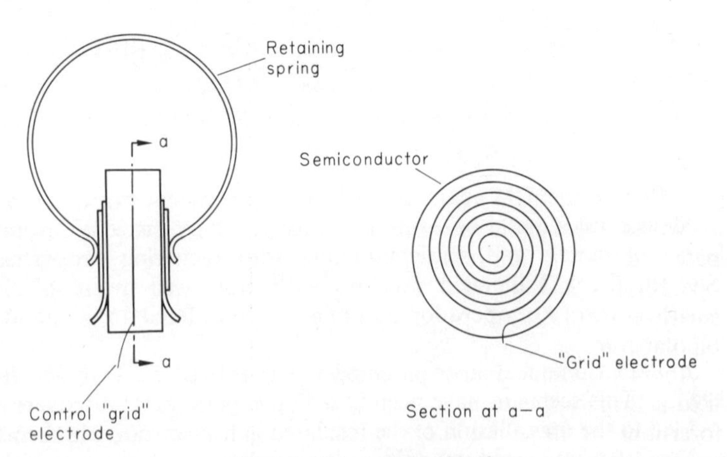

Fig. 1.5. Weber's FET of 1930

In 1936, G. Holst and W. van Geel of Philips, Eindhoven, applied for a U.S. patent for yet another device controlling currents in a rectifier. This patent is interesting in that it indicates that the device postulated by Weber had been built and found unsatisfactory. They therefore proposed a new arrangement. Their device as drawn in Fig. 1.6 consists of a metal separated from a semiconductor (such as selenium) by an insulating layer in which is embedded a control electrode made from a semiconductor. The emissive property of this semiconductor is described as being vital to the performance of the device.

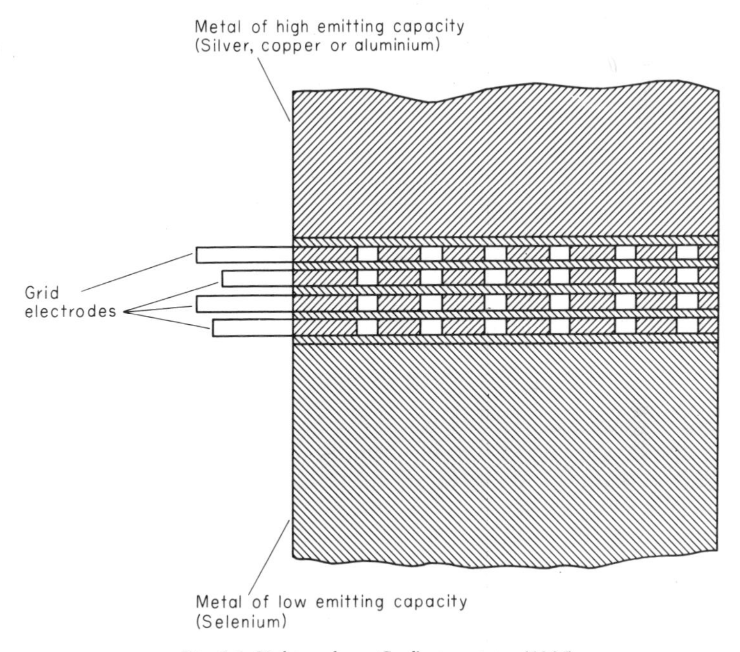

Fig. 1.6. Holst and van Geel's transistor (1936)

No real theory is given nor are any characteristics shown, but the patent gives the impression that the device was constructed. Its operation would depend very largely on the thickness of the insulating layer. If it was perforated then it would have been a kind of bipolar transistor.

About the same time, the famous German engineer O. Heil had filed a British patent (1934) which could be said to describe a true forerunner of the field-effect transistor. Again it is not clear whether a successful device was built.

In 1938, R. Hilsch and R. W. Pohl published a paper in Germany describing a three-electrode device using a crystal of potassium bromide. Such a device exhibits ionic as well as electronic conduction and although a central electrode could control the flow of current between dissimilar electrodes placed on the crystal, the bandwidth was of the order of 1 Hz only. This was an interesting paper, in that it had a valid theoretical exposition and many experimental results.

A. Glaser, W. Koch and H. Voigt applied for a U.S. patent in 1939 for yet another device very similar to that proposed by van Geel except that now the insulator was modified to produce more conduction under certain circumstances. This modification would now be described as doping a crystalline material. It is an interesting

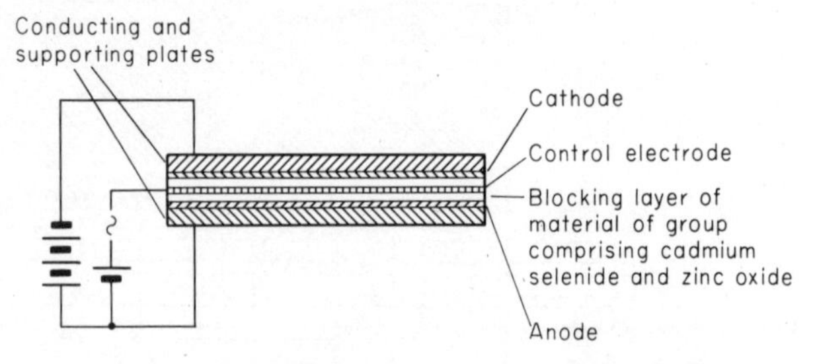

Fig. 1.7. Glaser, Koch and Voigt's transistor of 1939

example of the application of scientific discovery. The insulators or semiconductors proposed for the central layer between the layers of a conventional rectifier were cadmium selenide or zinc oxide. The device could be called a transistor for it had three layers, two of which were definitely semiconductors as shown in Fig. 1.7. Presumably the war prevented further exploitation.

In 1943 and 1945 W. van Geel applied for U.S. patents for blocking layer cells with one or more grids embedded in the blocking layers. This work obviously followed on from his previous invention. The first patent in fact was for a cell with an improved geometry and was somewhat as shown in Fig. 1.8. There are three layers and van Geel found many combinations of materials which had the appropriate qualities, but he recommended that both anode and cathode should be semiconductors though one could be a metal. The blocking layer in which the control grid was embedded was an organic insulator such as polystyrene or Canada Balsam, whilst the control electrode was a semiconductor with resistivity usually intermediate between that of anode and cathode. The insulating layer had to be very thin as conduction had to take place through it.

Its performance could not have been very good and could not match up to the well-developed thermionic valve.

The 1945 patent contained further modifications to the patent of 1943. Though the device depended on emission of charged particles or carriers into an insulator by a semiconductor, the theory of operation was obscure and it is probable that the proposals were based on experimental work.

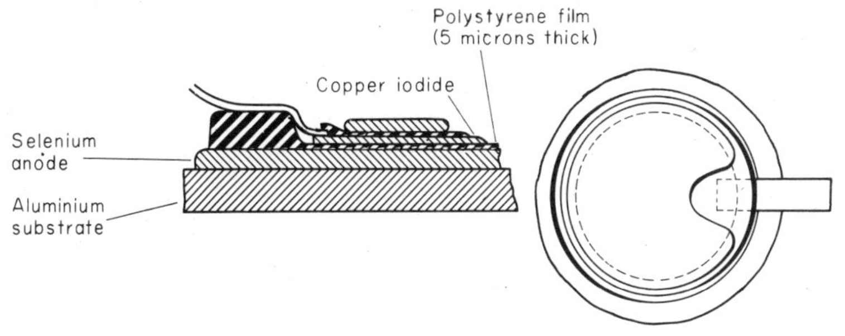

Fig. 1.8. A transistor proposed by van Geel (1943)

1.3. The modern FET

It is to the genius of William Shockley that is owed the re-invention of the FET from which the modern development of this branch of electronics stems. The paper published[1] by him in 1952 may be said to be all-important in this context and in it he refers to the observation of a field effect by Pearson and Shockley[2] in 1948, but also explains that due to surface effects it was much smaller than expected. The theory of surface states which had been proposed by Bardeen suggested that the additional charge induced by the controlling field would be 'trapped' in an immobile condition, and could not, therefore, contribute to conduction. Shockley's paper of 1952 proposed the use of a reverse-biased rectifying junction, similar to that of Lillienfeld's patent of 1925, as the means for overcoming these undesired surface effects, and went on to describe practical transistors fabricated in germanium which operated on this principle.

Subsequent development was continuous, with commercial devices fabricated by the alloy process appearing two years later.

The earliest type of structure was described about 1952. The *p–n* junctions were typically produced by alloying, which resulted in a rather limited useable channel cross-section due to the hemispherical profile of the junction. The distance between source and drain had to be kept fairly large to facilitate handling and give adequate physical strength. The result was an FET having a relatively high pinch-off voltage and a high gate voltage for drain current cut-off. Some improvement in the latter was possible if planar gate-channel junctions were used, but the essentially high voltage characteristic remained.

A later, rather more favourable geometry, is shown in Fig. 1.9. Here the channel is in the form of an annular ring between the central source 'pip' and the outer drain ring. These two ohmic contacts were usually formed by fusing metal contacts of suitably doped gold to the silicon chip, and could thus be used to mask the regions beneath from an etchant. This acid etching agent was used to remove the exposed silicon and thus decrease the effective width of the channel, with consequent reduction in the device pinch-off voltage.

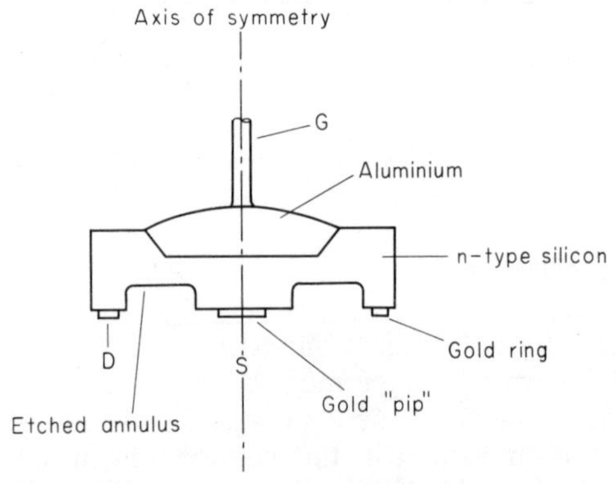

Fig. 1.9. An alloyed junction FET

Quite a different facrication technique was used in the Tecnetron, an early field-effect transistor specifically designed for high frequency use. Here a small bar of semiconductor had ohmic contacts made to its ends and (after electrochemical etching) a metal layer was electrolytically deposited on the bar, rather in the way that metal-film emitter and collector electrodes were deposited on a surface barrier bipolar transistor. If the metal film is correctly deposited, rectifying action is obtained between it and the underlying semiconductor, and it may thus be used as a gate. The technique lent itself to the production of devices with very small physical dimensions, by the standards of the late 1950s, and hence small interelectrode capacitances were made possible. As will be realised from consideration of the equivalent circuit, these determine the upper limit on the frequencies at which the device has useful circuit properties.

However, drain current is small, because of the small cross-sectional area, and the device is far less robust and has a much shorter thermal time-constant than a channel of the same cross-sectional area embedded in a much larger block of silicon.

With the coming of silicon planar processes for the fabrication of bipolar transistors, after Atalla's famous work published in 1959, new and much improved double-diffused and epitaxial-diffused FET structures have been developed, and the solution to the problem of a satisfactory insulated gate device found. It will be convenient to postpone descriptions of these modern structures to Chapters 2 and 3, where the physical theory of modern field-effect transistors is developed.

1.4. Conclusions

When Shockley's epoch-making paper was published in 1952 there was already, unknown to him, a substantial body of published work on solid state field-effect amplifiers. However, progress had been baulked by lack of scientific knowledge and inadequate solid-state technology. Thus the re-invention by Shockley was effective because, above all, it was timely, coming as it did at a period when these problems were at last being successfully tackled. However, this is not to detract from the significance of the three very remarkable patents by Lillienfeld, all filed within a period of three years in the mid 1920s, when he invented the junction FET, the bipolar transistor, and the insulated gate FET. At that early date Lillienfeld conceived of the triad of devices which currently dominate electronics technology, and he must surely be regarded as the great precursor of modern electronics.

REFERENCES

1. Shockley, W. 'A unipolar field-effect transistor', *Proc. I.R.E.*, **40**, 1365–76 (November 1952).
2. Shockley, W. and Pearson, G. L. 'Modulation of conductance of thin films of semiconductors by surface charges', *Physical Review*, **74**, 232–233 (1948).

2

THE JUNCTION FET

2.1. Theory of operation

The junction field-effect transistor is essentially a semiconductor element, the resistance of which is controlled by applying an electric field perpendicular to the direction of current flow. It is known as a unipolar device because current is carried only by one type of carrier, the majority carrier of the bulk semiconductor, which drifts along the element under the action of the electric field applied between the ends of this element. This is in contrast to the bipolar transistor, where the current in the active region is due to minority carriers diffusing in a relatively field free region.

The controlling electric field in the junction FET results from reverse biasing a *p–n* junction or junctions. Fig. 2.1 shows a hypothetical block device of the type considered by Dacey and Ross[1]. This consists of a slab of *n*-type crystal with ohmic contacts made at either end, together with two *p*-type regions known as *gates* which define the *channel*. Consider the case when both gates are short-circuited to the left-hand contact. If a positive potential V_{DS} is applied to the right-hand contact a current of electrons will flow along the crystal from the left-hand contact, or *source* to the right-hand contact, or *drain*. Because the crystal has a finite resistivity there will be a steady rise in potential from the source to the drain and since the source and the gates are at the same potential this increasing positive potential will appear across the *p–n* junctions bounding the gates causing them to be reverse biased. When a *p–n* junction is reverse biased, a depletion or space-charge region develops on either side of the junction, the thickness of which is a function of the relative dopings of the two types of material and the applied voltage. The depletion region has a low conductivity since there are no free carriers within it other than those generated thermally.

Returning to Fig. 2.1 the thickness of the depletion region will be greatest near the drain end of the gates since here the reverse bias is larger than at the source end. This means that the current flowing will be confined to the wedge-shaped region of the semiconductor, that is, the channel. If the drain voltage is increased, the depletion regions extend further into the bulk, so that the channel is narrowed, and the source-to-drain resistance increases until a point is reached when the depletion regions adjacent to either gate meet. This is the position represented in Fig. 2.1, the depletion regions being indicated.

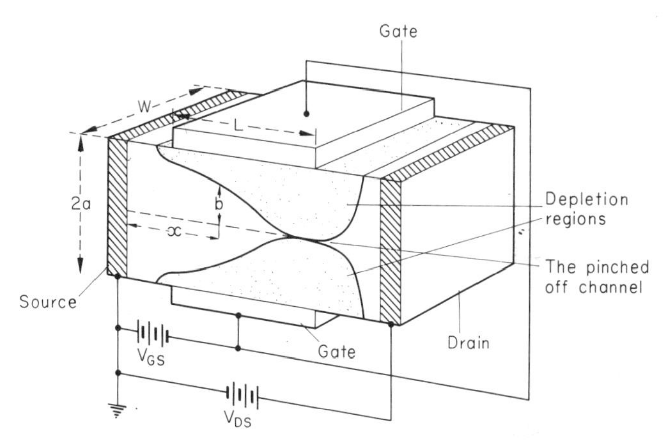

Fig. 2.1. Schematic diagram of FET formed in an n-*type bar of semiconductor with* p-*type gate regions.* (After Dacey and Ross[1])

The channel is here said to be pinched off. Further increase in drain voltage causes only a slight increase in drain current since the additional voltage drop now occurs across the space charge region near the drain electrode. The pinched-off channel also tends to spread back towards the source giving rise to a longer pinched-off region. The upper curve of Fig. 2.2 shows the relevant drain characteristics.

Consider now the effect of applying a bias voltage to the gate. If the gate is biased negatively with respect to the source then the depletion regions will be thicker for a given value of V_{DS} than when there is no bias. This means that pinch-off and saturation of I_D will occur at lower values of V_{DS} and I_D. Fig. 2.2 shows typical drain characteristics using the value of V_{GS} as a parameter. The similarity between these characteristics and those of the pentode valve are striking and the junction FET may therefore be used in a similar way to the pentode to produce amplification.

The theory of operation of an ideal simple block junction FET was worked out by Shockley[2] in 1952 and modified by many subsequent authors. Shockley's model will first be considered, after which it will be shown how it can be modified to give a working model of a practical device operating at high field intensities and high frequencies.

Consider again the block device of Fig. 2.1 where the channel defined by the shaded depletion region has a thickness 2b at a distance x from the source electrode. W is the width of the channel

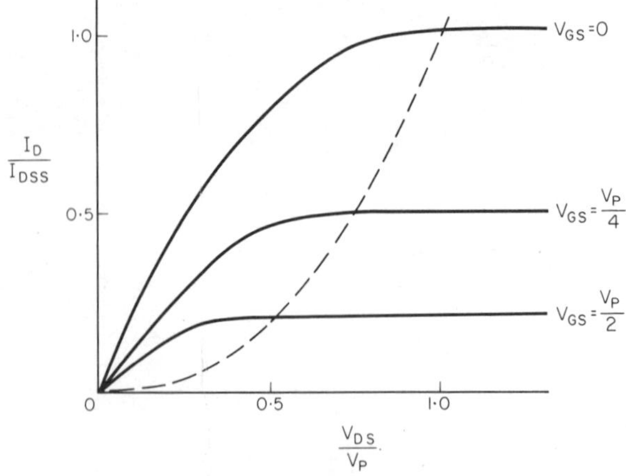

Fig. 2.2. Typical normalised drain characteristics of a junction FET showing the locus of the pinch-off points

and L is the gate length. The n-type bar of semiconductor has a conductivity σ_0 corresponding to a donor charge density ρ_0, which is assumed to be very much less than the acceptor charge density in the gate p-type regions so that the depletion regions will be much narrower in the gate than in the channel. The actual junction Shockley considered to be a step junction: that is, there is an abrupt change in carrier concentration in crossing from the p-type gate to the n-type channel.

Considering the case when the source is grounded and the gate is also at source potential, then if a small positive value of V_{DS} is applied, a current of electrons will flow along the channel from the source to the drain, and for *small* values of V_{DS} the value of the channel resistance R_0 will simply be[1]

$$R_0 = \frac{L}{2aW\sigma_0} \tag{2.1}$$

As the value of V_{DS} is increased, more current flows in the channel and the IR drop along the length of the channel will be such that the portions of the gate-channel p–n junction furthest from the source will be biased most heavily in the reverse direction. This will mean that a wedge-shaped depletion region (or space charge region) will form as indicated in Fig. 2.1. The current between source and drain will, of course, flow only in the channel. At a specific value of V_{DS} the channel will become pinched off near the drain by the two depletion regions associated with gates 1 and 2 coalescing. Hereafter, there will be only a minimal increase in channel current with further increase of V_{DS} but the channel shape will be further modified by the pinched-off length of channel extending back towards the source.

The value of V_{DS} which causes the channel to be pinched-off is known as the pinch-off voltage and will henceforth be denoted by V_P.

The value of V_P may be shown to be[2]

$$V_P = \frac{\rho_0 a^2}{2K} \tag{2.2}$$

Here K is the dielectric constant

$$K = 1{\cdot}42 \times 10^{-12} \text{ farads/cm for germanium}$$

$$K = 1{\cdot}06 \times 10^{-12} \text{ farads/cm for silicon}$$

If now the gate is considered to be negatively biased with respect to the channel, then the magnitude of the IR drop along the channel necessary to produce pinch-off will be smaller, since part of the reverse bias at the gate channel junction is now being provided by V_{GS}. This gives rise to a lower value of saturation current in the channel occurring at a lower value of V_{DS}. Shockley[2] has shown that if the wedge-shaped channel narrows sufficiently slowly (so that $\partial^2 V/\partial X^2 \to 0$ within the channel) then for drain voltages V_{DS} less than the value at which current saturation occurs the current in the channel per unit width (in the Z direction)

$$I_D = \frac{1}{L}[J(V_{DS} - V_{GS}) - J(-V_{GS})] \tag{2.3}$$

when the function J is given by

$$J(x) = 2\sigma x[1 - (2/3)(2xK/\rho a^2)^{\frac{1}{2}}] \tag{2.4}$$

which may be written from (2.2) as

$$J(x) = g_0 x[1 - (2/3)(x/V_P)^{\frac{1}{2}}] \tag{2.5}$$

where

$$g_0 = 2\sigma_0 a = 2\rho_0 \mu_n a \ (\mu_n \text{ is carrier mobility}) \qquad (2.6)$$

is the conductance of a unit square of the n-type channel region 2a thick at zero bias of both gate and drain.

Above the saturation value of V_{DS} of course, the channel current I_D will remain almost constant.

The drain voltage-drain current characteristic obtained from Shockley's gradual channel approximation (equations 2.3 and 2.4) are plotted in a normalised form in Fig. 2.2 for three values of the gate voltage V_{GS}. The locus of the saturation or pinch-off points is shown also in Fig. 2.2 and it may easily be shown that this curve is simply the curve for $V_{GS} = 0$ rotated through 180°.

From these characteristics the transconductance of the simple block device may be obtained since

$$g_{fs} = \left(\frac{\partial I_D}{\partial V_{GS}}\right)_{V_{DS} = \text{constant}} \qquad (2.7)$$

$$= (2\sigma_0 a/LV_P^{\frac{1}{2}}) \left[(V_{DS} - V_{GS})^{\frac{1}{2}} - (-V_G)^{\frac{1}{2}}\right] \qquad (2.8)$$

If operation in the pinch-off region is considered so that

$$V_{DS} - V_{GS} = V_P \qquad (2.9)$$

then g_{fs} becomes g_{fsG} given by

$$g_{fsG} = (2\sigma_0 a/L) \left[1 - (-V_{GS}/V_P)^{\frac{1}{2}}\right] \qquad (2.10)$$

$$= g_{fs0} \left[1 - (-V_{GS}/V_P)^{\frac{1}{2}}\right]. \qquad (2.11)$$

Here

$$g_{fs0} = \frac{2\sigma_0 a}{L} \qquad (2.12)$$

is the maximum transconductance, and the saturation drain current for any gate voltage V_{GS} from (2.3) and (2.4) is

$$I_D = (g_{fs0} V_P/3) \left[1 + (V_{GS}/V_P)\left(3 - 2\sqrt{(-V_{GS}/V_P)}\right)\right] \qquad (2.13)$$

From this equation it is apparent that if

$$V_{GS} = V_P \qquad (2.14)$$

the drain current is completely cut off and that the maximum value of drain current will be obtained when there is zero bias on the gate, i.e. $V_{GS} = 0$.

In this case

$$I_{DSS} = g_{fs0} \frac{V_P}{3} \qquad (2.15)$$

The 3 in this equation originally derived by Shockley[2] arises directly by considering a uniformly doped channel with a step profile of impurities between channel and gate. If an impurity profile similar to that found in most modern diffused junction FETs is used the equation is modified to that of equation (2.28). It is this latter form of the equation that will be used in subsequent chapters where a square law transfer characteristic for the FET is assumed.

2.1.1 THE CUT-OFF FREQUENCY

In the early years junction transistors were very restricted in the range of frequency over which they would operate. Using the knowledge gained from making high frequency bipolar transistors the frequency response of junction FETs has been comparatively good from the beginning and it is now possible to operate them at frequencies up to 1 GHz with power outputs up to a few watts.

The frequency response of a junction FET is limited by two principal factors,

(*a*) the capacitive coupling between gate and channel contacts and
(*b*) the transit time of carriers along the channel.

Considering (*a*) first; in order to change the gate voltage the capacity of its *p–n* junction must be charged through the resistance of the channel. For a wedge-shaped channel pinched off at the drain end and open at the source, with $V_{GS} = 0$ the capacity/unit length across the channel is approximately:

$$C \simeq \frac{4KL}{a} \tag{2.16}$$

and this will be charged by a current flowing along roughly half the length of the channel, which has a resistance of

$$R = \frac{L}{2a\sigma_0} \tag{2.17}$$

The cut-off frequency f_c therefore is

$$f_c = \frac{1}{2\pi RC} = \frac{1}{2\pi}\left(\frac{a^2\sigma_0}{2L^2K}\right) \tag{2.18}$$

Considering the carrier transit time τ, if the mobility is assumed to be constant, Dacey and Ross[1] have shown that

$$\tau = \frac{3}{2}\left(\frac{L^2}{\mu_0 V_p}\right) = \frac{3KL^2}{a^2\sigma_0} \tag{2.19}$$

This means that the transit time differs by only a factor of about $\frac{3}{2}$ from the time constant estimated from RC above in equation (2.18). This indicates that the estimated cut off frequencies are essentially in agreement[1].

2.1.2 DEPARTURES FROM THE IDEAL DEVICE

So far an ideal structure has been considered; this has to be modified to make it applicable to practical devices. Consider the following factors (so far neglected):

(i) the series resistance present at the source and drain contacts within the semiconductor,
(ii) temperature effects,
(iii) changes in carrier mobility at high electric fields,
(iv) junction impurity profiles other than step junctions.

These factors have effects as follows:

(i) The effect of the small bridges of semiconductor between the source contacts and the channel will modify the transconductance of the device to an apparent value g'_{fs} where[1]

$$g'_{fs} = \frac{g_{fs}}{1 + r_s g_{fs}} \tag{2.20}$$

Here r_s is the source resistance which must be as small as possible to maximise g'_{fs}. The effect of a resistance at the drain contact r_D is less serious, the main effects being that power $I_D^2 r_D$ has to be dissipated and that a higher supply voltage is needed to overcome the voltage drop.

The source resistance, of course, will also affect the cut-off frequency by increasing R in equation (2.18).

(ii) *Temperature effects.* Three separate mechanisms can cause an FET to be temperature sensitive: variation of gate leakage current due to the generation of electron hole pairs by thermal processes; variation of carrier mobility in the channel; and variation of the contact potential. Of these, the latter two are the most significant but it has been shown (see Sevin[3]) that FETs have one bias point where the transconductance is independent of temperature and another where drain current for constant drain–source voltage is unaffected by changes in temperature. These current values are different, for p-channel devices having values of $V_P > 1\text{V}$; however, they may be related empirically by the equation[3]

$$I_{Dg} = \frac{1}{4} I_{Dq}$$

where I_{Dg} is bias current for zero temperature coefficient of g_{fs} and I_{Dq} for zero coefficient of the static characteristic. This result makes possible the design of an amplifier (at least, theoretically) in which either AC or DC gain is virtually independent of temperature. This concept is particularly important for DC amplifiers.

(iii) In both germanium and silicon there is a critical field strength E_c above which Ohm's Law fails. This is because the carrier mobility decreases as the half power of the applied field E for $E > E_c$ giving rise to an effective change in conductivity. For germanium

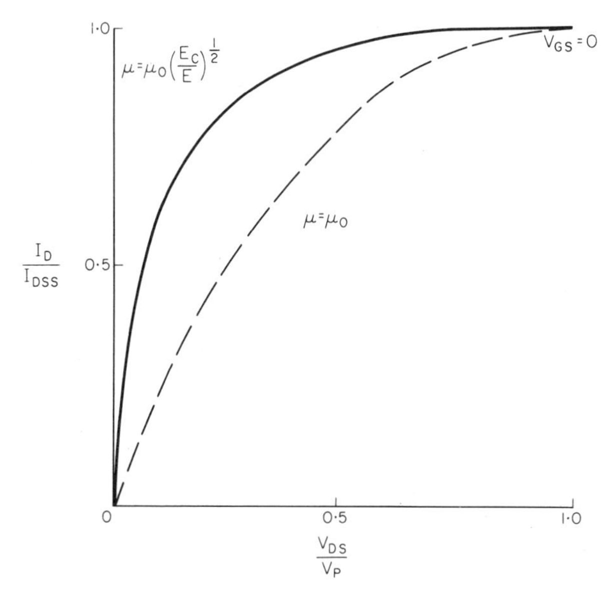

Fig. 2.3. Comparison of drain characteristics for the constant and square root mobility cases. (After Dacey and Ross[1])

$E_c \approx 1200$ V/cm a value that may easily be exceeded in a practical design of FET. It is therefore necessary to consider the effects of the square root mobility variation on the transistor characteristics.

Following a similar approach to that of Shockley[2] for the constant mobility case, Dacey and Ross[1] arrive at an equation for the characteristics of an n-channel device having a pinch-off voltage so high that the field in the channel $E > E_c$ over most of its length.

The pinch-off current for zero gate bias for the square root mobility case is

$$I_{DC} = 2\sigma_0 a (V_P E_{c/6L})^{\frac{1}{2}} \tag{2.21}$$

whereas for the constant mobility case it was

$$I_{DSS} = g_{fs0}\, V_P/3 = 2V_P\sigma_0 a/3L \tag{2.15}$$

The effect of non-constant mobility therefore is to reduce the maximum current in the ratio

$$\frac{I_{DC}}{I_{DSS}} = 1{\cdot}21\,(E_c L/V_P)^{\frac{1}{2}} \tag{2.22}$$

The drain characteristic for the square root mobility case has a steeper initial slope compared with the curve of Fig. 2.3 for constant mobility and a more gradual pinch-off. The device appears therefore to have a lower pinch-off voltage. The transconductance is also reduced; if an FET with a pinched-off drain and zero gate bias is considered, the ratio g_{fsc}/g_{fs0} is the same as the ratio of I_{DC}/I_{DS} giving a value of transconductance for the square root mobility case g_{fsc} of

$$g_{fsc} = 3I_{DC}/V_P \tag{2.23}$$

If the variation of g_{fsc} with gate voltage is studied it is found that there is initially a more rapid fall of g_{fsc} than is found for the constant mobility case. This suggests that if maximum transconductance is required gate bias must be held close to zero. This causes a serious limitation on the use of the FET for values of $E > E_c$.

If the variation of the current beyond pinch-off flowing in the FET is considered as a function of gate bias for constant and square root mobility cases the transfer characteristic shown in Fig. 2.4 results. It is seen that the current falls less rapidly with gate bias in the non-constant mobility case.

(iv) Depending on the method used for forming the junction in the FET structure the profile of impurities within the channel can vary greatly. Fig. 2.5 indicates some possible impurity profiles varying from the step or abrupt junction formed by alloying the gate to a uniformly doped semiconductor, to the exponential distribution. In the limit for the exponential distribution when $m \to \infty$ (see Fig. 2.5) the charge carriers are concentrated in the centre of the channel giving a profile that has been called a spike profile.

Considering the transfer curves for the two extreme distributions gives equations for drain circuit of the form

$$I_D = I_{DSS}\left[1 - 3\frac{V_{GS}}{V_P} + 2\left(\frac{V_{GS}}{V_P}\right)^{\frac{3}{2}}\right] \tag{2.24}$$

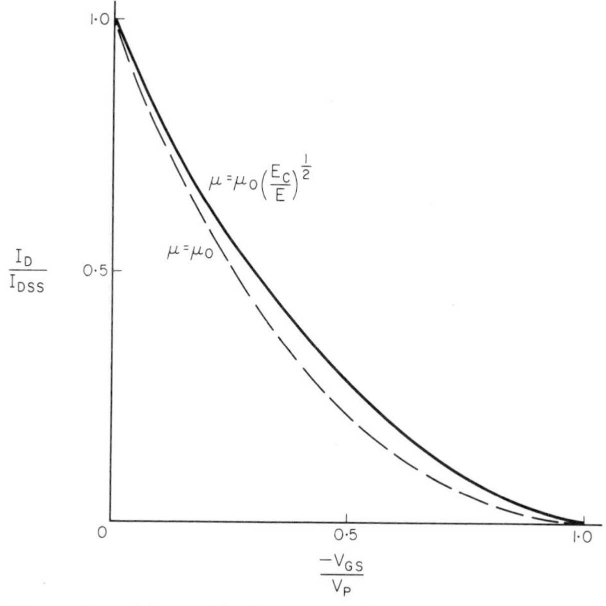

Fig. 2.4. Comparison of the transfer characteristic for constant and square root mobility cases. (Dacey and Ross[1])

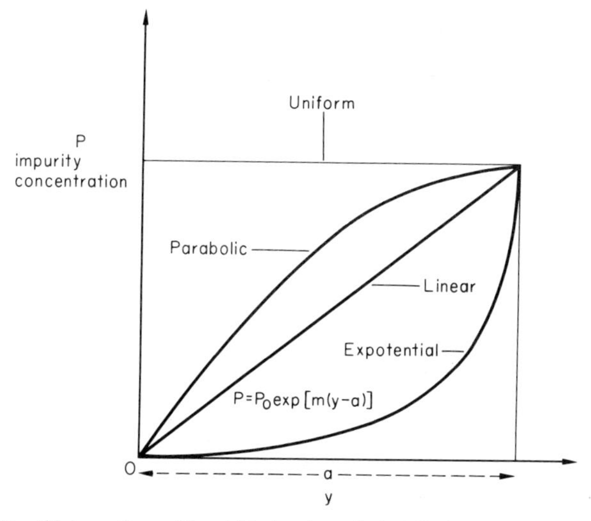

Fig. 2.5. Possible impurity profiles within the channel of a FET having a geometry similar to that of Fig. 2.1. The uniform distribution would apply for the step or abrupt junction and the exponential distribution would apply for diffused junction devices.

for the step junction and

$$I_D = I_{DSS}\left(1 - \frac{V_{GS}}{V_P}\right)^2 \tag{2.25}$$

for the spike profile.

When both curves are plotted as in Fig. 2.6 it is obvious that there is only a narrow range of possible transfer characteristics for FETs regardless of the methods used for forming the *p–n* junction. The relatively simple square law equation (2.25) gives a good approximation for all FETs having narrow channels and hence low pinch-off voltages.

The impurity profile of most FETs whose junctions are formed by diffusion will obviously lie between that of the uniform and spike profiles. The transfer characteristic therefore approaches even closer to that of the square law curve shown in Fig. 2.6. Cobbold

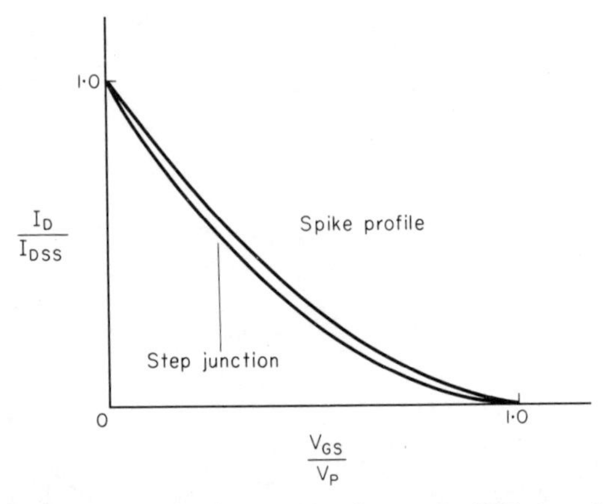

Fig. 2.6. Comparison of transfer characteristics for junction FETs assuming first a step junction profile, equation (2.24), and secondly spike profile, equation (2.25). (Sevin[3])

and Trofimenkoff[4] have considered in detail the effects of linear and graded junctions of various geometries in the performance and design of FETs. They show that for the abrupt junction considered by Shockley, that:

$$\frac{g_{fs}}{g_{fs0}} = 1 - \left(\frac{V_{GS}}{V_P}\right)^{\frac{1}{2}} \tag{2.26}$$

whereas if a device is considered having a linearly graded impurity distribution

$$\frac{g_{fs}}{g_{fs0}} = 1 - \left(\frac{V_{GS}}{V_P}\right)^{\frac{2}{3}} \tag{2.27}$$

The case of a diffused device would be intermediate between these. Because these transconductance characteristics are so similar it can be assumed to a first approximation for any diffusion profiles that the variation of g_{fs} with gate voltage will be similar to that for a step junction.

In order to describe the transfer curve for any particular device values of V_P and I_{DSS} must be specified. If the square law approximation to the curve is assumed to be accurate, the difficulty in measuring V_P because the curve changes so slowly at low values of drain current can be overcome, the only measurements that need to be made are I_{DSS} and g_{fs0} for then

$$V_P = \frac{2I_{DSS}}{g_{fs0}} \tag{2.28}$$

The 2 in this equation occurs by consideration of the channel impurity profile and the fact that the square law equation of the transfer curve is used. This is in contrast to Shockley's original equation for pinch-off current which assumes a uniform channel and leads to equation (2.15). This is straightforward since both I_{DSS} and g_{fs0} are measured at the same bias point.

2.2. Summary of design considerations

Before considering some of the many methods used for making junction FETs it is worthwhile to sum up some of the more important design considerations based on the foregoing theory.

(*a*) The pinch-off voltage depends on only one dimensional factor—the channel thickness—and in fact for most geometries it has been recorded (Warner[5]) that $V_P \propto a^2$. V_P is also proportional to channel doping so that a heavily doped thin channel may be pinched-off with the lowest gate voltage. This also gives a high value of transconductance.

(*b*) Gate voltage is most effectively used when the depletion region is formed largely in the channel, depletion layers spreading into the gate regions serving no useful purpose. For the *n*-channel device we have been considering, a p^+–n–p^+ structure should therefore be aimed at. In order not to conflict with (*a*) above, which required a high concentration of impurities in the channel, a compromise where gate and channel dopings are roughly equal is not unusual.

(*c*) g_{fs} is obviously increased by having a high carrier mobility. For this reason *n*-channel devices are to be preferred to *p*-channel if high g_{fs} is important. Note that mobilities are significantly higher for germanium than silicon and are higher still for one type of carrier or another in certain intermetallic semiconductors. (See Table 2.1.)

Table 2.1. CARRIER MOBILITIES IN ELEMENTAL AND INTERMETALLIC SEMICONDUCTORS

Semiconductor	*Forbidden Energy Gap (0°K, eV)*	*Electron Mobility $cm^2/Volt$-sec. (300°K)*	*Hole Mobility $cm^2/Volt$-sec. (300°K)*
Germanium	0·75	3800	1800
Silicon	1·165	1300	500
Gallium phosphide	2·4	1000	
Gallium arsenide	1·58	8500	400
Indium phosphide	1·3	4600	700
Indium antimonide	0·23	70,000	1000
Cadmium sulphide	2·4	200	

(*d*) The range of field allowable within the channel is limited by the decrease of mobility for values of $E > E_c$. This limits the increase in g_{fs} and cut-off frequency that can be obtained by increasing the field within the channel. Optimum performance when the limitations caused by power dissipation within the device are also considered, lead to operation at fields $E \simeq E_c$.

(*e*) *Cut-off frequency* is very dependent on the mobility of the charge carriers and if high f_c's are to be obtained the high mobility of carriers in germanium can give rise to cut-off frequencies up to 1 GHz. Similarly devices using gallium arsenide would show even higher values of f_c.

(*f*) g_{fs}, pinch-off current and cut-off frequency are all functions of the aspect ratio of the channel a/L. This needs to be maximised and may typically be ~20–100, in practical designs. Further increase of a/L leads to an undue increase in parasitic capacitances since the minimum dimension L is limited by the processing procedures.

2.3. Fabrication technology

In Chapter 1 some of the earlier attempts at making junction FETs have been discussed, here attention will be focused on more recent devices made by processes compatible with those currently in use in the production of monolithic integrated circuits.

Two distinct methods are considered; (a) those using only the diffusion process and (b) those using the epitaxial process as well as diffusion.

2.3.1 DIFFUSED PLANAR FETS

There are two major possibilities for all diffused devices, in the one shown in section in Fig. 2.7 the *p*-type gate regions are simultaneously diffused into an *n*-type slice leaving a narrow *n*-type channel sandwiched between them. Source and drain contacts are made subsequently by an n^+ diffusion into the top surface of the slice. The snag here is obviously the difficulty in controlling the channel thickness, this depends on the original wafer having its

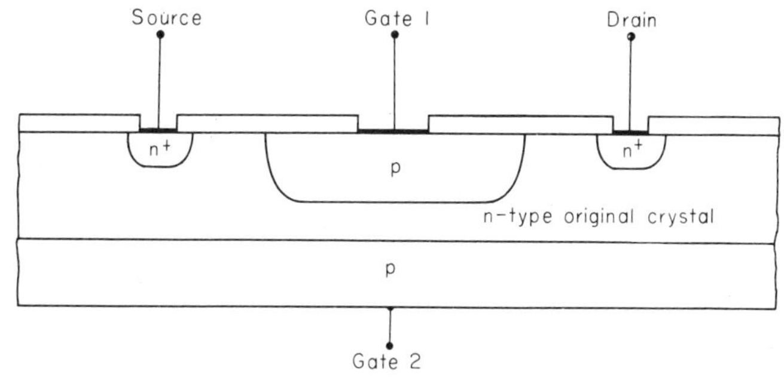

Fig. 2.7. A diffused junction FET where both gates are simultaneously formed by diffusion from either side of an n*-type silicon wafer.* n^+ *source and drain contacts are formed at a second diffusion stage*

thickness accurately specified and the diffusion profile being carefully calculated. The alternative process which has been widely used both for FETs and bipolar transistors is that known as the double diffused process. A *p*-channel device made this way is shown in section in Fig. 2.8.

Consider in more detail how this structure is arrived at using conventional planar technology. A silicon dioxide layer is first formed on an *n*-type slice by passing steam over the slice when it is heated to an elevated temperature in a diffusion type furnace. The oxide is then removed selectively by photolithographic techniques ready for the first diffusion step which will define the lower limit of the channel and the source and drain regions. This diffusion to form a *p*-type region is carried out in two stages using BBr_3 or BCl_3 as a source. In the first stage a boron rich glass is formed on the top surface of the exposed silicon and a heavily doped *p*-type layer formed in the silicon. In the second stage the boron is driven in to a prescribed depth of a few microns, and the oxide layer is allowed to reform on the top surface. Following a second photolithographic processing to cut out openings for the gate diffusion the wafers are

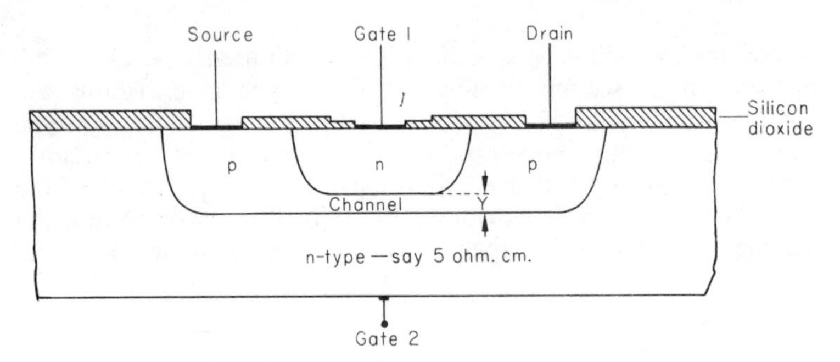

Fig. 2.8(a). A double diffused p*-channel junction FET. The channel is defined by the first* p*-diffusion. Typically the dimension* Y *may be 0·5μ*

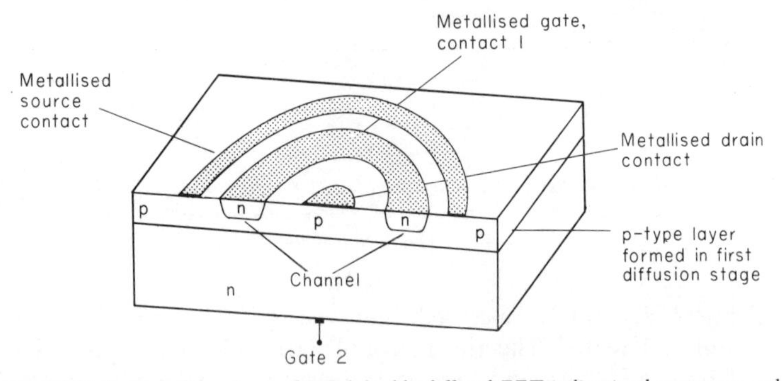

Fig. 2.8(b). Section through a p*-channel double-diffused FET indicating how source and drain contacts are isolated from each other by virtue of the diffused gate region*

removed to a second set of furnaces for the *n*-type diffusion which forms the gate region. Here phosphorous is diffused from one of a number of possible sources such as $POCL_3$, P_2O_5 etc. Obviously the depth of this second diffusion is all important as it defines the channel geometry of the device. The depth of the diffusion can be calculated in advance if the impurity concentrations within the original wafer are known, together with the surface concentrations of boron and phosphorous in the two stages of diffusion. This

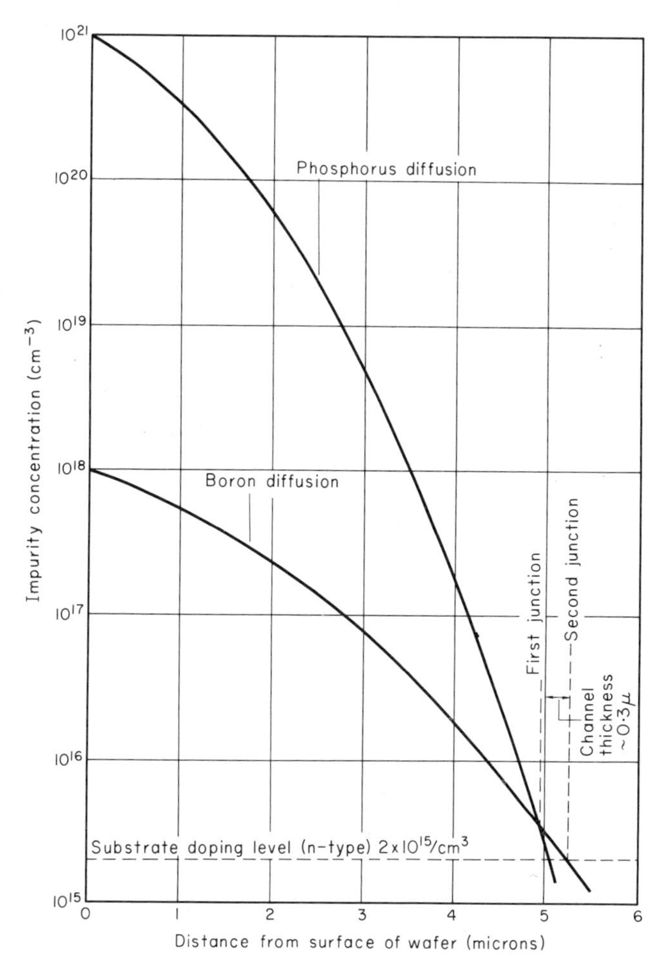

Fig. 2.9. Typical junction profile for a p-*channel junction FET*

enables a profile of the junctions to be drawn showing the channel thickness (Fig. 2.9).

The last stages in the construction involves the making of suitable contact areas to the source, drain and two gates. This is done by evaporating a film of aluminium or gold into the top surface of the wafer and selectively etching away the unwanted conducting film by a further photolithographic stage. An additional diffusion stage makes it possible to have a number of electrically isolated FETs of this kind on the same silicon slice.

2.3.2 EPITAXIAL PLANAR DIFFUSED FETS

One of the difficulties in making double diffused structures is the accurate control necessary in the diffusion stages to create a channel

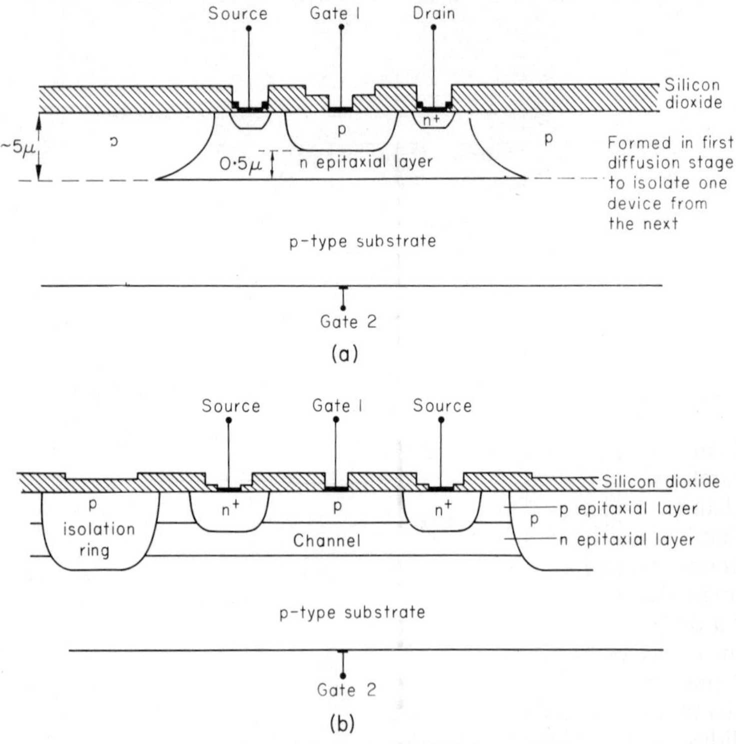

Fig. 2.10. Epitaxial planar diffused junction FETs.
(a) using a single n*-type epitaxial layer on a* p*-type substrate.*
(b) using a p*-type substrate having* n*- and* p*-type epitaxial layers.*
In both cases, isolation between one device and the next uses a deep p*-diffusion stage*

of known geometry and conductivity, bearing in mind that the first junction will not remain static during the second diffusion stage. This can be overcome using the epitaxial process to give a channel in a region of well-defined sheet resistance while using the diffusion process to define the device geometry. Again two types of device may be considered one using a *p*-type substrate having a single *n*-type epitaxial layer and the second having two epitaxial layers, *n* and *p* grown on a *p* type substrate. Schematic drawings of *n* channel devices made by both methods are shown in Fig. 2.10.

For the device of Fig. 2.10(*a*) the first stage in the processing is to carry out a deep *p*-type isolation diffusion to isolate one device

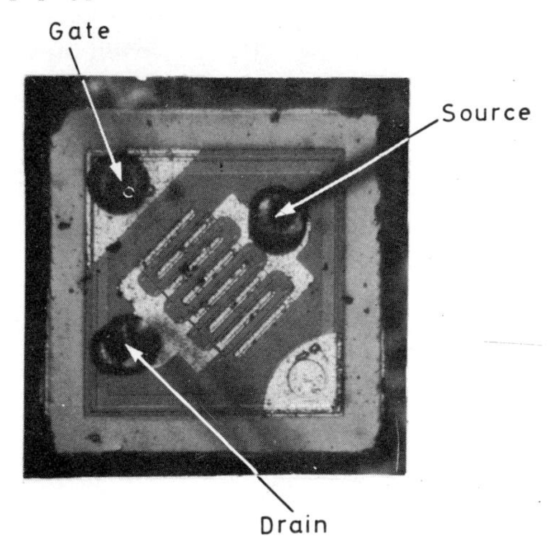

Fig. 2.11(a). Texas Instruments junction FET Type 2N3874 (Magnification × 50)

from another on the wafer, the gate geometry is then defined by a carefully controlled *p*-type diffusion which also determines the channel thickness. Finally an n^+ diffusion is carried out to ensure that low resistance source and drain contacts are possible when the drain, source and gate contact areas are metallised. Typically if the original epitaxial layer is about 4 μ thick the channel would be about 2 μ deep giving a pinch-off voltage ~4 volts in 1 Ω-cm silicon. For the device of Fig. 2.10(*b*) the channel sheet resistance is completely defined by the *n*-type epitaxial layer so that only the geometry of the channel in the plane of the junction needs to be determined by diffusion. Two diffusion stages would be used, the first a deep *p*-type diffusion for isolation and termination of the *n*-type channel and secondly a phosphorus diffusion into the *n* layer for defining the top gate and channel lateral geometry. This diffusion stage also connects the source and drain contacts to the channel ends, its depth is non-critical since the channel sheet resistivity is already defined by the growth of the *n*-type epitaxial layer. Again assuming a channel of 1 Ω-cm *n*-type silicon ~2 μ thick a pinch-off voltage of 4 V would be obtained.

Figs. 2.11(*a* and *b*) illustrate chip photographs of two commercial types of FET. It is worth noting the topology of these devices,

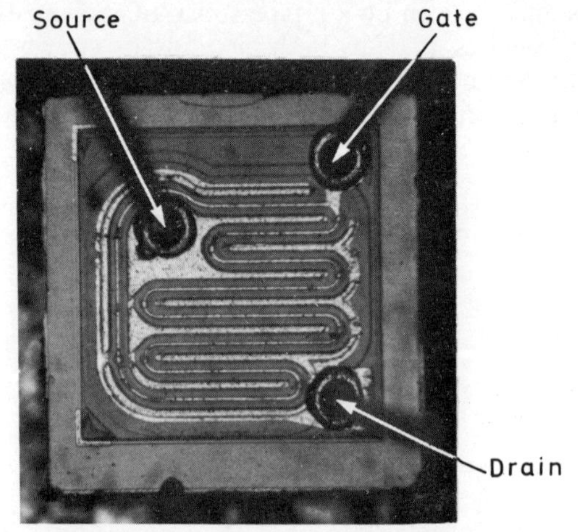

Fig. 2.11(b). Texas Instruments junction FET Type TIXS 80 (Magnification × 50)

particularly the way the upper gate electrode must be of such a geometry that it completely surrounds the drain or source electrode. This makes it impossible for current to flow from source to drain around the edges of the gate electrode.

2.3.3 GERMANIUM FETS

It has been mentioned that germanium offers the advantage of a higher carrier mobility and consequently the possibility of making devices to operate at higher frequencies than is possible with silicon.

Manufacture of germanium junction FETs can be carried out using the epitaxial diffused planar process in an analogous manner to that used for silicon. The most significant change in the processing is that the silicon dioxide used for the photolithographic processing has to be deposited on the germanium by a process such as the pyrolysis of tetraethyl-ortho-silicate. This adds to the cost of making the device by adding additional stages in the manufacture.

Devices using gallium arsenide are discussed in Chapter 3.

REFERENCES

1. Dacey, G. C. and Ross, I. M. 'Unipolar field-effect transistor', *Proc. I.R.E.*, **41**, 970–9 (August 1953); 'The field-effect transistor', *B.S.T.J.*, **34**, 1149–89 (November 1955).
2. Shockley, W. 'A Unipolar field-effect transistor', *Proc. I.R.E.*, **40**, 1365–76 (November 1952).
3. Sevin, L. J. *Field Effect Transistors*, New York, McGraw-Hill (1965).
4. Cobbold, R. S. C. and Trofimenkoff, F. H. 'Theory and application of the field-effect transistor', *Proc. I.E.E.*, **111**, 1981–92 (December 1964).
5. Warner, R. M. *Integrated Circuits—Design Principles and Fabrication*, New York, McGraw-Hill (1965).

3

THE INSULATED-GATE FET (MOST)

3.1. Introduction

The metal oxide semiconductor field-effect transistor (MOSFET or MOST) which is the more usual form of insulated gate FET or IGFET differs from the junction FET already considered by having its gate electrode isolated from the channel by a thin layer of insulating material, usually silicon dioxide. The conductivity of the channel which is formed in the layer of semiconductor immediately next to the insulator is modulated by charges induced within it by the potential difference applied across the insulator. Because of trapping effects taking place at the oxide-semiconductor interface, in practice two types of MOSFET are of most significance.

(*a*) *p*-type units having *p*-type channels, source and drain contacts. These units are usually enhancement mode devices, i.e. they are normally off and only conduct when a negative bias is applied to the gate.

(*b*) *n*-type units having *n*-type channels, source and drain contacts. These units are usually depletion mode devices in that they normally conduct when zero bias is applied to the gate and the drain current is depleted as a negative bias is applied to the gate.

Of course, *p*-channel depletion MOSTs and *n*-channel enhancement are also available but are much less common.

Since the *p*-channel enhancement MOST is of great importance from a circuit design standpoint, it is this type we shall now consider in some detail.

Consider the *p*-channel MOST shown in Fig. 3.1. This is formed on a high resistivity *n*-type substrate which has two *p*-type regions diffused into the surface to form the source and drain. A thin layer of silicon dioxide covers the silicon between these *p*-type regions and on top of this is evaporated a metal gate electrode. Application of a negative bias between the gate electrode and the substrate causes a *p*-type conductive path known as the channel to form between the source and drain enabling a current to flow between them. This current can be controlled by varying the bias applied to the gate electrode. Let us consider this further.

Using the MOST structure of Fig. 3.1 with source and drain at ground potential, and neglecting any effects caused by surface states and the difference in work function of the semiconductor and

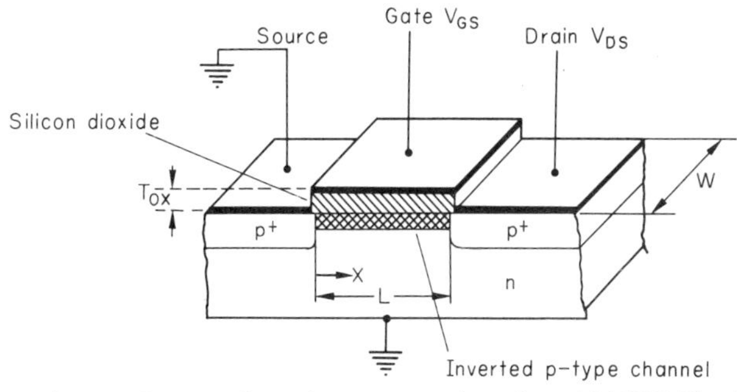

Fig. 3.1. Schematic drawing of an enhancement mode p-channel *MOST. The channel is formed by applying a negative bias between gate and substrate*

oxide, consider the effect of a steadily increasing negative potential applied between the gate and the substrate. It will give rise to a variation of the energy bands in the semiconductor and cause a change of the charge distribution with distance in a direction normal to the surface: this is shown in Fig. 3.2 for three values of applied bias.

In Fig. 3.2(*a*) where a positive gate potential V_G is applied, electrons, the majority carriers, are attracted to the region adjacent to the oxide-semiconductor interface and their resulting charge distribution is shown in an idealised form. In this case it is said there is an *accumulation* of majority carriers next to the surface. If a small negative bias is applied, then electrons will be repelled from the region next to the semiconductor surface leaving a space charge region occupied by uncompensated ionised donors. Because there are so few carriers in this region it is called a depletion region and the depth of this region obviously increases as the bias is increased (Fig. 3.2(*b*)). Further increase in V_G will attract a number of minority carriers (holes in this case) to the interface so that while some of the charge on the semiconductor consists of the charge of the ionised donors another part consists of holes in the inverted layer next to the surface. As V_G further increases a larger fraction of the charge will be contributed by the charge within the inversion layer.

Surface states can cause either an accumulation or a depletion of majority carriers at the surface. This in turn will cause the voltage at which an inversion layer is formed in the MOS device to vary. Because the surface states have time constants which can vary over a very wide range they affect the response of the space charge to rapid changes in V_G. The trapped immobile charge on the surface of the semiconductor will also reduce the differential transconductance of the MOS device. These surface states vitally affect

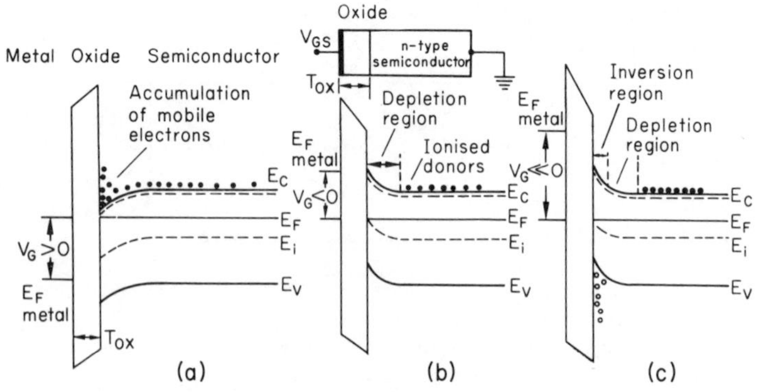

Fig. 3.2. Energy bands of an MOS structure showing the distribution of minority and majority carriers for the three different bias conditions.

(*a*) accumulation *of majority carriers near surface*
(*b*) depletion *of majority carriers from surface layer*
(*c*) inversion. *An accumulation of minority carriers has formed near surface*

the operation of the device; the fact that they are neglected in the following derivation of the device equations causes discrepancies to occur between measured and calculated parameters of the MOS device. These discrepancies may be minimised in practice if the magnitude of the surface trapped charge is always much less than the charge applied to the gate. This is the case for all modern devices.

The density of surface states on silicon is found to be significantly reduced when a thermally grown silicon dioxide layer is added to the chemically etched crystal. This surface passivation process of course forms the basis of the planar technology and has led among other things to the successful epoxy encapsulation of transistors and integrated circuits.

In the present context, the reduction of the number of surface states allows the electric field produced by the gate potential to penetrate the silicon surface and to modulate its conductivity. The mechanism which causes this improvement in the surface properties of silicon remains obscure but it is thought that the smooth transition from silicon to silicon dioxide eliminates unsaturated or dangling bonds which are commonly associated with surface states, in addition the hyperclean conditions under which the oxide is grown minimises the possibility of surface contamination of the silicon surface from the atmosphere.

Considering some quantitative data about the surface state densities that are acceptable. If we have an applied field across the silicon dioxide of 10^6 V/cm, this terminates in a charge on the silicon surface of about 2×10^{12} electrons/cm^2. Since the breakdown strength of the oxide is only 5×10^6 V/cm a surface density of states of less than 5×10^{11}/cm^2 is necessary if an acceptable device is to be made. Present day surface state densities lower than 2×10^{11}/cm^2 are obtainable with current technology. It is worth mentioning here that the $1/f$ noise associated with surface devices has been attributed to the transfer of carriers in and out of surface states so that if this is to be further reduced even lower surface state densities are desirable. This is discussed further in Chapter 4.

Consider now the effect on the device we have considered of applying a potential difference between drain and source. If the value of V_{DS} is low compared with the gate voltage V_{GS}, then assuming this is large enough to create an inverted channel there will be a linear drop of potential along the channel and the drain current I_D will be linearly dependent on V_{DS}. As V_{DS} is increased, however, the potential drop along the channel increases until a point is reached near the drain where the field across the oxide, which of course is weakened by the flow of drain current, is not sufficient for the inversion layer to be sustained. The channel is said then to be pinched-off in a manner analogous to that of the junction FET and the drain current that flows tends to saturate. Fig. 3.3 indicates that the channel is widest at the source tapering to zero thickness at the point of pinch-off. Further increase of V_{DS} causes the length of the channel to be shortened and the device is driven further into saturation. In this case the depletion region which is widest at the drain end of the channel tends to become still wider and may punch through to the source if V_{DS} rises high enough.

The voltage across the gate oxide at which the inverted channel is pinched-off is defined as the *threshold or pinch-off voltage* V_{th}. It is the voltage of course necessary to set up the invasion layer in the first instance. (In this chapter V_{th} will be used to describe the threshold voltage of the enhancement mode *p*-channel MOST).

Fig. 3.4 shows a typical set of drain characteristics for a *p*-channel

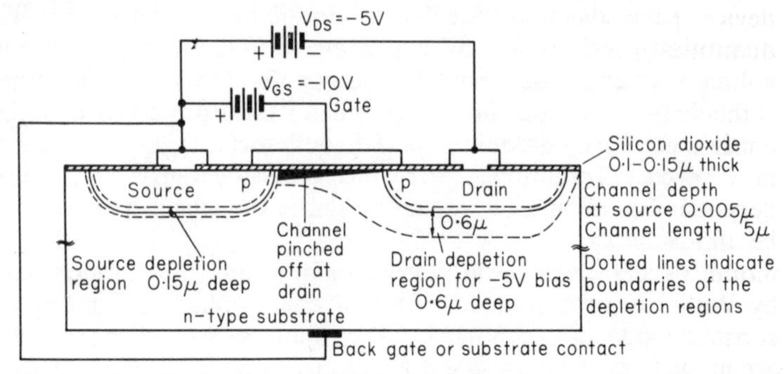

Fig. 3.3. A p*-channel enhancement MOST biased so that the channel is just pinched off.* $V_{th} = 5$ *V. Drawing not to scale. Approximate lengths of some of the more important dimensions are indicated*

enhancement MOST. The current is saturated beyond the saturation point, the locus of which is marked on the curves. The similarity between these characteristics and those of a junction FET may be noted.

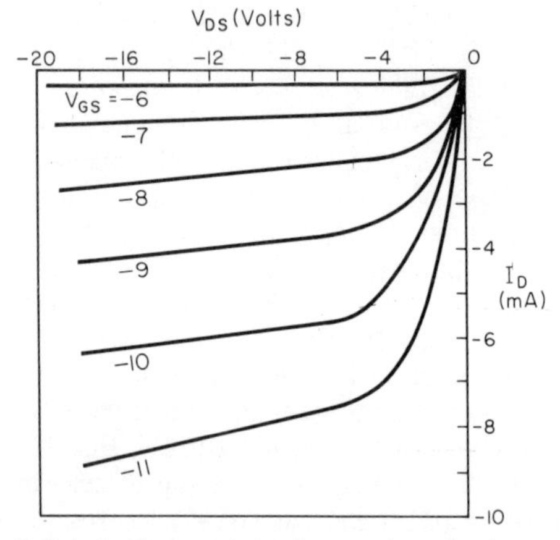

Fig. 3.4. Typical drain characteristics for a p*-channel enhancement MOST*

3.2. Mathematical analysis of the MOST

A large number of papers[1] have attempted to analyse more or less completely the operation of the MOST. Here a relatively simple model of the device will be considered and a number of simplifying assumptions introduced to arrive at equations for the more important device parameters. For example, one of the most important quantities it is necessary to know in order to work out the current-voltage characteristics, transconductance or the frequency response is the drift mobility of the carriers moving in the channel. We have considered briefly the effects of high electric fields on drift mobility in Chapter 2 and here again the device performance will be dependent on whether it can be assumed that mobility is constant, i.e. the carrier velocity is proportional to the field or whether the mobility is proportional to $1/\sqrt{E}$. This has been considered in detail by Wallmark and Johnson[2], here we will assume the mobility is constant with changing drift field. The following other assumptions are made:

(1) The dielectric is assumed to be a perfect insulator having a thickness large compared to the channel depth. This means that leakage currents are neglected and that the voltage drop in the semiconductor may be neglected compared with the voltage appearing across the oxide.
(2) The substrate is lightly and uniformly doped so that its gating action on the channel may be neglected.
(3) A gradual channel approximation is assumed so that the variation of channel thickness along the channel length is small.
(4) Effects caused by source and drain resistances within the bulk of the semiconductor are neglected.
(5) The consequencies of surface traps are not directly considered.

3.2.1 THE TRIODE REGION OF THE CHARACTERISTICS

We consider first the portion of the characteristics before saturation sets in, this is sometimes known as the triode region. Following Candler and Jordan[3] and using Fig. 3.1 as a model:

If $V_0(x)$ is the gate-channel DC voltage and $V_0(x) > V_{th}$ then there will be a mobile surface charge density σ C/cm^2 in the channel given by

$$\sigma = \frac{\varepsilon_{ox}}{T_{ox}}[V_0(x) - V_{th}] \tag{3.1}$$

when T_{ox} and ε_{ox} are the thickness and dielectric constant of the gate oxide measured in centimeters and Farads/cm respectively. If μ_p is the mobility of carriers (holes) within the channel and W is the device width (cm) measured in the direction transverse to the current flow then the channel resistance/unit length is

$$R = \frac{T_{ox}}{\mu_p \varepsilon_{ox} W[V_0(x) - V_{th}]} \tag{3.2}$$

Using the assumption that T_{ox} is great enough for it to be assumed that most of the gate-channel voltage V_0 is dropped across the oxide gives the gate-channel capacitance per unit length as

$$C = \frac{\varepsilon_{ox} W}{T_{ox}} \quad (3.3)$$

The DC drain current I_D will be

$$I_D = -\frac{1}{R}\frac{dV_0}{dx} \quad (3.4)$$

so that integrating (3.4) from a value $(V_0)_S$ at the source to $(V_0)_D$ at the drain gives

$$I_0 = -\frac{\mu_p \varepsilon_{ox} W}{2LT_{ox}}[((V_0)_D - V_{th})^2 - ((V_0)_S - V_{th}))^2] \quad (3.5)$$

$$= -\frac{\mu_p \varepsilon_{ox} W}{2LT_{ox}} V_{DS}[V_{DS} - 2(V_{GS} - V_{th})] \quad (3.6)$$

where L is the channel length in cm and V_{GS} and V_{DS} are gate and drain voltages.

Equation (3.6) may be written as

$$I_D = -\beta[(V_{GS} - V_{th}) V_{DS} - \tfrac{1}{2}V_{DS}^2 \quad (3.7)$$

for $$|V_{DS}| < |V_{GS} - V_{th}|$$

where $$\beta = \frac{W\varepsilon_{ox}\mu_p}{LT_{ox}} = \frac{C\mu_p}{L^2} \quad (3.8)$$

The transconductance of this device

$$g_{fs} = \left(\frac{\partial(I_D)}{\partial V_{GS}}\right)_{V_{DS}}$$

$$= -\frac{\mu_p \varepsilon_{ox} W}{LT_{ox}} V_{DS} = -\beta V_{DS} \quad (3.9)$$

for the triode region $V_{DS} < |V_{GS} - V_{th}|$

This simple dependence of g_{fs} on drain voltage leaving g_{fs} independent of V_{GS} is only true for low values of gate voltage. A fuller analysis where the variations of carrier mobility with field are considered shows that g_{fs} falls off with increase in gate voltage.

The drain or output conductance of the device can also be found from equation (3.7) by differentiation

$$\frac{\partial I_D}{\partial V_{DS}} = -\beta(V_{GS} - V_{th}) + \beta V_{DS} \tag{3.10}$$

This leads to a value of incremental drain resistance for the triode region if $V_{DS} \to 0$ which is

$$r_{dt} = \frac{1}{-\beta(V_{GS} - V_{th})} \tag{3.11}$$

Since the transconductance may be looked on as being the gain parameter of the device and this is $g_{fs} = -\beta V_{DS}$ we see that the

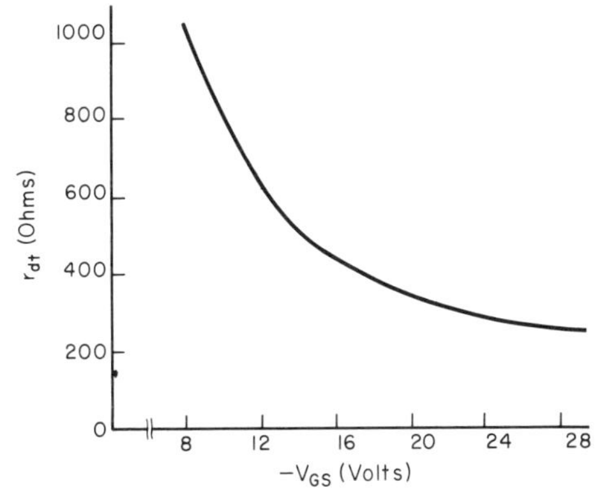

Fig. 3.5. Graph showing the variation of the incremental drain resistance (r_{dt}) with gate voltage in the triode region of the characteristics for a Texas Instruments Type TIXS 11 device $|V_{DS}| < 50\,mV$.

higher the gain the lower the value of r_{dt}. If the FET is being used therefore in the triode region as a switch its on resistance will be determined by V_{GS} and by the parameter β. Fig. 3.5 gives a typical variation of r_{dt} against V_G.

3.2.2 THE SATURATION REGION

Consider now the operation of the MOST in its saturation region. As the drain voltage is increased from zero, drain current increases almost linearly at first and then tends to level out at the saturation value as V_{DS} becomes large. This is caused by the field across the oxide being insufficient to maintain the mobile charges at the drain end of the channel so that the channel becomes pinched-off. The boundary between the triode and saturation region can be determined if the charge in the channel at the drain end is allowed to become zero. Then

$$V_{DS} = V_{GS} - V_{th} \tag{3.12}$$

V_{th}, the threshold voltage can be given a value if we consider the charging of the condenser which is formed between the gate and semiconductor. If the trapped charge on the semiconductor surface is Q_{SS} and Q_D is the bulk charge associated with the depletion region of the semiconductor, then since C is the gate-channel capacitance per unit length (3.3)

$$V_{th} = -\frac{Q_{SS} + Q_D}{C} \tag{3.13}$$

since V_{th} must be just sufficient to neutralise the charges Q_{SS} and Q_D.

If $|V_{DS}| > |V_{GS} - V_{th}|$ and the MOST is operating in the saturation region, to a first approximation the drain current becomes independent of V_{DS} and the characteristics for different values of V_{GS} give a series of lines parallel to the V_{DS} axis (See Fig. 3.4). This implies an infinite output impedance which of course is not met with in practice, this is considered further in the next section.

Substituting $V_{DS} = V_{GS} - V_{th}$ into equation (3.7) gives the saturation current as a function of V_{GS} so that

$$I_D = -\frac{\beta}{2}(V_{GS} - V_{th})^2 \tag{3.14}$$

This is valid for $|V_{DS}| > |V_{GS} - V_{th}|$; it is obvious that there is a square law dependence of drain current on gate voltage, this square law relationship has been shown experimentally to apply over a wide range of currents and for devices of various geometries.

The accurate square law transfer curve obtained from equation (3.14) which is similar to that found for the junction FET gives the MOST and the junction FET a series of unique applications. Some of these are discussed later. (See also Sevin[4].)

In most devices the error introduced when V_{DS} is made equal to V_{GS} for the saturation region is small enough to be neglected. In this case the saturation current $I_{DS} \approx -(\beta/2)V_{GS}^2$

$$= -\frac{W\varepsilon_{ox}\mu_p V_{GS}^2}{LT_{ox}} \tag{3.15}$$

This is the same as saying that the device has a zero threshold voltage, a fact that will be true if the surface doping of the semiconductor is very low.

The transconductance

$$g_{fs} = \left(\frac{\partial I_D}{\partial V_{GS}}\right)_{V_{DS}}$$

obtained by differentiating 3.14 gives

$$\frac{\partial(I_D)}{\partial V_{GS}} = -\beta(V_{GS} - V_{th}) \approx -\beta V_{GS} \approx \frac{\mu_p \varepsilon_{ox} W}{T_{ox} L} V_{GS} \quad (3.16)$$

g_{fs} may therefore be increased by either an increase of gate bias V_{GS} or by maximising the width to length ratio W/L of the channel.

The characteristic we have so far considered in terms of a triode and a saturation region represented by equation (3.7) and (3.14) where the boundary between the two equations is a parabola

$$I_D = \frac{\beta(V_{DS} - V_{th})^2}{2}, \quad (3.17)$$

is that of an idealised device. In particular, the fact that I_D increases slowly after saturation has been reached has been neglected. Let us consider what happens in this region to give the MOST a finite output impedance.

Consider the model of a MOST shown in Fig. 3.6 which has a variable drain voltage applied sufficient to drive the device into its saturation region of operation. The channel has been pinched off so that the depletion region is shown extending into the region normally occupied by the channel. The voltage drop across the channel will be $V_{GS} - V_{th}$ and this is almost independent of the value of V_{DS} so that for a drain potential of V_{DS}, a voltage of $V_{DS} - (V_{GS} - V_{th})$ must be dropped across the depletion region existing between the end of the

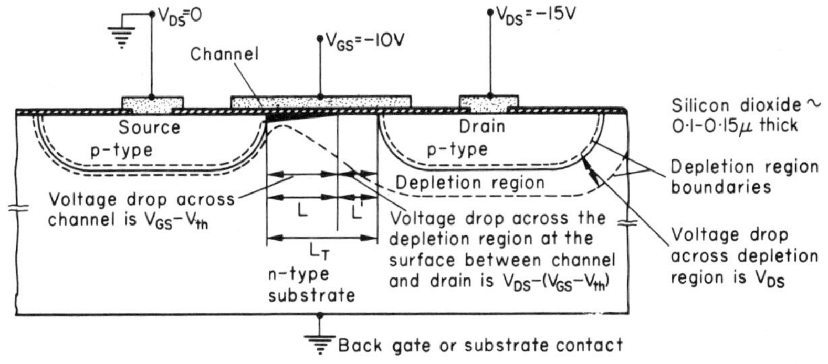

Fig. 3.6. Model of p-channel *enhancement MOST operating in the saturation region of characteristics* $V_{th} = 5$ *V. Compare this with Fig. 3.3 where the channel is only just pinched off*

channel and the gate. As V_D increases, the length of this depleted region say L' must increase thereby shortening the length of the channel, the resistance of which consequently decreases. If this is so then as the voltage across the channel remains constant the current through it must rise slowly as V_{DS} increases due to the shortening of the channel length, this increase in current with accompanying increase in output voltage results in the MOST having a finite output impedance which may be some tens of thousands of ohms in a typical device. This effect where the channel length is modulated by the drain voltage is an example of positive feedback, it is analogous to the Early effect in bipolar transistors where the modulation of the base width with collector voltage gives rise to a finite output impedance.

Equation (3.14) may be modified to take channel length modulation into account by writing

$$I_D = I_{DP}\frac{L_T}{L_T - L'} \tag{3.18}$$

where

$$I_{DP} = -\frac{\beta}{2}(V_{GS} - V_{th})^2\Big|_{V_{DS}=V_{GS}-V_{th}} \tag{3.19}$$

is the drain current at the pinch-off point and L_T is the channel length at pinch-off ($L_T = L + L'$).

3.2.3 TRANSCONDUCTANCE

The variation of I_D with V_{DS} just considered in the saturation region gives rise to a change in the value of the transconductance or gain of the MOST, given by equation (3.16). In fact the effective value of β increases slightly with drain voltage, β also decreases with increasing gate voltage. For many practical purposes, however, equations (3.7) and (3.14) can be used for designing circuits. This is analogous to designing bipolar circuits without regard for the dependence of current gain on operating point.

If it is desired to maximise the transconductance or gain of the MOST by either increasing V_{GS} or increasing the width: length ration W/L of the channel there are limitations in both cases. Considering these two possibilities in turn, an increase of V_{GS} over a wide range gives rise to a linear change in g_{fs} but excessive gate voltage of course leads to breakdown of the gate oxide and consequent failure of the device. The limiting factor therefore is the breakdown strength of the oxide (5×10^6 V/cm for SiO_2). There are, or course, possibilities of using alternative insulating layers which have higher breakdown strengths. A great deal of work using silicon nitride is currently taking place and MOSTs using composite layers of silicon dioxide and silicon nitride are now marketed. (See also section 3.2.2.).

Changes in the W/L ratio can obviously only be undertaken at the device design stage. Values of β in a typical integrated circuit logic MOST may be about 20 $\mu A/V^2$, but values between 0·1 and 500 $\mu A/V^2$ are possible by varying W and L. The dimensions of the channel are also important when considering the frequency response of the device. It has been demonstrated by Johnson and Rose[5] that the carrier transit time in a charge controlled device such as a bipolar or field-effect transistor is a good measure of the speed of the device. For an ideal FET the transit time τ_r is given by the ratio of C_{in}/g_{fs} which provides a useful figure of merit. If a constant carrier mobility is assumed in the channel an approximate value of τ_r may be simply arrived at:

$$\tau_r = \frac{L}{v_d} \text{ where } v_d \text{ is the average drift velocity} \tag{3.20}$$

$$= \frac{L^2}{\mu V_{DS}} \text{ where } V_{DS}/L \text{ is the average drift field} \tag{3.21}$$

If then a transit time of 10^{-9} seconds is required for a drain voltage of 10 volts, the channel length must not exceed 15 microns if a reasonable value of carrier mobility near the surface of 225 cm^2/V-sec is assumed (Wallmark[1]). A linear channel dimension as small as this is quite feasible with the present state of the integrated circuit technology where line widths down to one micron can be achieved with photolithographic techniques.

It may be noted that the gain bandwidth product C_{in}/g_{fs} increases as L is decreased. However, both C and g_{fs} increase together as the channel width is increased or the oxide thickness is thinned leaving their ratio unchanged. The advantage in changing these dimensions is that it leads to devices having lower impedance levels so that they can deliver more current at fixed drain and gate voltages. There are snags in fabrication here, however, if these changes are extended too far as the probability of making a transistor with a faulty gate is enhanced, there being a greater chance that the gate electrode will cover a pin-hole in the oxide. The lower impedance level is obtained therefore at a sacrifice of transistor yield and reliability.

3.2.4 DEPLETION MODE DEVICES

So far we have been considering a specific type of device namely a *p*-channel enhancement MOST. It is, however, possible by doping the surface layer of the semiconductor to form a depletion type

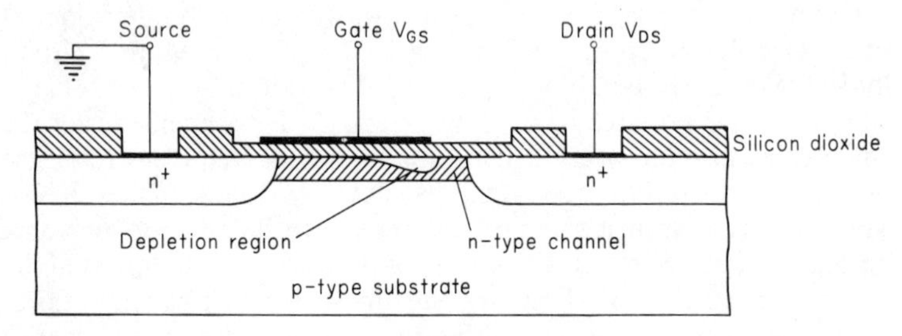

Fig. 3.7(a). An n*-channel depletion mode MOST with offset gate geometry to reduce feedback capacitance between gate and drain. The depletion region is shown diagrammatically for a negative bias applied to the gate*

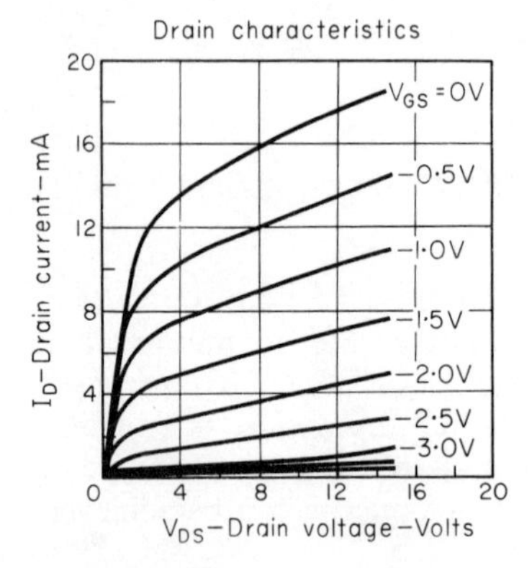

Fig. 3.7(b). Drain characteristics of an n*-channel depletion mode MOST*

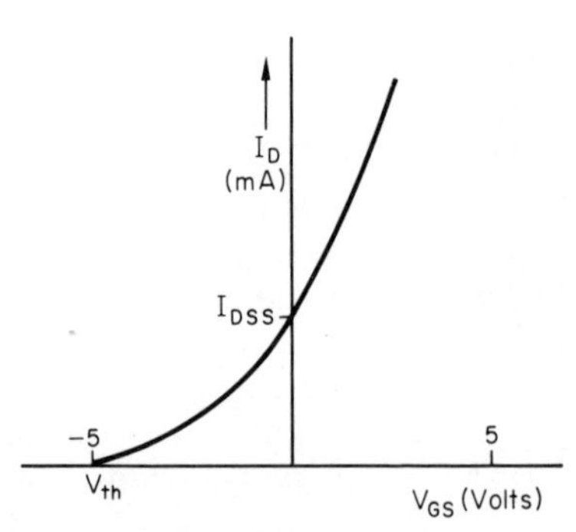

Fig. 3.7(c). Transfer characteristic for an n*-channel depletion mode MOST*

MOST having characteristics very similar to those of the junction FET. Fig. 3.7 shows a schematic diagram of a depletion MOST and gives typical drain and transfer characteristics for an *n*-channel device. From the transfer characteristic Fig. 3.7(*c*) it can be seen that there is a natural built-in reference point namely the drain current at zero gate voltage. This current I_{DSS}, can easily be measured and can be related to the other parameters of the device.

Considering again the equation for the saturation drain current of a MOST, equation (3.14)

$$I_D = -\frac{\beta}{2}(V_{GS} - V_{th})^2 \quad (3.14)$$

For zero gate voltage

$$I_D\big|_{V_{GS}=0} = I_{DSS} = -\frac{\beta}{2}V_{th}^2 \quad (3.22)$$

So that substituting in (3.14).

$$I_D = I_{DSS}\left(1 - \frac{V_{GS}}{V_{th}}\right)^2$$

The slope of the transfer curve is the transconductance of the MOST. At $V_{GS} = 0$ we have

$$\left.\frac{\partial I_D}{\partial V_{GS}}\right|_{V_{GS}=0} = g_{fs0} = -\frac{\beta}{2}(2V_{th})$$

or

$$g_{fs0} = \frac{2I_{DSS}}{-V_{th}}$$

where g_{fs0} is the transconductance of the depletion MOST measured at I_{DSS}.

The advantage in using I_{DSS} in calculations is that it is an easily and accurately measured parameter. V_{th} on the other hand is not so easy to estimate, it is often specified in data sheets as the gate voltage required to produce a drain current of 10 μA, this value may differ from the true value by 10%.

3.2.5 SUMMARY OF EQUATIONS

The important equations for the MOST have been summarised by Crawford[6] in the form of a table where the polarities of the applied voltages are indicated for *n* and *p* channel devices for both modes of operation.

For all MOSTs:

For the triode region where $|V_{DS}| < |V_{GS} - V_{th}|$

$$I_D = -\beta[(V_{GS} - V_{th})V_{DS} - \tfrac{1}{2}V_{DS}^2] \quad (3.7)$$

For the saturation region where $|V_{DS}| \; \mathrm{I} \; |V_{GS} - V_{th}|$

$$I_D = -\frac{\beta}{2}(V_{GS} - V_{th})^2 \tag{3.14}$$

Both these equations must be modified to take account of source and drain resistances, of course I_D is given by equation (3.18) when the variation of saturation current with drain current is considered.

Considering now the polarity of the applied potentials and currents for different types of MOST.

For n-*channel devices*

Electrons are majority carriers so by convention

$$\beta = \frac{W \varepsilon_{ox} \mu_e}{L T_{ox}} \text{ will be } < 0 \text{ (since } \mu_e < 0)$$

$$(V_{GS} - V_{th}) > 0$$

$$V_{DS} \text{ and } I_D > 0$$

For depletion mode operation $V_{th} < 0$
For enhancement mode operation $V_{th} > 0$
$V_{GS} > 0$

For p-*channel devices*

Positive holes are majority carriers so $\beta > 0$

$$(V_{GS} - V_{th}) < 0$$

$$V_{DS} \text{ and } I_D < 0$$

For depletion mode operation $V_{th} > 0$
and for enhancement mode operation $V_{th} < 0$
$V_{GS} < 0$

3.3. MOST device technology

Consider the fabrication of a typical integrated *p*-channel enhancement MOST for use in digital circuit applications where direct-coupled inversion will be possible without the need for level shifting between stages. The device will be required to show a very low drain current at zero gate bias with the current increasing rapidly with increasing negative gate voltage. An important design feature must be that the gate electrode must cover the entire channel, overlapping to some extent both the source and drain regions. If this is not the case then any channel region left exposed contributes a high series resistance to the device since there are few mobile carriers in the channel at zero bias. The overlap of the gate with the source and drain obviously results in an enhanced gate-source capacitance C_{GS} and gate drain capacitance C_{GD} unless precautions are taken to thicken the oxide in the overlapping regions. Since the feedback capacitance C_{GD} is amplified by the Miller effect when reflected at the

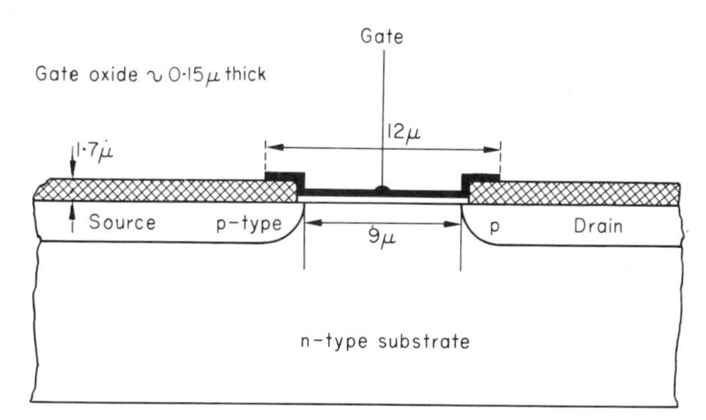

Fig. 3.8. Typical p-*channel enhancement MOST*

input this parasitic capacitance poses a limitation to the speed at which the transistor can operate. This may to some extent be taken care of in the design of the MOST by increasing the thickness of the oxide in the overlap region.

A typical geometry of a MOST designed as a low level chopper is shown in Fig. 3.8. Here the channel width is ~9 μ and the overlap at source and drain ~$1\frac{1}{2}$ μ. The values of C_{GS} and C_{GD} are typically at least 3·5 and 2·0 pF respectively. Consider the stages in manufacture of a MOST similar to that shown in diagramatically in Fig. 3.9.

A clean chemically polished *n*-type wafer of about 10 Ω-cm resistivity has its surface oxidised in a diffusion furnace through which steam is passed. Openings for the source and gate areas are then made using the conventional photolithographic process, and the source and gate simultaneously formed by diffusing boron through the openings. The oxide in the area between source and drain is then stripped chemically and a thin gate oxide ~ 0·15 μ.

thick grown in a dry oxygen atmosphere. The final stage is the metallisation of the source, drain and gate contacts. It is obvious from this that there are fewer stages in the manufacture of a MOST than in making a junction FET or bipolar device, this means that the chip cost should be substantially less and the yield correspondingly greater for the MOS device. The main potential drawback of the

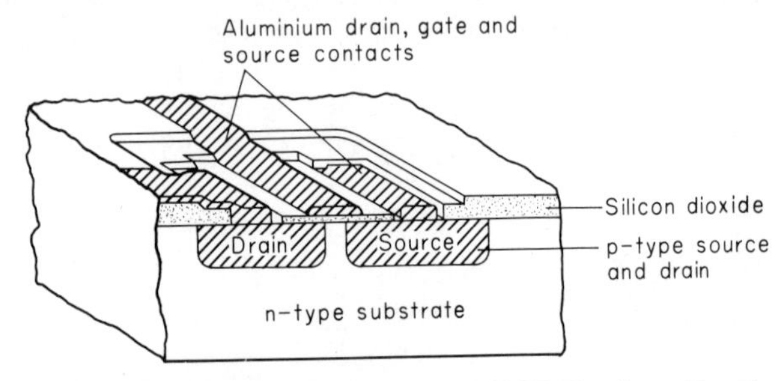

Fig. 3.9. Section through a p-channel *enhancement MOST. The channel length would be typically* 5 μ

MOST is the fact that the current path lies along the silicon–silicon dioxide interface, this region is known to be the least stable part of the planar device and precautions have to be taken to stabilise the oxide in order to give a stable set of characteristics. We will consider this a little later.

An alternative approach to the making of MOSTs which has been extensively used for making arrays of devices having a ladder geometry is shown diagramatically in Fig. 3.10. The FETs are arranged in tandem so that the drain of one FET forms the source of the adjoining one and so on. This arrangement greatly simplifies the interconnection of many transistors on the slice and reduces the capacitance at each node. For these devices the deposition of

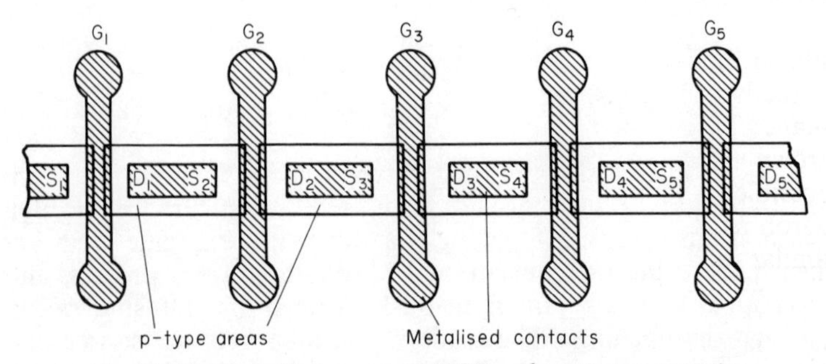

Fig. 3.10(a). The ladder geometry MOST—schematic representation

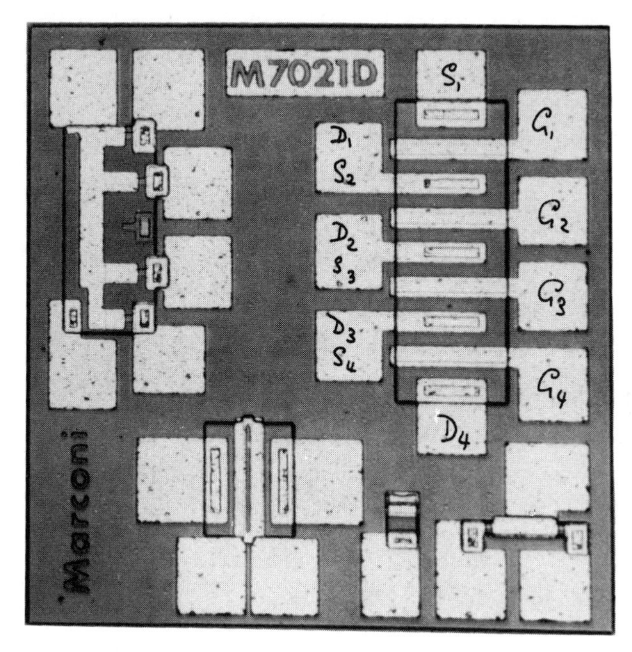

Fig. 3.10(b). The ladder geometry MOST. Marconi-Elliott Microelectronics MOST circuit Type E6015D showing an array of ladder MOSTs at top right

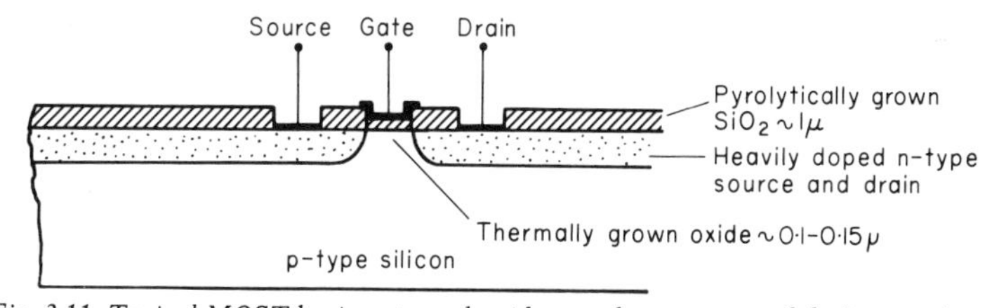

Fig. 3.11. Typical MOST having stepped oxide to reduce source and drain capacitance

silicon dioxide over the slice may be achieved by the pyrolysis of tetraethyl orthosilicate (TEOS), or by the controlled oxidation of silane. This gives an oxide similar in structure to the thermally grown oxide already described but has the advantage that it is possible to lay down doped oxide layers containing phosphorus or boron as the need arises[7]. The stages in manufacture for a device similar to Fig. 3.11 are:

(1) A clean chemically polished *p*-type wafer has an *n*-type oxide layer deposited over the entire slice area to a depth of at least one micron.

(2) The oxide is coated with photo resist, a pattern exposed on to it and developed to leave rectangular blocks of resist over the source and drain regions. The remainder of the oxide is then removed chemically and finally the remainder of the resist is removed.

(3) The cleaned wafers are placed in a diffusion furnace so that the phosphorus in the doped oxide diffuses into the slice to form *n*-type source and drain regions. While the wafers are still in the diffusion furnace a thin thermally grown oxide film is formed over the exposed silicon surface to a depth of say 0·1–0·15 μ.

(4) Using photo resist again holes are opened in the oxide to make contacts for source and drain. These contacts together with the gate electrodes are made by evaporating a film of gold or aluminium in the conventional manner.

It may be noted from Fig. 3.11 that the oxide in the overlapping regions between source and gate and drain and gate is appreciably thicker than the gate oxide. This means that the capacitances C_{GD} and C_{GS} are substantially reduced thereby relaxing the otherwise stringent tolerance on the degree of overlap that can be allowed particularly in the gate-drain region.

A disadvantage of this method of production of MOSTs using pyrolitically grown doped oxides is that surface concentrations of electrically active donor atoms in the silicon is limited to a value of say 10^{20} atoms/cc. This gives rise in the finished device to larger values of series source and drain resistance than can be obtained when the N^+ source and drain regions are diffused by conventional means; then surface concentrates as high as 10^{21} atoms/cc are achieved. In power transistors where heat dissipation has to be kept to a minimum obviously thermally grown oxides are used. Fig. 3.12(*a*) shows such a transistor in which the gate electrode snakes between the thin source and drain electrodes producing a channel having a large W/L ratio; in this way a high transconductance is achieved. Fig. 3.12(*b*) shows typical drain characteristics for this device, it can be seen that drain currents of 50 mA are possible for applied gate voltages $\sim$20 V.

3.3.1 DEPLETION TYPE DEVICES

It is possible to make depletion MOSTs by methods similar to those used for enhancement types. The main difference is that the source, drain and channel will all be of the same conductivity type resulting in a flow of a large drain current for zero gate bias. In order that complete pinch-off of the channel may be obtained with moderately low gate voltages the doping in the channel region is very much lower than that of the drain or source region. A typical depletion MOST having an offset gate geometry together with a set of characteristics was shown in Fig. 3.7. This type of transistor finds application as a small signal linear amplifier over a wide range of frequencies. Note that unlike its predecessor, the thermionic valve, the gate when biased positively for an n-channel device draws no current.

The offset gate geometry of the depletion MOST of Fig. 3.7 is designed so that the gate overlaps the source but not the drain region because the feedback capacitance is low. This gives a device

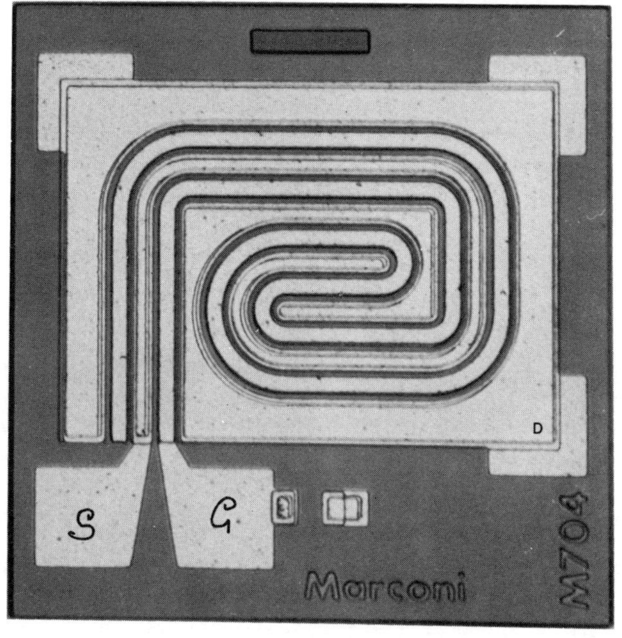

Fig. 3.12(a). The power MOST. Marconi-Elliott Type E6019 p-*channel enhancement MOST*

having an improved high frequency performance. Of course the unmodulated region of the channel near the drain adds a series resistance in the saturation region, so effectively increasing the drain voltage at which saturation occurs, this however has little effect on the circuit gain provided the additional resistance is small compared with the output resistance of the device. Of course, if there was also a series resistance at the source this would have a more dramatic effect on the circuit performance when the reflected value of this resistance is considered at the output.

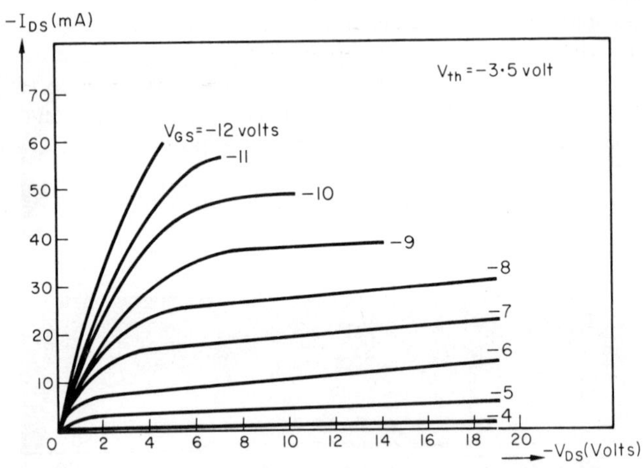

Fig. 3.12(b). The power MOST. Typical drain characteristics for the Type E6019 MOST at 25°C

3.3.2 OXIDE STABILISATION AND CHARACTERISTIC DRIFT

It has previously been mentioned that under normal operating conditions the gate oxide in a MOST is subjected to extremely high fields $\sim 10^6$ V/cm, a value close to the dielectric breakdown strength of silicon dioxide. It was found in all early MOSTs that the characteristics showed a marked instability which was directly attributable to the migration of ions, particularly sodium ions, within the oxide resulting in a built in electrostatic field. The effect manifested itself in a number of ways, from the production standpoint successive batches of similar MOSTs had threshold voltages varying from perhaps −8 V to −20 V or more, and individual units when subjected to heat treatment or prolonged biasing showed V_{th} drifting by as much as −40 V. Fig. 3.13(*a*) shows a typical drift in characteristics for an unstabilised MOST. The effect is most significant in enhancement mode MOSTs for with depletion mode devices near zero gate voltages are used with correspondingly lower electric field stresses within the oxide.

The problem of drift was very largely overcome by the discovery that exposure of the gate oxide to a phosphorous ambient at a temperature of about 850°C has a stabilising effect. In practice about the top third of the oxide is converted into a phosphorus glass onto which subsequently the gate electrode is evaporated. Devices made in this way will be subject to characteristic drift of $\ll\frac{1}{2}$ V. It is also possible to produce stable MOSTs by paying particular attention to the conditions under which the gate oxide is grown. Elimination of

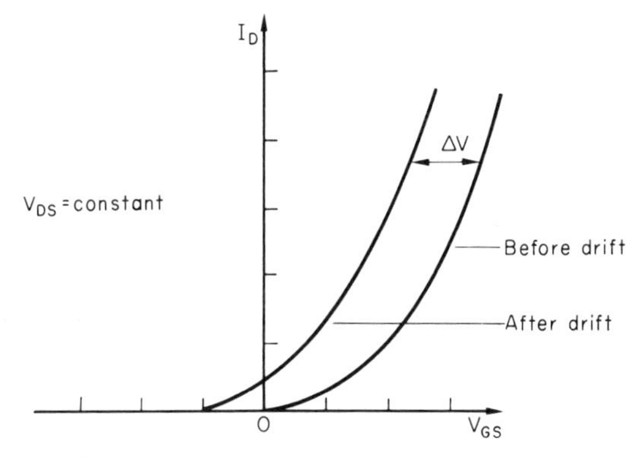

Fig. 3.13(a). The effect of using a non-stabilised gate oxide on the transfer characteristic of an n-*channel MOST subject to temperature cycling or prolonged storage. In an extreme case,* ΔV *may be as large as 5 V. Using a well-stabilised oxide* ΔV $\sim$ *0·1 V*

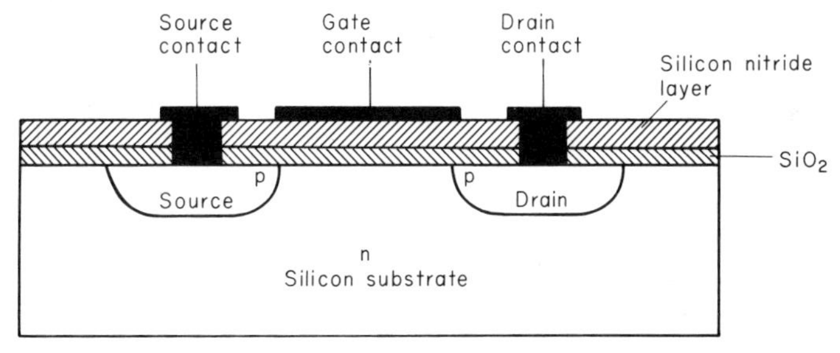

Fig. 3.13(b). Schematic diagram of a Motorola MOST which uses a silicon nitride layer to passivate the silicon dioxide

contamination at this stage is the method used by many manufacturers to produce stable devices.

An alternative approach is to use silicon nitride as the gate insulator since this eliminates the ion migration problem encountered with silicon dioxide. Unfortunately silicon nitride alone has not proved satisfactory since sufficiently low surface state charge densities have not been attainable, however a promising approach adopted by a number of manufacturers is to use a composite layer of silicon dioxide and silicon nitride. In this case the oxide layer gives rise to a low surface charge and the nitride minimises the migration of ions. Fig. 3.13(*b*) shows a schematic diagram of a Motorola MOSFET for which it is claimed that even when subjected to reverse bias at 200°C for 1000 hours the characteristics remain stable.

If enhancement MOSTs are to be made having values of threshold voltage less than the value presently available further effort has to be made to reduce the surface charge on the silicon oxide interface. Ideally, for the *n*-channel device a negative rather than a positive surface charge is desirable. This would enable currently available types of MOST to have lower values of V_{th} and to operate at higher speed without the necessity for going to smaller channel dimensions. (See section 3.6).

3.3.3 COMPLEMENTARY TRANSISTORS

As it is possible to make MOSTs having either *n*- or *p*-type channels it is possible also using presently available planar epitaxial techniques to make circuits containing complementary pairs of MOSTs, a feature which has application for a wide range of circuits. Unfortunately, there are difficulties in manufacture, the most important of which is that *p*-channel devices normally operate in the enhancement mode whereas *n*-channel devices normally operate in the depletion mode. This difficulty may be overcome using different doping levels in the *p*- and *n*-type silicon to give comparable threshold voltages in the two types of MOST. In the typical complementary MOSTs shown in Fig. 3.14 the *p*-channel MOST is fabricated in 1–3 Ω-cm *n*-type silicon and the *n*-channel MOST in 0·2–0·5 Ω-cm *p*-type silicon. The characteristic curves of Fig. 3.15 were obtained by White and Cricchi[8] for devices having channel lengths ~12 μ and a channel width ~500 μ with a gate oxide ~0·1 μ thick. The *p*-type region of the substrate was formed by etching a groove in the *n*-type epitaxial slice and regrowing epitaxially *p*-type silicon within the groove. Only in this complex fashion could regions having sufficiently well controlled resistivity for complementary transistors be formed.

Two other disadvantages of complementary MOSTs follow from what has already been said. The first is that they occupy a very much larger area on the silicon slice than arrays of *n*- or *p*-channel devices which do not require electrical isolation from each other. This decreases the packing density of devices, one of the principle advantages of MOS logic circuitry over its bipolar counterpart. The second is that the cost per device is significantly increased for

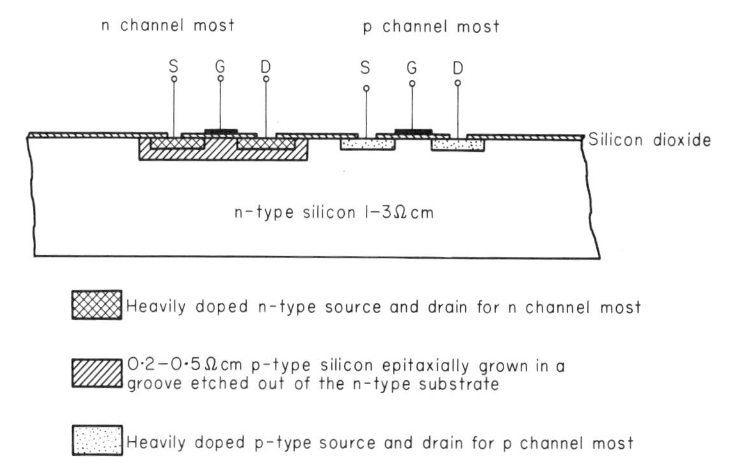

Fig. 3.14(a). Complementary MOSTs formed in an n-*type silicon wafer. Channel length is typically 12 μ and the channel width 500 μ*

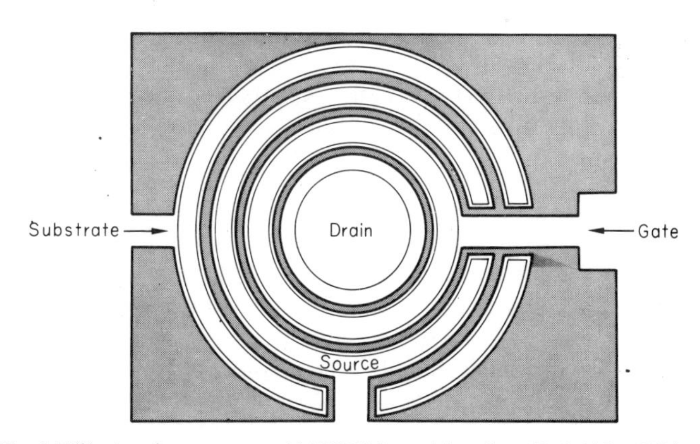

Fig. 3.14(b). Annular geometry of MOST (channel length ≈ 12 μ). (After White and Cricchi[8])

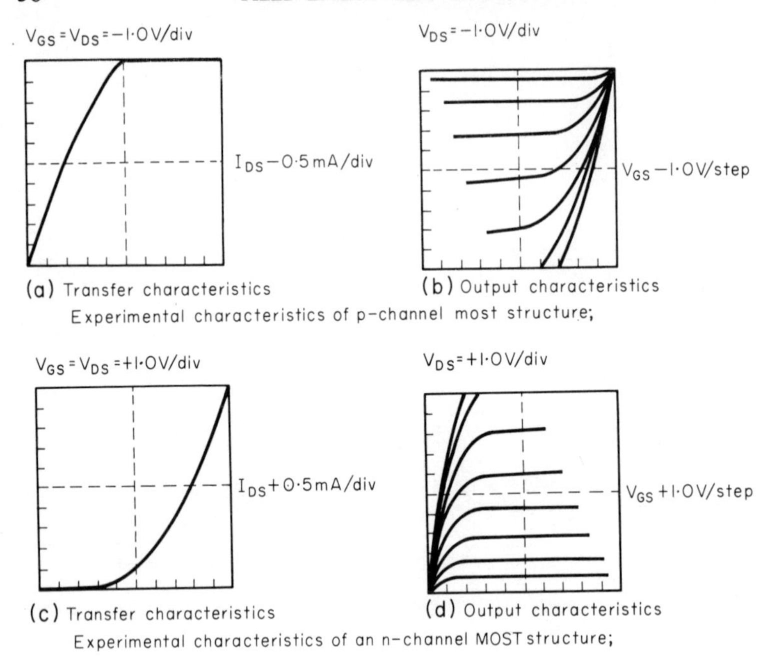

Fig. 3.15. Characteristics of complementary n- *and* p-*channel MOSTs.* (After White and Cricchi[8])

complementary MOSTs because of the increased number of manufacturing stages and the consequent reduction in overall yield.

3.3.4 GATE PROTECTION

Because the gate oxide is only ~0·1–0·15 μ thick on typical MOSTs of all types the maximum gate voltage before dielectric breakdown occurs is ~50 V. Since the electrical leakage across the oxide is extremely small it is possible for the gate-channel capacitor to be accidentally charged to a voltage which exceeds the breakdown voltage causing a destructive breakdown to occur. In order to prevent this it is common practice to build into the MOST some form of gate protection. This can take the form of a diode between gate and substrate that will conduct at a voltage slightly lower than the breakdown voltage of the oxide, or alternatively a second enhancement MOST having a high value of V_{th} may be placed in shunt between gate and channel. This additional MOST

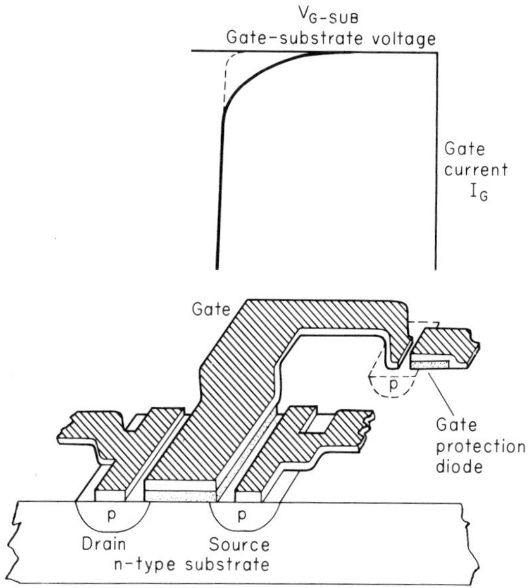

Fig. 3.16(a). Schematic drawing of diode gate protection system for p-*channel enhancement MOST* (Plessey).
Inset shows typical diode breakdown curve

S
Thick oxide most
G
Gate
protection
diode
Thick oxide most
D

Fig. 3.16(b). Equivalent circuit of MOST with diode gate protection

protects the gate oxide of the first MOST since it will conduct before breakdown occurs. It does not need protection itself since its gate oxide is significantly thicker. Fig. 3.16(*a*) shows a sketch of a typical Plessey MOST incorporating diode gate protection. An equivalent circuit of a MOST protected in this manner is shown in Fig. 3.16(*b*). Here the two additional MOSTs have thick gate oxides and are formed between the source and drain diffused areas of the protected MOST and the protection diode diffused area. The gate is formed by the aluminium connecting the gate protection diode to the gate of the MOST.

3.4. Thin film transistors

The MOSTs so far considered have all had as their basic starting point a substrate of single crystal silicon or other semiconductor. During recent years much work has been undertaken to develop thin film transistors (TFTs) where the active region of the device is a thin polycrystalline film of semiconductor laid down by evaporation, or some epitaxial technique, onto an insulating substrate of ceramic, glass or perhaps single crystal sapphire.

The main difficulty in making TFTs is that of making and reproducing transistors having reasonable performance and stable characteristics. Fig. 3.17 shows three types of TFT, all due to Weimer[9], in each of which the thickness of the films has been greatly exaggerated for the sake of showing the structure more clearly. The metal electrodes would normally be $\sim$0·05 μ thick, the insulators from 0·02–0·20 μ and the semiconductor from 0·05–1 μ thick. Most present day TFTs use cadmium sulphide as the semiconductor though tellurium and certain selenide films may also be used. The insulator most commonly used is silicon monoxide with magnesium fluoride as an alternative. Metallisation for the electrodes may be either gold or aluminium.

All evaporated devices such as those of Fig. 3.17 may have channel lengths as small as 10 μ with widths of 250–2500 μ. Dimensions as small as these can only be achieved by elaborate masking techniques within the evaporation chamber employing photo etched masks, wire masks, or combinations of both types.

The most critical of the thin films to be deposited and treated is of course the semiconductor film, this is deposited in polycrystalline form by one of several possible methods of which evaporation under vacuum is by far the most important. The choice of semiconductor is governed by a number of factors. For good high frequency response the carrier mobility should be as high as possible, to a limited extent this can be compensated for by reducing the channel length but this causes the device cost to rise since the yield falls. The high mobility must be accompanied by a carrier concentration that is not too large to be modulated by the gate potential, compare the other FETs where relatively high resistivity channels are preferred. The charge

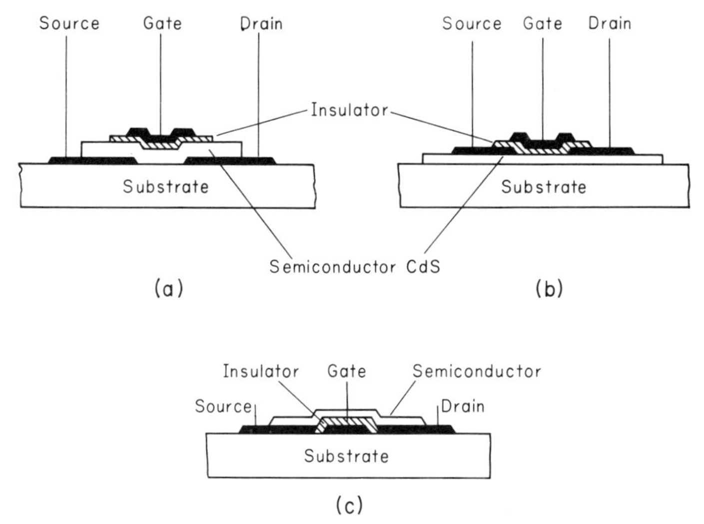

Fig. 3.17. Three typical thin film transistor structures using cadmium sulphide as the semiconductor.
- (*a*) *a staggered electrode structure.*
- (*b*) *a coplanar structure having the advantage that all precision masking is done after the semiconductor layer is formed.*
- (*c*) *an inverted form of* (b) *where the semiconductor is deposited last: this gives the best control of the semiconductor film.* (All after Weimer *et al*[9])

that the gate can control of course is determined by the dielectric constant of the insulator and its breakdown strength. The two factors of requiring high carrier mobility and resistivity make intermetallic semiconductors advantageous for TFTs. Silicon however should also be considered a suitable material though so far its potential is far from realised.

3.4.1 PERFORMANCE OF TFTS

Data is only readily available for cadmium sulphide units of the type illustrated in Fig. 3.17. They have been shown by Weimer[9] to operate in both enhancement and depletion modes with characteristics similar to those found in more conventional FETs. Fig. 3.18 shows typical characteristics for both modes of operation for TFTs having channels 10 μ long and 0·25 cm wide. High values of transconductance up to 25,000 μ mhos are possible coupled with input resistances up to 10^9 Ω. For the characteristics illustrated the dynamic input impedance (the slope of the saturated portion of the curve) is ~40 000 Ω and the voltage amplification factor ~160.

As may be expected with devices of this kind stability of the TFT characteristics is a major problem since there are high densities of traps near the surface of the semiconductor. These traps of

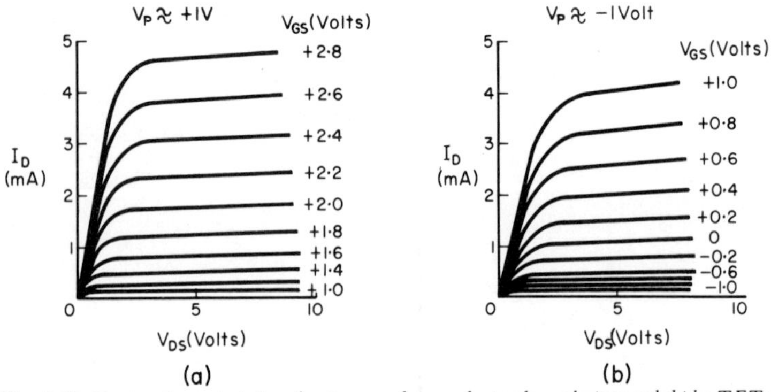

Fig. 3.18. Drain characteristics for two coplanar electrode cadmium sulphide TFTs: (a) *an enhancement mode unit,* (b) *a depletion mode unit.*
V_p *is the pinch-off or threshold voltage and is the gate voltage necessary for the pinch off or onset of drain current.* (Weimer et al.[9])

course are slowly emptying and filling during the operation of the TFT and give rise to hysteresis effects on the characteristics. The avoidance of these slow surface states is a materials processing problem that is being actively studied at the present time and on the success of this work the future of the TFT to some extent depends.

Notwithstanding this deficiency, integrated arrays of TFTs have been built by Weimer and others to form flip flops as well as more complex arrays for solid state image scanning devices that incorporate evaporated photo sensitive elements[10]. (See Chapter 13)

The potential advantage of this form of circuitry is largely economic for it should be possible to form complex thin film circuits comprising active and passive components on relatively cheap substrates; both types of component would be made simultaneously within the vacuum chamber and really large area substrates may be used. Indeed it is possible to conceive of using flexible substrates of some inert insulating material which would enable thin film circuits to be made by a continuous process. This could lead to an enormous reduction in unit cost.

3.4.2 OTHER THIN FILM CIRCUITS

Mention has already been made to the use of sapphire substrates. In this case silane SiH_4 is decomposed at a temperature of about 1000°C to grow a film of elemental silicon on the sapphire. This film may be monocrystalline if care is taken to cut the sapphire crystal along the correct crystal plane, otherwise a homogeneous

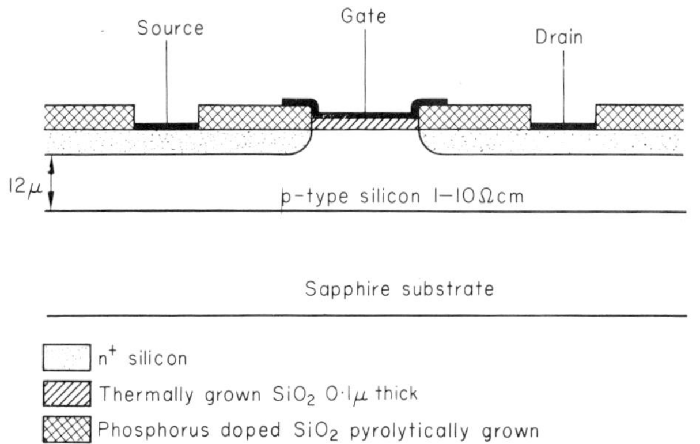

Fig. 3.19(a). MOS TFT formed by growing silicon epitaxially on a sapphire substrate. (After Mueller and Robinson[11])

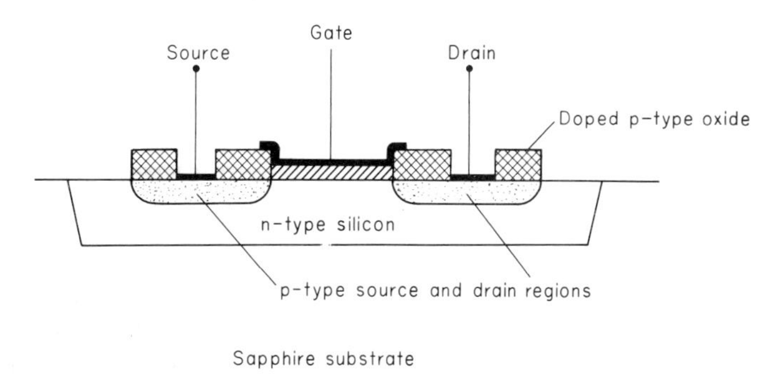

Fig. 3.19(b). Possible TFT formed by growing silicon epitaxially in a groove or circular depression in a sapphire substrate. This method can also be used for making junction devices and has the advantage that individual devices are electrically isolated from each other by the substrate

polycrystalline layer results. In either event if the silicon is etched to form small islands insulated from each other by the sapphire it is possible using conventional diffusion and oxidation techniques to form field-effect or bipolar transistors of both polarities. These transistors are isolated from each other by the substrate and are easily interconnected using conventional metallisation techniques. Fig. 3.19 shows two examples of sapphire devices in the second of

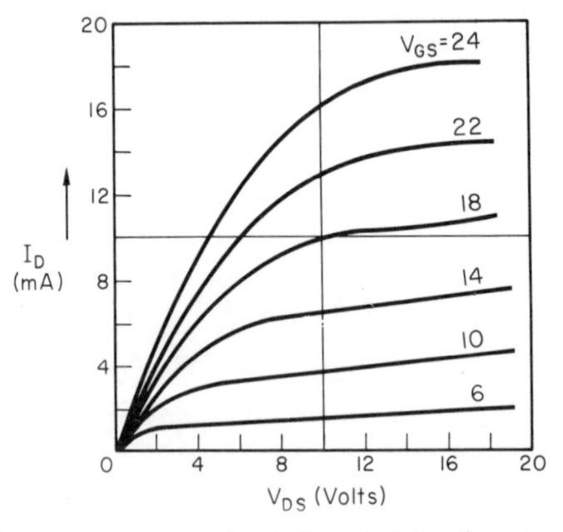

Fig. 3.19(c). Drain current versus voltage characteristics for a transistor of the type shown in (a). *The source drain spacing was* ~ 10μ *and the active width* ~ $120\ \mu$. *The maximum transductance was* ~ $100\ \mu$ *mho*[11]

which a planar geometry is achieved by cutting small depressions in the original substrate and filling with epitaxially grown silicon. The operating frequency of thin film transistors made by techniques similar to these have been reported as being as high as 4 GHz with a possible doubling of this being achieved in the near future.

3.5. The Shottky barrier gate FET

As already indicated, the high frequency performance of an FET is limited by the dimensions of the channel and the mobility of the carriers within the channel. A semiconductor which offers an electron mobility five times higher than silicon is gallium arsenide but it has proved very difficult to produce a conventional junction FET or MOSFET with this material despite the large research effort so far expended. However, an FET proposed by Mead[12] in 1966 shows considerable promise for use at microwave frequencies both as a discrete component and as part of an integrated circuit.

This is an *n*-channel junction gate FET having a gate formed by a metal-semiconductor Shottky barrier diode. A schematic diagram of an FET of this type is shown in Fig. 3.20 (after Hooper and Lehrer[13]).

The barrier gate is formed by a thin evaporated metal film which makes good contact with the *n*-type semiconducting GaAs. Using this technique, very much shorter gate lengths may be achieved than using conventional diffusion techniques. The source and drain contacts are also evaporated onto the top surface of the GaAs and

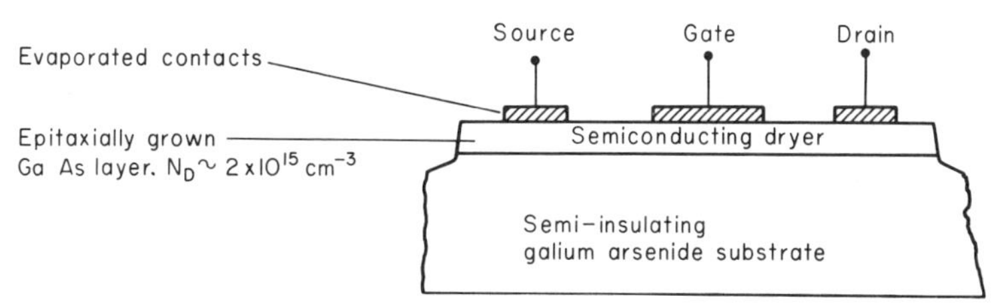

Fig. 3.20. Section through Schottky barrier gate FET

the geometry of the source, drain and gate regions fashioned using standard photolithographic techniques. The substrate for this device is made of semi-insulating GaAs, this has all the crystallographic properties of semiconducting GaAs but a resistivity of about 10^8 Ω cm. On top of this substrate a thin epitaxial layer of *n*-type GaAs is grown prior to the evaporation of the electrodes.

Fig. 3.21(*a*)[13] shows the source-drain characteristics of an FET of the type shown in Fig. 3.20 and Fig. 3.21(*b*) plots the transconductance as a function of gate voltage. The maximum transconductance at 2 GHz is ~20 millimhos, this is roughly a factor of 5 higher than would be attainable for a silicon FET of the same geometry.

An advantage of the Shottky barrier gate over a conventional junction FET is that the need to form a junction in the relatively wide gap semiconductor is removed with the result that the high reverse leakage current associated with the junctions in these materials is avoided. Compared with the MOST, the Shottky barrier gate offers the advantage that the bias dependence of the barrier depletion layer will not be affected by surface states so getting round the problem of gating devices in materials where conventional MOS techniques do not work. Also of course the instability and drift problems consequent upon using a thin oxide layer will be eliminated.

Because of the insulating properties of the GaAs substrate this forms an ideal substrate material for isolating one active device from another to form an integrated circuit containing GaAs FETs. A discrete FET capable in a modified form of being used in an integrated circuit is shown in the photograph of Fig. 3.22. In this device the channel length is ~3 μ and the channel to source/drain separation is also 3 μ. In order to minimise parasitic capacitances

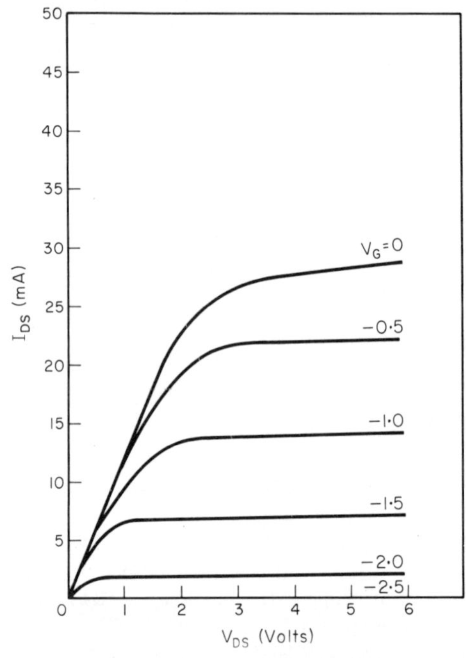

Fig. 3.21(a). Source–drain characteristics of Schottky Barrier gate GaAs FET. (After Hooper and Lehrer[13])

associated with the gate bonding pad of this FET it has been removed onto the semi-insulating substrate so that it then forms a negligible fraction of the total input capacitance. This particular Plessey prototype has a gain bandwidth product ($g_m/2\pi C_{in}$) of about 2·5 GHz. Fig. 3.23 shows the admittance parameters as a function of frequency for a device similar in construction to that of Fig. 3.22, but with a 12 μ channel length and an aspect ratio, gate width/length of 30:1; it is encapsulated in a T05 header which, of course, degrades the input capacity. This FET has a gain bandwidth product of 350 MHz and a power gain of 25 dB at 300 MHz. Work to produce a similar FET operating at frequencies up to 10 GHz using channel lengths ~2 μ is well advanced. This may be made

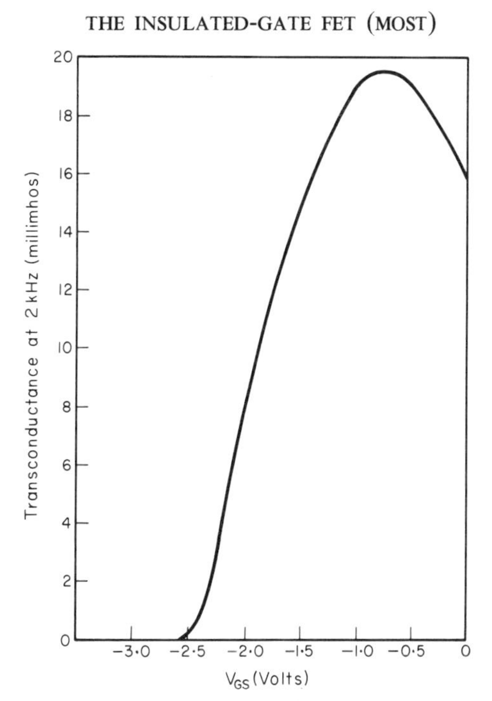

Fig. 3.21(b). Transconductance versus gate voltage curve at 2 kHz with $V_{DS} = +3$ *V for the same GaAs transistor* (After Hooper and Lehrer[13])

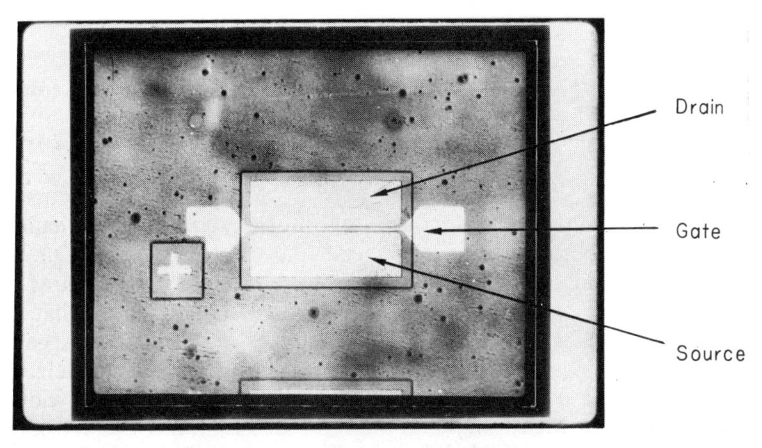

Fig. 3.22. Chip photograph of Plessey prototype GaAs FET having a gain bandwidth product of ~ 2·5 GHz. (Courtesy The Plessey Co.)

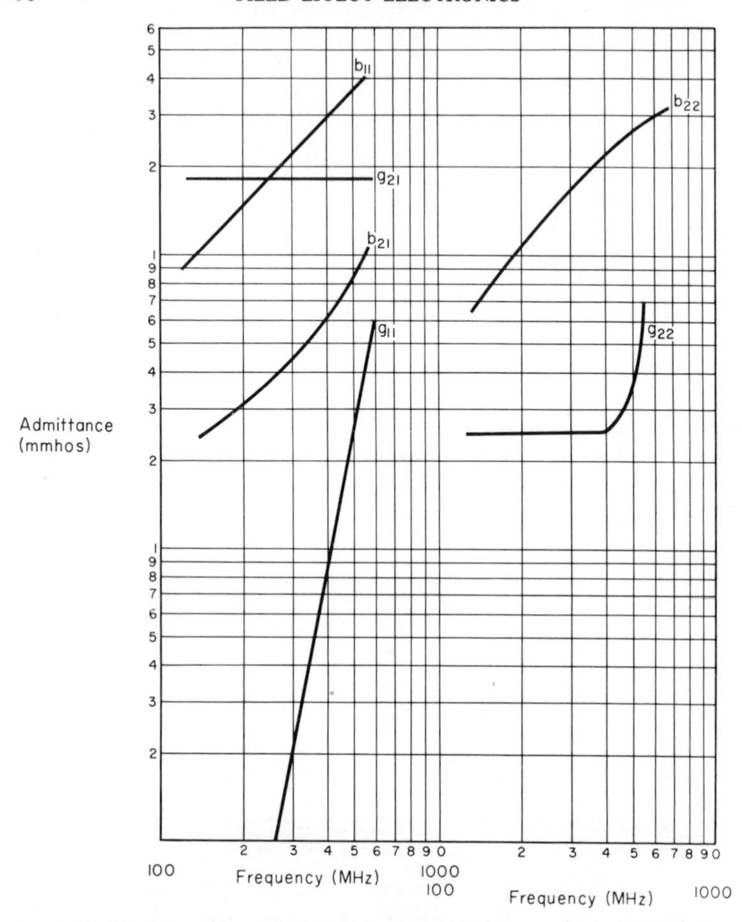

Fig. 3.23. CS admittance parameters for a prototype Plessey GaAs FET similar in construction to that shown in Fig. 3.22

using either conventional photolithographic techniques or by using an electron beam process to expose the photoresist; this latter process should lead to devices having closer tolerances and still shorter channel lengths.

A silicon Schottky barrier FET reported by a number of I.B.M. workers[14] uses a high resistivity *n*-type substrate (10–50 kΩ-cm) with a thin 0·2–0·5 *n*-type epitaxial layer of resistivity 0·05–0·2 Ω-cm grown on. The device shown in plan in Fig. 3.24(*a*) has a gate width of only 1 μ which typically gives a value of $g_{fs} \sim 42$ mA/V/mm gate length. The performance of the device is limited by the bonding pad

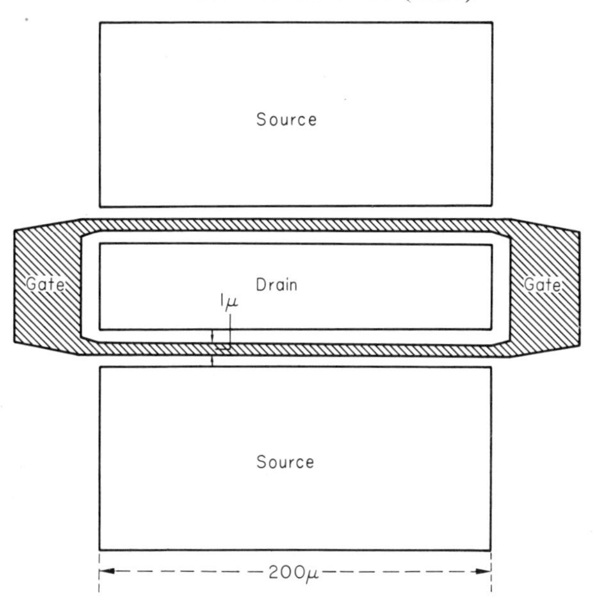

Fig. 3.24(a). Schematic diagram of a silicon Schottky gate FET. The gate is formed by successive evaporation of layers of Au, Cr and Nl (After Drangeid et al.[14])

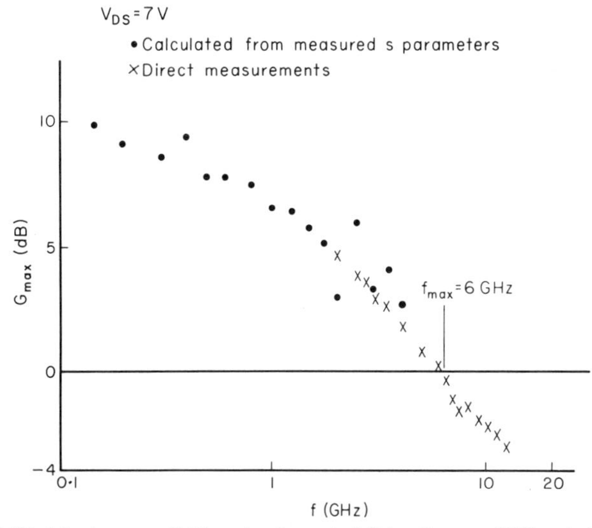

Fig. 3.24(b). Maximum available gain of a typical Schottky gate FET as in (a). (After Drangeid et al. [14])

arrangements but a maximum frequency of operation of 8 GHz has been achieved with a G_{max} of 5 dB at 4 GHz. Fig. 3.24(*b*) shows a curve of maximum available gain G_{max} vs. frequency for a transistor of this type. Considerable improvement in performance should follow if the contact and bonding problems can be overcome.

3.6. Silicon gate transistors

One of the important factors limiting the speed of conventional MOS transistors is the capacitance that exists between the gate and the source and drain. This is caused by the overlap of the gate with the source and drain diffused regions that is essential if misalignment is not to occur and give rise to a portion of the channel not covered by the gate (for example, see Fig. 3.8). During recent times a great deal of work has been done to develop auto-registration techniques which enable the source, drain and gate to be exactly located relative to each other without the need for overlap. The most important of these techniques involves laying down a layer of polycrystalline silicon on top of the gate oxide layer giving rise to what is commonly called the silicon gate MOS transistor[15]. This will now be described and its advantages reviewed.

Fig. 3.25 shows a series of stages in the making of the silicon gate device. After the gate oxide has been grown, polycrystalline silicon is deposited over the entire slice. By a conventional photolithographic stage openings are cut in the silicon and SiO_2 to allow the diffusion of the source and drain areas. The source and drain are then diffused and the polycrystalline silicon is converted into a low resistivity *p*-type layer. Because the SiO_2 is protected by the silicon layer no degradation of the oxide occurs during this high temperature stage and since the diffusion is a short one there is little lateral diffusion of the *p*-type impurity to give rise to any gate capacitance. The final stages of preparation are the growth or deposition of SiO_2 (or possibly Si_3N_4) over the entire slice and the making of aluminium contacts to the source, drain and gate.

There are a number of other benefits that this process brings besides an increase in speed of at least a factor of two. First, there is an increased packing density of devices on the clip since there is no longer a need to allow an overlap at the source and drain. This means shorter channels can be used without having to increase the quality of the masks and alignment equipment. Secondly due to the replacement of the aluminium contact by one of low resistivity silicon there is a reduction in the threshold voltage of the MOST. This is due to the lower work function of the silicon and the fact that the fixed surface state charge Q_{ss} tends to be reduced. In practice values of

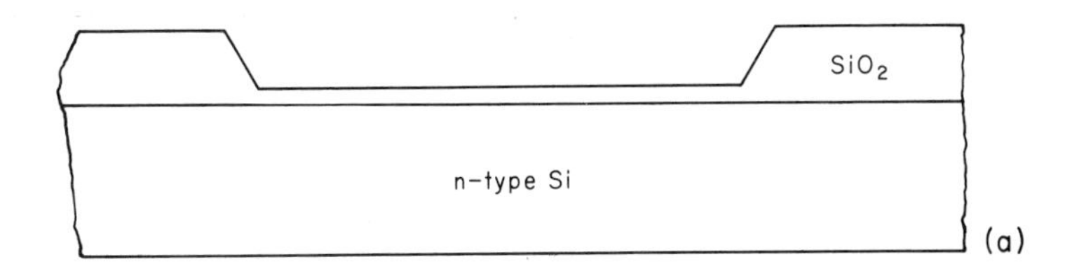

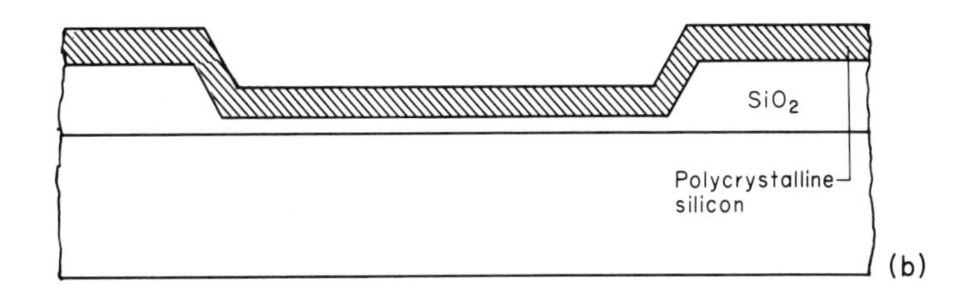

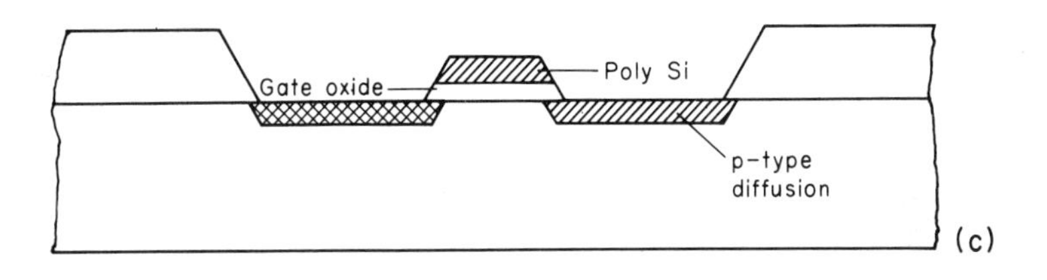

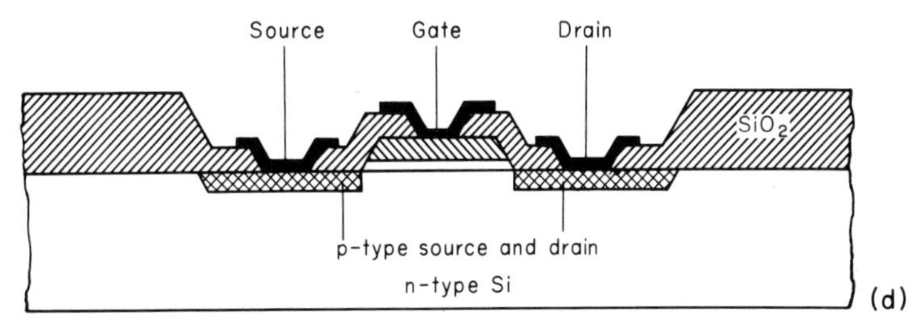

Fig. 3.25. Stages in the making of a silicon gate MOS transistor

(a) n-*type slice with gate oxide.*

(b) *Polyscrystalline silicon layer grown over complete slice.*

(c) *Source and drain diffusions completed after first openings in silicon and* SiO_2 *layers.*

(d) *Completed device passivated with a further layer of* SiO_2 *and with contacts made to sound drain and gate*

V_{th} at least 1 volt lower than that of conventional metal gate devices are found. Typically $V_{th} \simeq 1{\cdot}5$ V for a silicon gate device; this compares favourably with devices using SiO_2 and Si_3N_4 layers. It has been found that there is no sacrifice in reliability in using this process, accelerated life tests indicate very low drifts in the characteristics.

A further possible advantage of the silicon gate process is that since the silicon film that allows the gate oxide to be heated to high temperature without its properties being adversely affected the possibility exists for incorporating bipolar transistors into integrated circuits containing MOSTs. These bipolar transistors having shallow junctions would be formed after the MOS transistors were completed. This opens up exciting possibilities for a new hybrid technology to exploit the high speed of bipolar circuits and the high packing densities of the MOS technology.

REFERENCES

1. Hofstein, S. R. and Heiman, F. P. 'The silicon insulated-gate field-effect transistor', *Proc. I.E.E.E.*, **51**, 1190–1202 (September 1963).
   Storm, H. F. 'Field-effect transistor (FET) Bibliography', *I.E.E.E. Trans. on Electron Devices, E-D 14*, 710–7 (October 1967).
2. Wallmark, J. T. and Johnson, H. *Field Effect Transistors*', Prentice Hall (1966).
3. Candler, D. B. and Jordan, A. G. 'A small-signal, high-frequency analysis of the insulated-gate field-effect transistor', *International Journal of Electronics*, **19**, 181–196 (August 1965).
4. Sevin, L. J. *Field Effect Transistors*, New York, McGraw-Hill (1965).
5. Johnson, E. O. and Rose, A. 'Simple general analysis of amplifier devices', *Proc. I.R.E.*, **47**, 407–18 (March 1959).
6. Crawford, R. H. *MOSFET in Circuit Design*, New York, McGraw-Hill (1967).
7. Scott, J. and Ohmstead, J. 'A solid to solid diffusion technique', *R.C.A. Review*, **26**, 357–68 (September 1965).
8. White, M. H. and Cricchi, J. R. 'Complementary MOS transistors', *Solid State Electron.*, **9**, 991–1008 (October 1966).
9. Weimer, P. K. , Shallcross, F. V. and Borkan, H. 'Coplanar electrode insulated-gate thin-film transistors', *R.C.A. Review*, **24**, 661–675 (December 1963).
   Weimer, P. K. 'A *p*-type tellurium thin-film transistor', *Proc. I.E.E.E.*, **52**, 608–9 (May 1964).
10. Weimer, P. K., Borkan, H., Sadisiv, G., Meray-Hovath, L. and Shallcross, F. V. 'Integrated circuits incorporating thin-film active and passive elements', *Proc. I.E.E.E.*, **52**, 1479–86 (December 1964).
11. Mueller, C. W. and Robinson, P. H. 'Grown film silicon transistors on sapphire', *Proc. I.E.E.E.*, **52**, 1487–90 (December 1964).
12. Mead, G. A. 'Schottky barrier-gate field-effect transistor', *Proc. I.E.E.E.*, **54**, 307–8 (February 1966).
13. Hooper, W. W. and Lehrer, W. I. 'An epitaxial GaAs field-effect transistor', *Proc. I.E.E.E.*, **55**, 1237–8 (July 1967).
14. Drangeid, K. E., *et al.* 'Microwave silicon Schottky-barrier field-effect transistor', *Electronics Letters*, **4**, 362–3 (August 1968).
15. Faggin, E. and Klein, T. 'A faster generation of MOS devices', *Electronics*, **42**, 88–94 (September 1968).

4

NOISE IN FIELD EFFECT DEVICES

4.1. Introduction

An FET of either the junction or insulated gate type will generate noise in a manner similar to that found in any other electrical component.

Before considering the extent of the noise generated in FETs by the various noise generating mechanisms, the methods by which the noise properties of any device or circuit can be characterised will be studied. This will be done by taking as an example the noise in a linear amplifier.

4.1.1 SIGNAL-TO-NOISE RATIO AND NOISE FIGURE

Consider a signal applied to the input of a noisy infinite impedance linear amplifier of voltage gain A from a source having a purely resistive impedance R_G, Fig. 4.1(*a*). The signal will always be accompanied by some noise, the extent of which is defined as the *input signal-to-noise ratio.* As the amplifier also is noisy the output signal will contain a larger noise component so that the output signal to noise ratio will be smaller than that of the input. The two S/N ratios may be compared and used to define the *noise figure NF* of the amplifier

$$NF = \frac{\text{Input signal power/input noise power}}{\text{output signal power/output noise power}} \tag{4.1}$$

Because both the input signal power and the input noise power are dissipated within the same external input resistance R_G and the output signal power and noise power within the load resistance R_L, the power ratios may be replaced by voltage-squared ratios. Note that although the mean values of the noise components are zero the mean square values are finite: it is convenient therefore to rewrite equation (4.1) in terms of mean square voltages:

$$NF = \frac{\overline{v^2_{S(in)}}/\overline{v^2_{N(in)}}}{\overline{v^2_{S(out)}}/\overline{v^2_{N(out)}}} \tag{4.2}$$

where $\overline{v^2_{S(in)}}$ and $\overline{v^2_{S(out)}}$ are the mean square values of input and output signal voltages respectively and $\overline{v^2_{N(in)}}$ and $\overline{v^2_{N(out)}}$ are the mean square values of input and output noise voltages respectively.

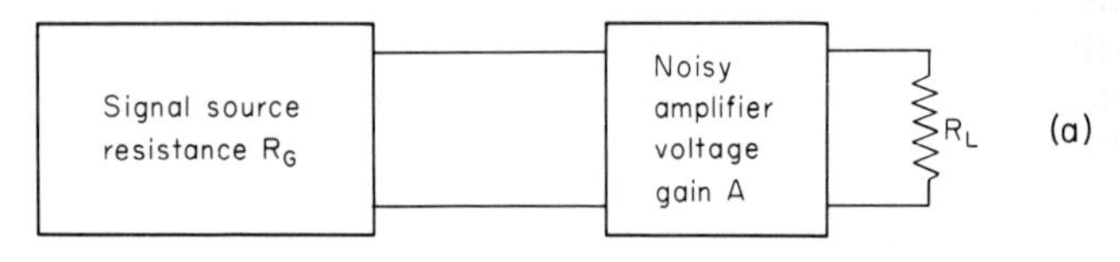

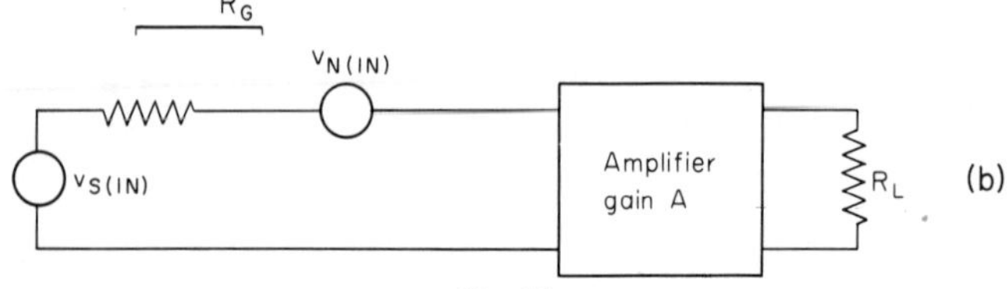

Fig. 4.1.
(a). Two port representation of a noisy amplifier supplied from a signal source of resistive impedance R_G
(b). Equivalent circuit of (a). The noise figure for this amplifier is:

$$NF = 1 + \frac{\overline{V_N^2}}{4A^2R_GkT\Delta f}$$

where $\overline{V_N^2}$ is the noise generated within the amplifier

Since the ratio $\overline{v^2_{S(out)}}/\overline{v^2_{S(in)}}$ is the voltage gain of the amplifier, equation (4.2) may be rewritten

$$NF = \overline{v^2_{N(out)}}/A^2\overline{v^2_{N(in)}} \tag{4.3}$$

The term $\overline{v^2_{N(in)}}$ will be the Johnson noise generated in the resistance R_G which has a mean square value

$$\overline{v^2_{N(in)}} = 4R_GkT\,\Delta f \tag{4.4}$$

where k is Boltzmann's constant, T is the absolute temperature and Δf is the noise bandwidth. See the equivalent circuit of Fig. 4.1(*b*).

The noise at the output will consist of the sum of the amplified Johnson noise and the noise generated within the amplifier, say v_N.

Thus, since there is no correlation between the two noise generators the output noise is

$$\overline{v_{\mathrm{N(out)}}^2} = A^2 4 R_G kT\,\Delta f + \overline{v_{\mathrm{N}}^2} \tag{4.5}$$

using equations (4.4) and (4.5) and substituting in equation (4.3) the noise factor becomes:

$$NF = 1 + \frac{\overline{v_{\mathrm{N}}^2}}{4A^2 R_G kT\,\Delta f} \tag{4.6}$$

Noise factor is often defined specifically as this ratio expressed in dBs. If it is not expressed in dBs it is then referred to as the noise figure.

In equation (4.5) the only quantity not known is v_{N}, the noise generated within the amplifier. As will be discussed later in this

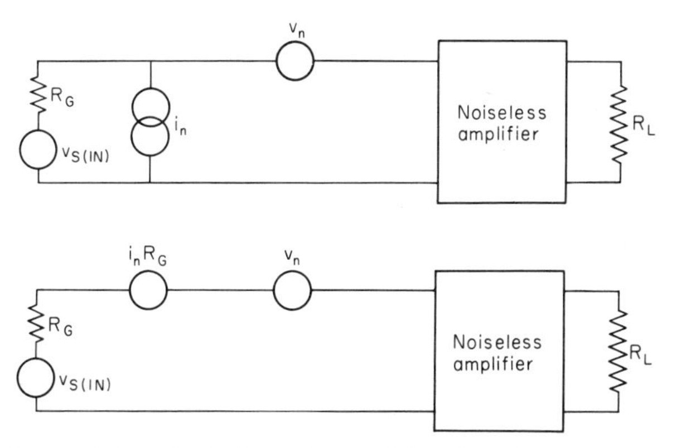

Fig. 4.2. Two equivalent circuits for a noisy amplifier using voltage and current noise generators at the input to represent the noise characteristics of the amplifier

chapter several phenomena contribute to v_{N} which make it frequency dependent. One convenient method of describing v_{N} is to replace the noisy amplifier in Fig. 4.1 by a hypothetical noiseless amplifier preceded by two equivalent noise generators, one a voltage generator v_n and the other a current generator i_n. This is shown in Fig. 4.2. For this circuit:

$$v_N = A(v_n + i_n R_G) \tag{4.7}$$

since for an FET amplifier $R_G \ll R_{\mathrm{in}}$.

Rewriting in terms of mean square quantities in place of the instantaneous values of equation (4.7):

$$\overline{v_N^2} = A^2\overline{(v_n + i_n R_G)^2} = A^2(\overline{v_n^2} + \overline{i^2}R_G + 2\gamma R_G(\overline{v_n^2}\,\overline{i_n^2})^{\frac{1}{2}}) \tag{4.8}$$

where γ is a correlation coefficient giving a measure of the correlation between the two noise sources v_n and i_n. For complete correlation $\gamma = 1$ and for zero correlation $\gamma = 0$.

Using (4.8) to substitute for $\overline{v_N^2}$ in (4.6) gives an expression for noise figure in terms of v_n and i_n which is:

$$NF = 1 + \frac{\overline{v_n^2}}{4R_G kT\,\Delta f} + \frac{\overline{i_n^2} R_G}{4kT\,\Delta f} + \frac{2\gamma(\overline{v_n^2}\,\overline{i_n^2})^{\frac{1}{2}}}{4kT\,\Delta f} \tag{4.9}$$

The value of F in this equation is independent of the input resistance of the amplifier if it may be assumed that v_n and i_n represent *all* the noise sources within the amplifier.

The units used to indicate the strength of the noise signals can be simplified if the bandwidth over which each noise generator is measured is incorporated into the measurement itself giving $\overline{v_n^2}/\text{Hz}$ and $\overline{i_n^2}/\text{Hz}$. If then a true r.m.s. instrument is used for the measurements the directly measured voltage and current divided by Δf gives units of voltage and current per root-Hertz or in practical terms $\mu\text{V}/\sqrt{\text{Hz}}$ and $\text{pA}/\sqrt{\text{Hz}}$ for the noise generator amplitudes. These are the units normally quoted on data sheets for FET noise generator values. The symbols $\overline{v_n^2}/\text{Hz}$ and $\overline{i_n^2}/\text{Hz}$ are normally replaced by the simpler form: $\bar{e}_n^2$ and $\bar{i}_n^2$ so that (4.9) is written:

$$NF = 1 + \frac{1}{4kT}\left(\frac{\bar{e}_n^2}{R_G} + \bar{i}_n^2 R_G + 2\gamma \bar{e}_n \bar{i}_n\right) \tag{4.10}$$

For FETs in general the two noise generators $\bar{e}_n$ and $\bar{i}_n$ are caused by different noise generating processes; typically $\bar{i}_n$ corresponds to gate current noise and $\bar{e}_n$ to channel and 1/f noise. Between these there is no correlation, so that $\gamma = 0$. NF is then:

$$NF = 1 + \frac{\bar{e}_n^2 + \bar{i}_n^2 R_G^2}{4kTR_s} \tag{4.11}$$

In this equation the relative values of $\bar{e}_n$ and $\bar{i}_n$ are such that if $R_s < 10\ \text{M}\Omega$, $\bar{i}_n R_G^2 \ll \bar{e}_n^2$. Hence the equation

$$NF \simeq 1 + \frac{\bar{e}_n^2}{4kT\,R_G} \tag{4.12}$$

can in most cases be used for relating noise figure to $\bar{e}_n$.

The disadvantage with considering the noise figure of an amplifier or FET is that it relates only to a circuit having a specific value of R_G. The values of $\bar{e}_n$ and $\bar{i}_n$ on the other hand apply specifically to the amplifier or FET itself and are independent of R_G. This gives rise to an alternative method of specifying the noise in terms of equivalent noise resistances.

4.1.2 NOISE RESISTANCE AND OPTIMUM NOISE FIGURE

Series and parallel effective noise resistances R_{Nv} and R_{Ni} can be defined using the relation:

$$R_{Nv} = \frac{\overline{v_n^2}}{4kT\,\Delta f}, \qquad R_{Ni} = \frac{4kT\,\Delta f}{\overline{i_n^2}} \tag{4.13}$$

so that equation (4.9) becomes:

$$NF = 1 + R_{Nv}/R_G + R_G/R_{Ni} + 2\gamma\sqrt{(R_{Nv}/R_{Ni})} \tag{4.14}$$

(The numerical relationships themselves will be dealt with in section 8.5.1).

From equation (4.14) or (4.10) it is easy to see how F varies with changing R_G since obviously F becomes large for either low or high values of R_G. Differentiating (4.14) gives the optimum value of R_G for which F is a minimum, that is, for which the degradation of the signal passing through the amplifier is a minimum.

$$(R_G)_{\text{opt}} = \sqrt{(R_{Nv}R_{Ni})} = \sqrt{\frac{\overline{v_n^2}}{\overline{i_n^2}}}$$

$$\text{or } (R_G)_{\text{opt}} = \sqrt{\frac{\overline{e_n^2}}{\overline{i_n^2}}} \tag{4.15}$$

where the minimum noise figure is:

$$NF_{\min} = 1 + 2(1+\gamma)\sqrt{(R_{Nv}/R_{Ni})} = 1 + \frac{2(1+\gamma)}{4kT\,\Delta f}\sqrt{(\overline{v_n^2}\,\overline{i_n^2})} \tag{4.16}$$

$$\text{or, if } \gamma = 0 \quad NF_{\min} = 1 + \frac{2}{4kT}\overline{e_n}\,\overline{i_n} \tag{4.17}$$

According to Faulkner[1], the condition $R_{Ni} \gg R_G \gg R_{Nv}$ is that which R_G has to satisfy if a good noise figure is to be obtained. Since also a good noise figure of 1 dB ($=1{\cdot}26$) would be hard to distinguish from the best possible value of 0 dB, if $\gamma = 0$, such a noise figure may be achieved if $R_{Ni} \approx 60\ R_{Nv}$. This condition is usually satisfied for both bipolar and junction FETs, so that the noise performance of one of these devices can be completely specified by giving the minimum noise figure $NF_{\min}$ and the optimum source resistance $(R_G)_{\text{opt}}$. Equation (4.14) can be written in terms of these two quantities to determine the noise figure for any other value of R_G

$$(NF - 1) = (NF_{\min} - 1)\left[\frac{(R_G)_{\text{opt}}}{R_G} + \frac{R_G}{(R_G)_{\text{opt}}}\right] \tag{4.18}$$

Hence, as Faulkner[1] also points out, the low noise capability of a circuit or transistor over a given frequency range should be assessed from the value of its minimum noise figure rather than from the value of F obtained for any chosen value of R_G.

Since in practice the values of R_{Ni} and R_{Nv} or $\overline{e_n}$ and $\overline{i_n}$ are known as a function of frequency it is possible to evaluate the noise figure for any given value of generator resistance.

Fig. 4.3 illustrates typical noise voltage vs. frequency curves for two junction FETs.

The three processes which commonly give rise to noise in

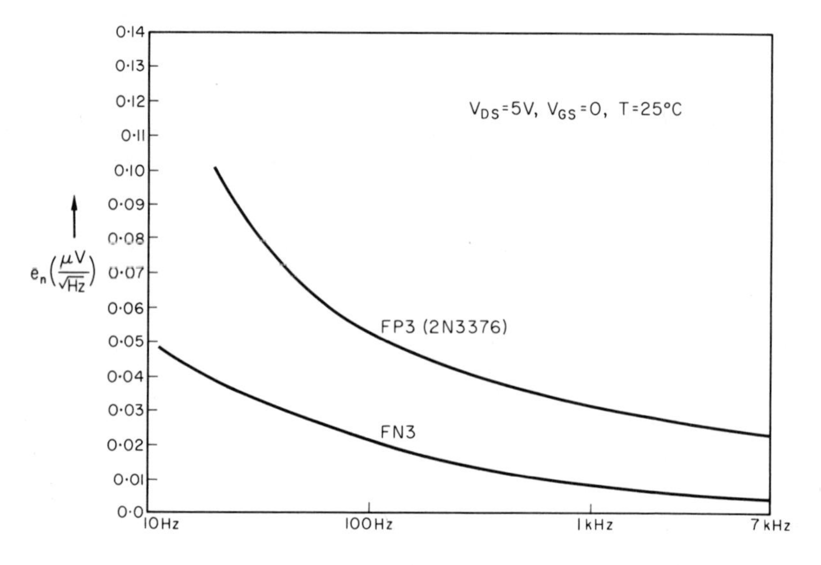

Fig. 4.3. FET noise voltage curves for Union Carbide p- *and* n-*channel junction FETs. Types FP3 and FN3*

electronic devices will now be considered and their relevance to junction and insulated gate FETs considered.

4.1.3 THERMAL NOISE

Thermal noise, sometimes also called Johnson noise is caused by the random thermal velocities of carriers. The motions of the carriers between collisions constitute currents and these will cause a fluctuating voltage to appear across the ends of any resistance. The average value of this voltage, of course, is zero but the mean square value for a noise bandwidth Δf for resistance R is

$$\overline{V^2} = 4kT\Delta f . R \tag{4.19}$$

This source of noise will be present for good and poor conductors in contrast to the next noise mechanism, shot noise, which only manifests itself when the conductor is carrying current.

4.1.4 SHOT NOISE

This is caused by the discrete nature of the charge carriers which means that any so called direct current flowing in a semiconductor is made up of carriers that are continually recombining and dissociating. This means that instead of a perfectly smooth flow of current through the device the carriers tend to move in a more jerky fashion. The mean square fluctuation in current for a noise bandwidth Δf is given by

$$\overline{i^2} = 2qI\ \Delta f \tag{4.20}$$

where I is the direct current flowing and q is the carrier charge.

It is found that both thermal and shot noise are essentially white, that is, the power spectrum of the noise is uniform over the range of frequencies normally of interest for the operation of FETs.

4.1.5 SURFACE OR $1/f$ NOISE

Within all types of semiconductor device there exists a source of noise which has a power spectrum that is inversely proportional to frequency and which extends over an unlimited bandwidth. This is known as $1/f$ noise, it is caused by the presence in the devices, particularly at the surface, of various forms of crystalline defect. Because of its frequency characteristic, at all but the lowest frequencies, thermally generated white noise tends to mask the presence of $1/f$ noise. In MOSTs as we shall see, because of the important role played by the silicon-oxide interface in the operation of the device, $1/f$ noise tends to be important over a wider frequency range than in planar bipolar and junction field-effect devices. This is because in these devices the surfaces are deliberately passivated with SiO_2 to reduce the role played by surface states, and the active region of the devices are well removed from the surface.

Since the time average noise power at frequency f can be written for this type of noise as $\overline{P}(f) = kf^{-1}$ where k is a constant; considering the total noise power P over a decade of frequency from f_0 to $10f_0$:

$$P = \int_{f_0}^{10f_0} \overline{P}(f)\,\mathrm{d}f = k\ln 10. \tag{4.21}$$

Provided then that the $1/f$ law is assumed to be valid over the range of frequencies considered, the noise power steadily increases with decreasing frequency at a rate 3 dBs per octave, having equal power for each decade of frequency. As the other sources of noise give outputs proportional to bandwidth they become insignificant compared with $1/f$ noise at the lowest frequencies.

4.2. Noise mechanisms in junction FETs

The principle sources of noise in junction FETs are thermal noise generated in the channel and in the bulk resistances at source and drain, and noise caused by the capacitive coupling between channel and gate, giving rise to a capacitive noise current i_g flowing in the short circuited gate. This lätter effect is most significant at higher frequencies whereas thermal noise generation is dominant at lower frequencies.

Considering first thermal noise generation, van der Ziel[2] has shown that the noise generated in the channel may be represented by a current generator $\sqrt{(\overline{i_d^2})}$ where

$$\overline{i_d^2} = 4kTg_{fs0}\,\Delta f Q(x, y) \tag{4.22}$$

where the function $Q(x, y)$ has the value 1·00 when V_{DS} is zero and is always $>$ 0·5 for other bias conditions. Although the theory presented by Van der Ziel[2] does not hold beyond saturation, it is found experimentally that the current generator given by (4.22) does give a good approximation for the noise generated provided the field strength in the cut-off part of the channel is not too large and $Q(x, y)$ is $\approx$ 0·7.

Thinking in terms of effective noise resistance a series noise resistance R_{Nv} can be defined by writing

$$\overline{i_d^2} = 4kTR_{Nv}\,\Delta f g_{fs}^2 \tag{4.23}$$

This gives a value of

$$R_{Nv} = \frac{g_{fs0}}{g_{fs}^2} Q(x, y) \tag{4.24}$$

and this in the saturation region of the characteristic becomes

$$R_{Nv} \approx \frac{0 \cdot 7}{g_{fs0}} \tag{4.25}$$

At frequencies below 10 kHz there may be excess noise due to the presence of $1/f$ noise. This is caused by the fluctuation of charge in the recombination–generation centres of the depletion regions of

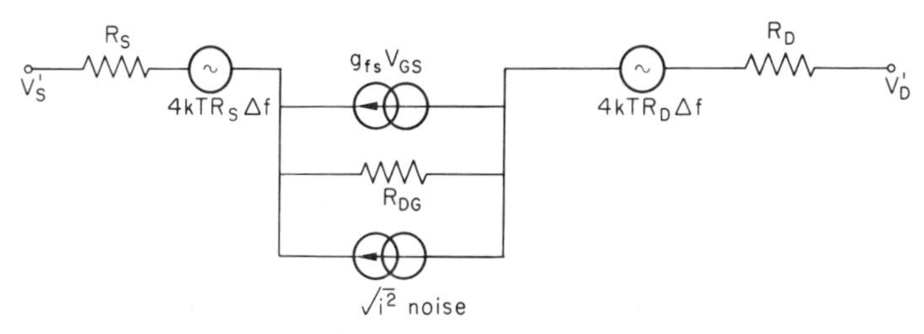

Fig. 4.4. A low frequency equivalent circuit of an FET taking thermal noise generated in the channel (Vi_2) *and at source and drain into account*

the device. In contrast to the MOST, $1/f$ noise can usually be neglected for present-day junction FETs at frequencies above about 100 Hz.

The thermal noise at the source and drain series resistances can be represented by voltage generators of the form

$$\left.\begin{aligned} \overline{v_S^2} &= 4kTR_S\Delta f \quad \text{at the source} \\ \overline{v_D^2} &= 4kTR_D\Delta f \quad \text{at the drain} \end{aligned}\right\} \tag{4.26}$$

A low frequency equivalent circuit of the FET taking thermal noise into account is shown in Fig. 4.4.

Considering next noise caused by the gate current, there will be shot noise contributed by the flow of electrons and holes comprising the gate current I_G. The noise current is given by Van der Ziel[2] as

$$\overline{i_g^2} = 2qI_G\Delta f \tag{4.27}$$

In an equivalent circuit this could be represented as a current generator between gate and source. Alternatively it may be looked on as a parallel noise resistance R_{Ni} at the input having a value (from 4.14).

$$R_{Ni} = 2kT/qI_G \tag{4.28}$$

Although this source of noise is predominant in the gate circuit at low frequencies, for which the theory of Van der Ziel specifically applies, as f is raised, because of the capacitative coupling, gate and channel thermal noise in the channel will increasingly affect the gate circuit. Van der Ziel and Ero[3] have shown that the noise current due to this cause at high frequencies is

$$\overline{i_g^2} = 4kTg_{11}\Delta f R(V_{GS}, V_{DS}) \tag{4.29}$$

where g_{11} is the input conductance of the FET and $R(V_{GS}, V_{DS}) = 1$ for $V_{GS} = 0$, and $R(V_{GS}, V_{DS})$ is slightly greater than unity for V_{GS}

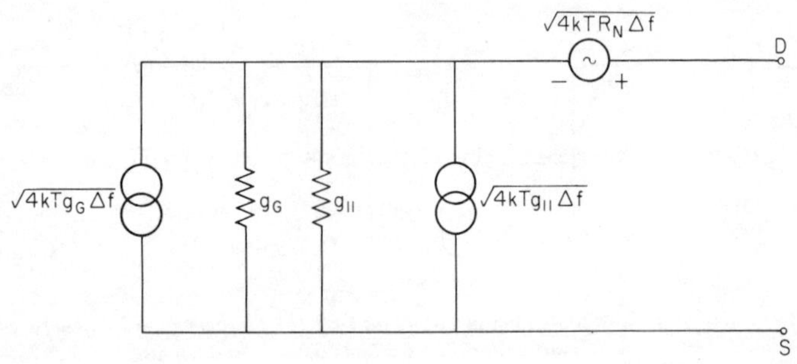

Fig. 4.5. Simple noise equivalent circuit of an FET according to Bruncke and van der Ziel[4]

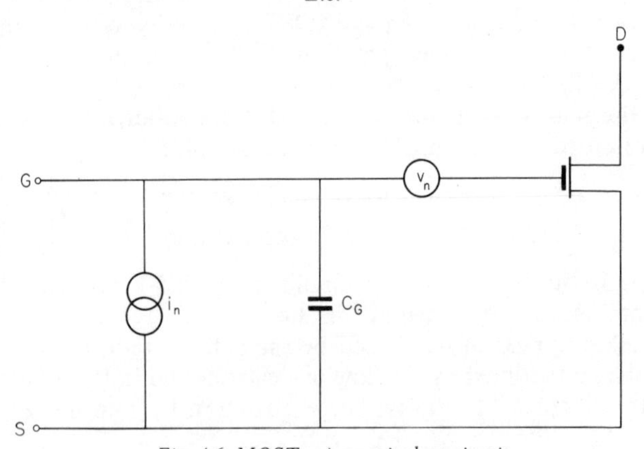

Fig. 4.6. MOST noise equivalent circuit

negative. This means that the noise generated can be approximated by that thermally generated in the full input conductance g_{11}. Experiments by Bruncke and Van der Ziel[4] have showed that this noise is partly correlated with the channel noise. Neglecting this correlation they showed that the noise equivalent circuit of the FET may be drawn as in Fig. 4.5 where for all practical purposes the device can be represented by the thermal noise of the input conductance g_{11} of the device and by an EMF $\sqrt{(4kTR_N\Delta f)}$ in series with the gate. Since for a frequency range of practical importance g_{11} varies as ω^2 and R_N the equivalent noise resistance is practically independent of frequency this makes the prediction of the noise performance of junction FETs at elevated frequencies relatively straightforward.

The noise figure of the equivalent circuit of Fig. 4.6 may be written as

$$NF = 1 + \frac{g_{11}}{g_G} + \frac{R_N}{g_G}(g_G + g_{11})^2 \tag{4.30}$$

($g_G = 1/R_G$ is the generator conductance) which gives a minimum value

$$NF_{\min} = 1 + 2R_N g_{11} + 2\sqrt{(R_N g_{11} + R_N^2 g_{11}^2)} \tag{4.31}$$

for an optimum value of generator conductance,

$$(g_G)_{\text{opt}} = \sqrt{\left(\frac{g_{11}}{R_N} + g_{11}^2\right)} \tag{4.32}$$

At frequencies where $R_N g_{11} \ll 1$ (4.31) and (4.32) become

$$\left.\begin{aligned} NF_{\min} &= 1 + 2\sqrt{(R_N g_{11})} \\ \text{for } g_G &= \sqrt{(g_{11}/R_N)} \end{aligned}\right\} \tag{4.33}$$

so that $NF_{\min}$ is comparatively small. As the frequency is increased $NF_{\min}$ rises because of the increase in g_{11}, however, over the working range of most FETs reasonably low noise figures may be obtained.

4.3. Noise mechanism in MOSTs

In contrast to the junction FET, low frequency noise generation mechanisms are of particular interest in the MOST when it is to be used as an amplifier and its extremely high input impedance exploited for applications such as electrometers or radiation detectors. The noise at low frequencies is primarily $1/f$ noise, this will tend to dominate thermal and shot noise at frequencies less than say 20 kHz in a typical MOST.

4.3.1 $1/f$ NOISE

Consider the origins of $1/f$ noise in the MOST. The charge on the gate electrode induces an equal charge of opposite polarity near the surface of the silicon substrate which can be considered to be divided between the charges in the surface states at the Si–SiO_2

interface, the carriers in the inverted channel and the ions in the depletion region adjacent to it. These charges which are all in a state of dynamic equilibrium will be subject to random fluctuations which can give rise to two separate effects leading to noise.

First fluctuations in the total surface charge as carriers move to and from the bulk of the semiconductor giving rise to similar fluctuations in gate charge and consequently a noise current in the gate circuit. Variations of charge between the depletion region, the inversion layer and the surface states can alter the instantaneous value of V_{GS} and so effect the channel conductance. This effect unlike the generation of thermal noise to be considered later, produces output noise independent of source impedance; it can be represented as a voltage generator in series with the gate electrode as in Fig. 4.6. Consider the movement of electrons and holes between the surface states and the bulk of the semiconductor; it is possible to think of these movements as having time constants.

$$\tau_n = C_t R_n, \qquad \tau_p = C_t R_p \tag{4.34}$$

where C_t is the capacitance/unit area associated with the depletion layer and R_n and R_p are resistances (Ω/cm^2) associated with the flow of carriers, both electrons and holes, from the bulk to the surface states. In practice for the p-channel MOST τ_p is extremely small $\sim 10^{-9}$ seconds and is of no importance in this case, the value of τ_n, for electrons however is $\sim \frac{1}{10}$ second typically. τ_n will vary with the bias voltage between gate and channel and consequently with position along the channel. The continuous distribution of time constants obtained in this way along the channel gives rise to the $1/f$ form of noise spectrum associated with the movement of carriers in and out of the surface states[5, 6].

From the circuit of Fig. 4.6 the noise figure is given by[5]

$$NF = 1 + \frac{\overline{i_n^2} R_G^2 + \overline{v_n^2}(1 + 4\pi^2 f^2 C_G^2 R_G^2)}{4kTR_G} \tag{4.35}$$

where C_G is the total input capacitance of the MOST.

If $\overline{v_n^2} = \dfrac{\text{constant}}{f}$ and $i_n^2 R_G^2$ is neglected

$$NF = 1 + \frac{\text{constant}}{4kTR_G}\left\{\frac{1}{f} + 4\pi^2 f C_G^2 R_G^2\right\} \tag{4.36}$$

which gives a minimum value at

$$f = \frac{1}{2\pi C_G R_G} \tag{4.37}$$

Experimentally measured noise figures fit the above equations well over a range of frequency 12 Hz–20 000 Hz[5]

Measurements with high source resistances ~1000 MΩ show that i_n is barely detectable and does not make a significant contribution to the total noise. Values of $\overline{i_n}$ measured with $R_G = 1000$ MΩ on a Plessey MTO-1 p-channel enhancement MOST gave a value

$\overline{i_n} \approx 4{\cdot}5 \times 10^{-15}$ A/$\sqrt{}$Hz at low frequencies (12–60 Hz)

$\overline{v_n}$ was typically 0·2 μ$V\sqrt{}$Hz at 1000 Hz for the same device.

4.3.2 THERMAL NOISE

An equivalent circuit for the FET discussed by Johnson[7] is shown in Fig. 4.7. In this circuit the principal contribution to the generator $\overline{i_d^2}$ will be thermally generated noise in the channel caused by

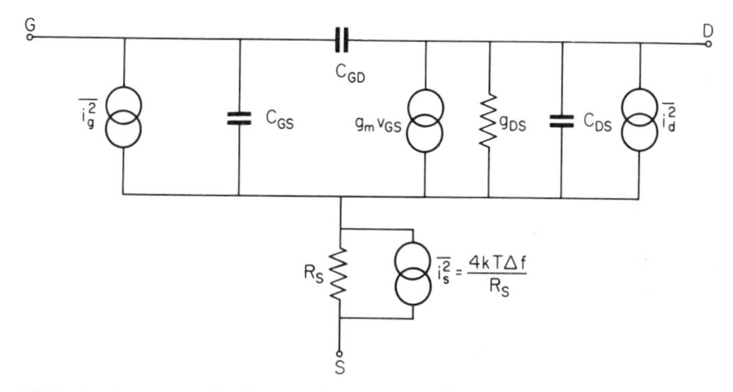

Fig. 4.7. Equivalent circuit for the FET due to Johnson[7] showing various noise generators

thermal fluctuations. The major contribution to $\overline{i_g^2}$ on the other hand will be $1/f$ noise at frequencies less than say 100 kHz, above this frequency thermal fluctuations in the channel coupled to the gate circuit via the gate-channel capacitance will become the dominant source of noise. As in the case of junction FETs at high frequencies a correlation exists between $\overline{i_d^2}$ and $\overline{i_g^2}$ which tends to minimise the total noise figure.

Thermal noise in MOSTs has been considered by many authors at both medium and high frequencies, the following is a summary of the most significant conclusions.

Using a gradual channel approximation Jordan and Jordan[6] considered the effects of thermal fluctuations in the channel by evaluating the noise generated in an elemental slice of channel perpendicular to the current flow and integrating over the channel length. In this case

$$\overline{i_n^2} = 4kT\Delta f \frac{\mu\varepsilon_{ox}}{LT} |V_{GS} - V_{th}| H(u) \tag{4.38}$$

where μ is surface mobility, L the channel length, ε_{ox} the permittivity of the oxide

$$u = \frac{|V_{DS}|}{|V_{GS} - V_{th}|} \quad \text{and} \quad H(u) = \frac{2}{3}\left[\frac{1-(1-u)^3}{u(2-u)}\right] \tag{4.39}$$

A plot $H(u)$ as a function of u as in Fig. 4.8 shows that for the range $0 < u < 1$ $H(u)$ varies between 1 and $\frac{2}{3}$ so that the thermal noise varies by the same amount in the range

$$0 < |V_{DS}| < |V_{GS} - V_{th}|$$

that is in the range from zero drain voltage to saturation of the drain characteristic (see Chapter 3). Although strictly speaking the equation only holds for the non-saturation region of the characteristics only

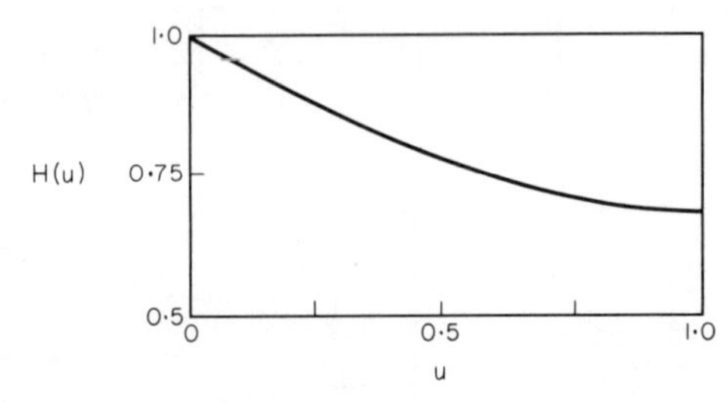

Fig. 4.8. Plot of the function H(u) *which appears in expression* (4.38) *for thermal noise in a MOST*[6] ($u = |V_{DS}|/|V_{GS} - V_{th}|$)

a small error is incurred if the noise current is assumed to remain constant at its value for $u = 1$ within the saturation region of the drain characteristics.

At saturation ($u = 1$) therefore:

$$\overline{i_d^2} = \tfrac{2}{3} . 4kT\Delta f\, g_{fs0} \tag{4.40}$$

and at $V_d = 0$[7]

$$\overline{i_d^2} = 4kT\,\Delta f\, g_{fs0}. \tag{4.41}$$

These results are similar to those of Shoji[8] who analysed the MOST by using a transmission line model for the channel.

Considering now the effect of thermal noise generation on the gate current, the noise generator $\overline{i_g^2}$ will be partly correlated with $\overline{i_d^2}$ at high frequencies because of the gate-channel capacitance so that the total output noise will not simply be the sum of the effects due to

$\overline{i_d^2}$ and $\overline{i_g^2}$. Johnson[7] has evaluated the contribution of thermal noise to the value of $\overline{i_g^2}$ by calculating the change in the charges on the gate to channel capacitor when there is a small voltage perturbation at a point within the channel

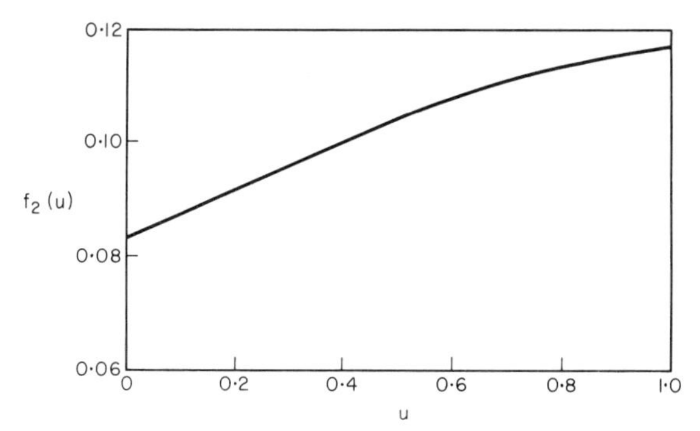

Fig. 4.9. Plot of the function f_2(u) *which appears in expression* (4.42) *for charge fluctuations occurring at the gate*[7]

The mean square charge fluctuation at the gate is shown to be

$$\overline{q^2} = 4kT\,\Delta f \frac{C^2}{g_{fs}} . f_2(u) \tag{4.42}$$

where $C = \varepsilon_{ox}\, L/T_{ox}$ is the gate to channel capacitance/unit width and $f_2(u)$ is a function plotted in Fig. 4.9. Since at saturation

$$u = |V_{DS}|/|V_{GS} - V_{th}| = 1 \tag{4.43}$$

$\overline{q^2}$ becomes

$$\overline{q^2} = 0{\cdot}12 \times 4kT\,\Delta f \frac{C^2}{g_{fs}} \tag{4.44}$$

and

$$\overline{i_g^2} = 0{\cdot}12 \times 4kT\,\Delta f \frac{\omega^2 C^2}{g_{fs}} \tag{4.45}$$

If $V_{DS} = 0$ then $u = 0$ and

$$\overline{i_g^2} = \frac{1}{12} \times 4kT\,\Delta f \frac{\omega^2 C^2}{g_{fs0}} \tag{4.46}$$

This may be compared with a value of

$$\overline{i_g^2} = \frac{64}{135} kT\,\Delta f \frac{\omega^2 C^2}{g_{fs0}} \tag{4.47}$$

obtained by Shoji[8] using a transmission line model of the FET at high frequencies.

The correlation factor for the drain and gate circuit noise generators is given by[8]:

$$\overline{i_g^* i_d} = \frac{4}{9} j\omega kTC \, . \Delta f \qquad (4.48)$$

for the saturation region.

A complex correlation coefficient

$$C = \frac{\overline{i_g^* i_d}}{\sqrt{\overline{i_g^2}\,\overline{i_d^2}}} \qquad (4.49)$$

can also be defined and its value in the saturation region shown to be[8]

$$C = 0{\cdot}395\, j \qquad (4.50)$$

Any correlation of gate and drain noise current generators will be only significant at high frequencies since the induced gate noise due to thermal effects is proportional to ω^2. For a practical device this usually means frequencies in excess of 100 Mc/s[7].

It is possible to evaluate a noise figure taking into account the correlation between $\overline{i_g^2}$ and $\overline{i_d^2}$ and this has been done by Shoji[8] using the transmission line model. A simpler expression for the minimum noise figure has been given by Johnson[7] for a MOST operating under saturation conditions, taking into account the correlation between $\overline{i_d^2}$ and $\overline{i_g^2}$, it is:

$$NF_{\text{min}} = 1 + 0{\cdot}52 \frac{\omega C}{g_{fs}} \qquad (5.51)$$

where C is gate to channel capacitance.

If C is taken typically as 10 pF and g_{fs} as 2000 μmhos, F_{min} at 100 Mc/s is $\sim$4·2 dB[7].

It is assumed, of course, here that the noise figure is calculated for an optimum source impedance. Any change of source impedance gives a changed value of F as discussed in section 4.1.

Because of the difficulty in specifying noise figure for a field-effect transistor many manufacturers use the alternative of giving values of the noise voltage $\overline{e_n}$ ($\mu V/\sqrt{\text{Hz}}$) at specified frequencies, $\overline{i_n}$ ($\text{pA}/\sqrt{\text{Hz}}$) may also be quoted for a specified bandwidth, it usually being implied that there is a complete lack of correlation between the two generators.

REFERENCES

1. Faulkner, E. A. 'The design of low-noise audio-frequency amplifiers', *J.I.E.R.E.*, **36**, 17–30 (July 1968).
2. Van der Ziel, A. 'Thermal noise in field-effect transistors', *Proc. I.R.E.*, **50**, 1808–12 (August 1962).
3. Van der Ziel, A. and Ero, J. W. 'Small-signal, high-frequency theory of field-effect transistors', *I.E.E.E. Trans. on Electron Devices*, **ED-11**, 128–35 (April 1964).

4. Bruncke, W. C. and Van der Ziel, A. 'Thermal noise in junction-gate field-effect transistors', *Ibid*, **ED-13**, 323–8 (March 1966).
5. Fry, P. W. 'Low-frequency noise measurements on the *p*-channel MOST', *Electronic Engineering* (*GB*), **38**, 650–3 (October 1966).
6. Jordan, A. G. and Jordan, N. A. 'Theory of noise in metal oxide semi-conductor devices', *I.E.E.E. Trans. on Electron Devices,* **ED-12**, 148–56 (March 1965).
7. Wallmark, J. T. and Johnson, H. *Field Effect Transistors,* Prentice Hall (1966).
8. Shoji, M. 'Thermal noise of MOS field-effect transistors', *I.E.E.E. Trans. on Electron Devices,* **ED-13**, 520–4 (June 1966).

5

FET CHARACTERISTICS

It will be clear from the preceding chapters that the junction FET will have different characteristics from those of the insulated gate FET in certain respects. In particular, the junction FET must by definition be a depletion type, for if more than a few tens of millivolts of forward bias are applied to the gate, serious conduction will occur, whereas with the IGFET it will not. It is therefore logical to divide this chapter into two sections, dealing respectively with the junction type and the insulated gate type (the IGFET).

5.1. The junction type

5.1.1 JUNCTION FET DC PARAMETERS

As an aid to the definition of various FET parameters, the drain characteristics as shown in Fig. 5.1 are very useful. However, as will be explained later, such a family of curves is not suitable for the initiation of circuit design, because the parameters for a given FET type have a spread which is too wide to depict on such a graph.

Fig. 5.2 shows the polarities of the direct voltages and currents which exist when the two types of FET are in operation. Normally, the drain characteristics for either type are drawn in the first quadrant (Fig. 5.1) and the channel type is indicated by the signs taken by the numerical values. Normally the convention that 'current-in' has a positive sign and 'current-out' has a negative sign is used, so that the relevant voltage directions are automatically designated. Hence, the values of I_D and V_{DS} will be positive for the *n*-channel type, while V_{GS} and I_G will be negative. The converse will be true for the *p*-channel type.

The first important parameter is the current which flows along the channel when $V_{GS} = 0$, that is, when the gate and source are short-

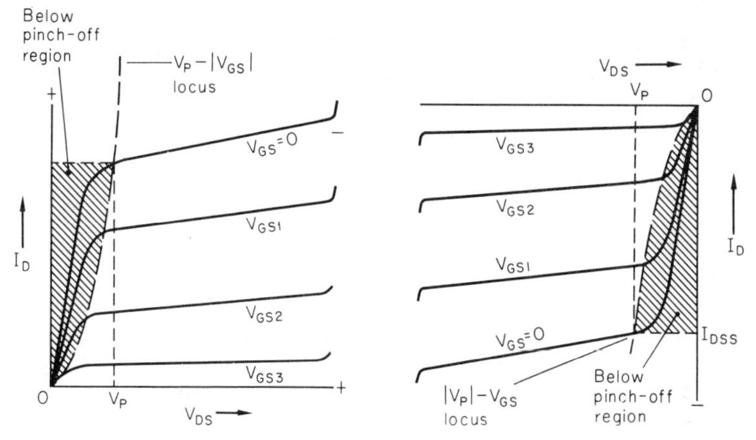

Fig. 5.1. (left) *Drain characteristics for an* n-*channel FET*
(right) *Drain characteristics for a* p-*channel FET*

circuited. This is given the symbol I_{DSS}, the subscripts referring to Drain current in common-Source connection with the gate Short-circuited. (In earlier references the symbol $I_{D(ON)}$ was used, but this is now obsolete in this context). Obviously, only the simple circuit of Fig. 5.3(*a*) is needed to measure I_{DSS}, though the value of V_{DS} must, of course, be large enough to ensure that saturation is just attained.

Originally, the pinch-off voltage referred to the rather arbitrary value of V_{DS} marked on Fig. 5.1, where the characteristic at $V_{GS} = 0$ bends over into the saturation region, and I_D becomes I_{DSS}. For other values of V_{GS}, saturation occurs at lower values of V_{DS}, and a pinch-off locus results where $|V_{DS}| = |V_P| - |V_{GS}|$, as is shown. However, when $V_{GS} = V_0$, the cut-off voltage, I_D is reduced almost to zero, and because $|V_0| = |V_P|$, the symbol V_P has come to refer also to the cut-off, or *gate pinch-off voltage.*

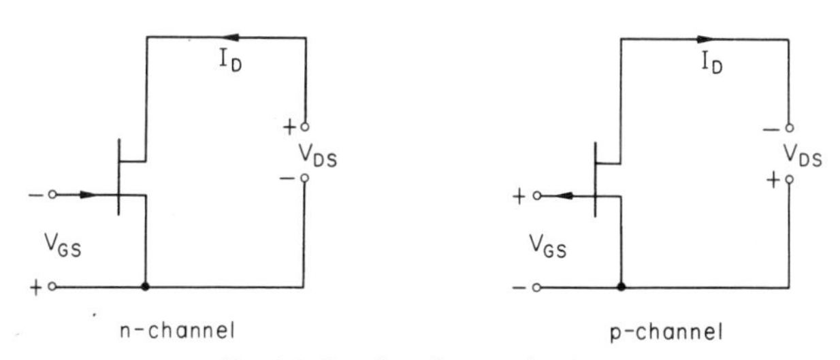

Fig. 5.2. Supply polarities for the FET

Most manufacturers rather arbitrarily quote the gate pinch-off voltage as that value of V_{GS} which reduces I_D to a very small value, typically 1 μA. The V_P so defined usually differs from that obtained from a more formal definition only by a few per cent, however. (The more formal definition, derived from the transconductance curve, is given later.)

When the FET is heavily cut off—that is, $|V_{GS}|$ exceeds $|V_P|$ by a volt or more—a small drain-to-gate leakage current flows. This is termed $I_{D(OFF)}$, and it may be measured by the simple circuit of Fig. 5.3(*b*).

The gate leakage current, being that of a reverse-biased diode, is markedly temperature sensitive. If the source and drain are short-circuited as in Fig. 5.4(*a*) the leakage current flowing is termed I_{GSS}, and a typical temperature dependence curve is given in Fig. 5.5. The gate leakages to the drain and source respectively can be measured

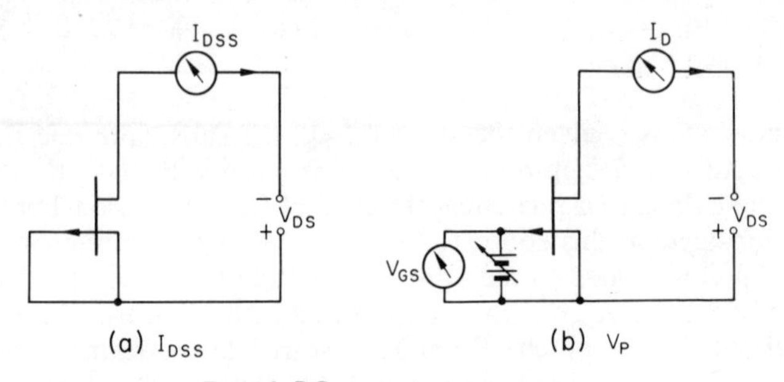

Fig. 5.3. DC parameter measurement

by means of the simple circuits of Fig. 5.4(*b*) and (*c*). Clearly, in the case of the gate-drain leakage (I_{GDO}), some contribution will be made by the current passing from the gate to the source end of the junction and up through the channel to the drain. Similarly, if the gate-source leakage (I_{GSO}) is measured, an analogous contribution will be made via the drain and channel. Consequently, the sum of I_{GDO} and I_{GSO} will be somewhat greater than I_{GSS}.

Again referring to Fig. 5.1, it will be seen that the characteristics rise sharply at the high V_{DS} ends. This sudden increase in current denotes that the channel breakdown voltage has been reached. The parameter BV_{DSS} is the maximum voltage which may be applied to the channel with the gate short-circuited to the source, and is quoted by the manufacturers as being slightly below the actual breakdown voltage of the 'worst' FET of the type number under consideration. Actually, the relevant phenomenon in this case is that of junction

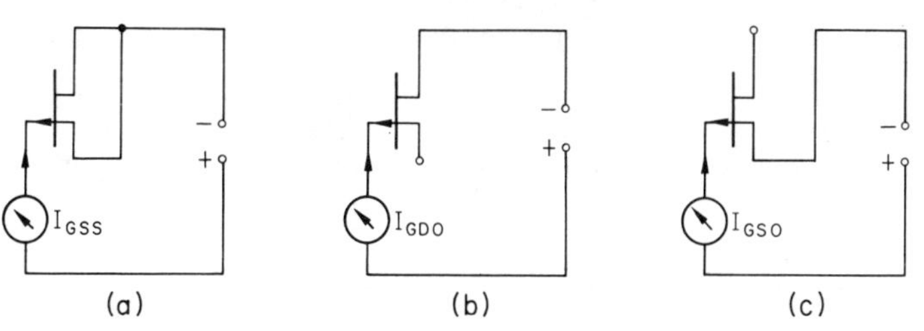

Fig. 5.4. Measurement of gate leakage current

avalanche breakdown, the gate being connected to the source, so that $BV_{DSS} = BV_{DGS}$. For any other (specified) condition, the quoted parameter would be BV_{DSX}, the limiting case being BV_{DSO}. Here, BV_{DSO} is likely to be greater than BV_{DSS}, there being no junction involved when the gate is floating.

From the circuit design point of view, perhaps the most important breakdown parameter is BV_{DGO}, the voltage at which the gate–drain junction will break down under open-circuited source conditions. Observing from the polarity diagrams (Fig. 5.2) that $V_{DG} = V_{DS} + V_{SG}$, it is clear that BV_{DGO} represents the limiting value of this sum.

The final breakdown parameter, BV_{GSO}, is not usually particularly important because its value will be much higher than V_P, and it is of course pointless to allow V_{GS} to rise much beyond V_P.

In order to give some idea of the magnitudes of the parameters listed above, it is convenient to tabulate some of them for a typical FET. This has been done in Table 5.1, which also shows some typical

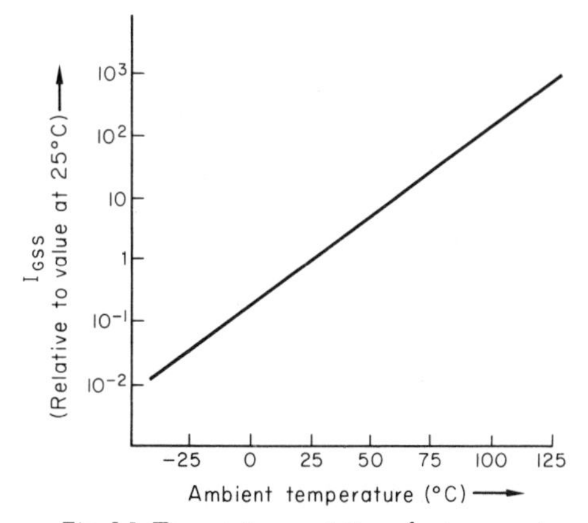

Fig. 5.5. Temperature variation of gate current

Table 5.1. SOME DC PARAMETERS FOR THE SILICONIX 2N2607 *p*-CHANNEL FET AT 25°C

I_{DSS} (mA)		V_P (volts)		I_{GSS} (nA)		$V_{BR(GDS)}$ (or BV_{DSS}) (volts)	
min	max	min	max	min	max	min	max
−0·30	−1·50	+1	+4	—	+3	−30	—

parameter spreads. Note in particular that the range of I_{DSS} for the 2N2607 is 5 to 1.

For practically all FETs operating in the saturation region the following expression[1] can be taken as approximately true:

$$I_D = I_{DSS}\left[1 - \frac{V_{GS}}{V_P}\right]^2 \tag{5.1}$$

The index of the term in brackets will actually lie somewhere between 1·9 and 2·1 for most modern FETs, though some references quote indices as low as 1·7.

Equation (5.1) can, in fact, be used as the basic bias circuit design equation for FETs, but its usefulness will here be demonstrated by employing it to derive a simple relationship between I_{DSS} and V_P, and also to introduce the small-signal parameters:

$$\frac{\mathrm{d}I_D}{\mathrm{d}V_{GS}} = g_{fs} = -\frac{2I_{DSS}}{V_P}\left[1 - \frac{V_{GS}}{V_P}\right] \tag{5.2}$$

where g_{fs} is the transconductance.
At $V_{GS} = 0$, this becomes,

$$g_{fs0} = -\frac{2I_{DSS}}{V_P} \tag{5.3}$$

If the transconductance curves for both *n*- and *p*-channel FETS are plotted according to equation (5.1), they will appear as shown in Fig. 5.6(*a*) and (*b*). From equation (5.2), the numerical value of g_{fs} is seen to be positive in both cases, corresponding to the positive slopes in the diagrams.

Fig. 5.6 also shows how the position of V_P may be defined (according to equation (5.3) if both g_{fs0} and I_{DSS} are known. Both g_{fs0} and I_{DSS} are in fact easy to measure.

5.1.2 JUNCTION FET SMALL-SIGNAL PARAMETERS

The transconductance g_{fs} is an important parameter in the design of FET amplifiers because it predicts how much control of the channel current will be exercised by the gate voltage. Fig. 5.6 shows that

g_{fs} varies quite widely depending upon the standing, or quiescent values of I_D and V_{GS} at which it is measured. Consequently, the conditions under which it is measured are always quoted in the manufacturers data. Also, because the stray capacitances associated with the FET can give rise to incorrect values for g_{fs}, it is always assumed that low frequency measurements only have been made. At higher frequencies, it is more correct to quote a complex transadmittance Y_{fs}.

The incremental channel resistance of a FET, r_{DS}, is seen from Fig. 5.1 to be a function of both V_{DS} and V_{GS}. When $|V_{DS}| < |V_P|$, the slope of the characteristic is determined largely by V_{GS}; hence the

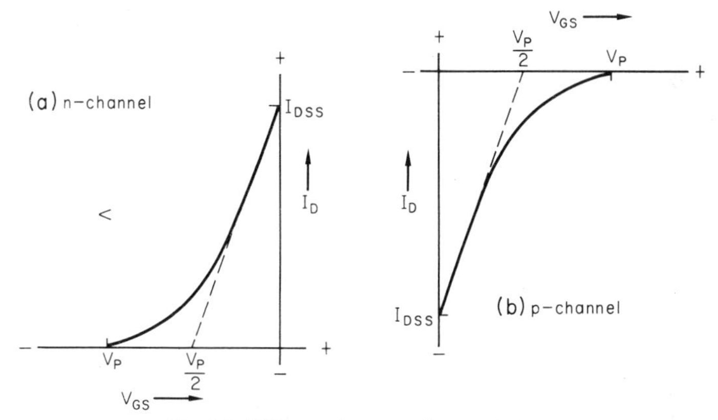

Fig. 5.6. FET transductance characteristics

FET becomes essentially a voltage controlled resistor. For switching applications r_{DS} is often required to be as low as possible when the FET is in the ON condition.

In the foregoing discussion it has been implicitly assumed that the FET has been connected with the gate as the input terminal and drain current as the output parameter. Having recognised this assumption, consider the equivalent circuit for a FET having enough drain–source voltage applied to it for the operating point to be in the pinched-off region, and with the gate as input terminal and the drain as output terminal (Fig. 5.7). Some of the parameters already discussed can be identified here: r_{DS} is seen to be ther incremental resistance presented by the channel, whilst the transconductance, g_{fs}, appears in the output current generator expression. Note that the numerical value of g_{fs} is always positive, so that the sense of this current generator is as shown, since the convention is that all currents entering the FET are to be considered positive.

The (very high value) gate leakage resistances, r_{GS} and r_{GD} are shunted by the capacitances, C_{GS} and C_{GD}, whilst r_{DS} is shunted by the channel capacitance C_{DS}. (Actually, C_{DS} is usually very much smaller than C_{GS} or C_{GD}, being largely the encapsulation, or header capacitance.) If the drain and source are short-circuited to AC, the measured input capacitance is termed C_{iss} (or C_{gss}) where

$$C_{iss} \simeq C_{GS} + C_{GD} \tag{5.4}$$

However, when the drain and source are not short-circuited, the actual input capacitance becomes much larger than this because of

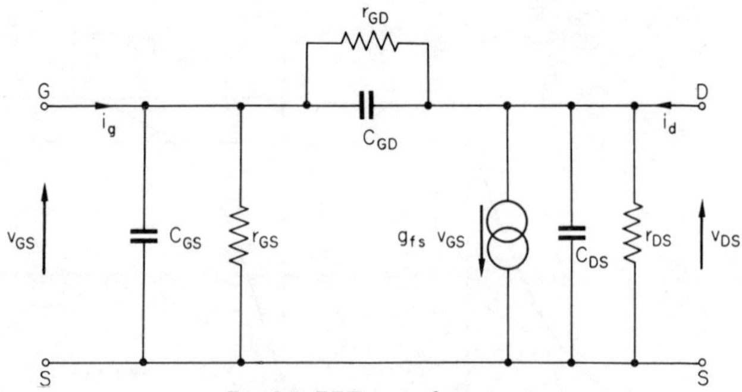

Fig. 5.7. FET equivalent circuit

the Miller Effect. This can be seen by referring to Fig. 5.8, where the FET is represented by an amplifier of voltage gain A_v across which C_{GD} is connected. Here, the voltage across C_{GD} is:

$$V_{in} - V_{out} = \frac{1}{C_{GD}} \int i \, dt$$

and since $V_{out} = A_v V_{in}$, this becomes:

$$V_{in}(1 - A_v) = \frac{1}{C_{GD}} \int i \, dt$$

or

$$V_{in} = \frac{1}{(1 - A_v) C_{GD}} \int i \, dt$$

This shows that the apparent input capacitance is given by:

$$C_{in} = C_{GS} + (1 - A_v) C_{GD} \tag{5.5a}$$

Clearly, $C_{in} > C_{iss}$, because A_v has a negative numerical value. Often, the symbol for the reverse transfer capacitance, C_{rss}, is used

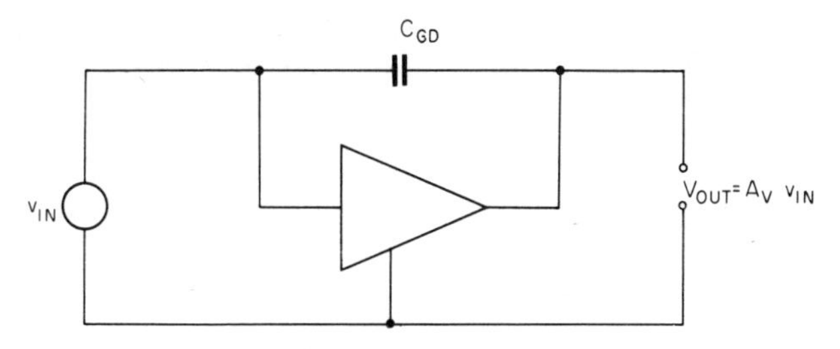

Fig. 5.8. Illustration of the Miller Effect

in place of C_{GD}; and if C_{DS} is much smaller than C_{rss} (as it usually is, being mainly the header capacitance), then $C_{rss} \simeq C_{oss}$. Hence, $C_{GS} \simeq C_{iss} - C_{rss}$ when equation (5.5a) becomes:

$$C_{\text{in}} \simeq C_{iss} - C_{rss} + (1 - A_v)\, C_{rss}$$

or

$$C_{\text{in}} \simeq C_{iss} - A_v\, C_{rss} \tag{5.5b}$$

Because these junction capacitances are functions of applied voltage, the conditions of measurement must be specified as

Table 5.2. SOME PARAMETERS FOR THE 2N3684 *n*-CHANNEL FET

V_P		I_{DSS}		g_{fso}		r_{DS}		C_{rss}		C_{iss}	
$V_{DS} = 20$ V, $I_{D(OFF)} = 1$ nA		$V_{DS} = 20$ V, $V_{GS} = 0$		$V_{DS} = 20$ V, $V_{GS} = 0$, $f = 1$ kHz		$V_{DS} = 0$, $V_{GS} = 0$		$V_{DS} = 20$ V, $V_{GS} = 0$, $f = 1$ kHz		$V_{DS} = 20$ V, $V_{GS} = 0$, $f = 1$ kHz	
min	max	min	max	min	max	min	max	min	max	min	max
−2	−5	2·5	7·5	2000	3000		600		1·2		4
volts		mA		μmho		ohms		pF		pF	

(Reproduced by permission of Siliconix Inc.)

demonstrated in Table 5.2. Note in particular that in this table, I_{DSS} is quoted not at $|V_{DS}| = |V_P|$, but at $V_{DS} = 20$ V. This is a common practice which, though resulting in an I_{DSS} slightly higher than for $|V_{DS}| = |V_P|$, does give a realistic value relevant to the actual working conditions of the device.

5.1.3 JUNCTION FET PARAMETER VARIATIONS

Like all semiconductor devices, the FET has characteristics which vary significantly with temperature. Prominent among these are variations in I_{GSS} (which doubles for each 10°C rise, approximately), V_P and I_{DSS}. These latter two parameters vary as the result of two effects:

(a) the contact potential between gate and channel decreases with temperature, so causing I_D to rise. (For most FETs, the rate of change is about −2·0 to −2·2 mV per °C.)
(b) carrier mobility decreases with temperature, so that I_D falls.

These two effects work in opposition, so that for a correctly chosen FET working under optimum conditions, perfect temperature stability of I_D should be attainable. This possibility has been

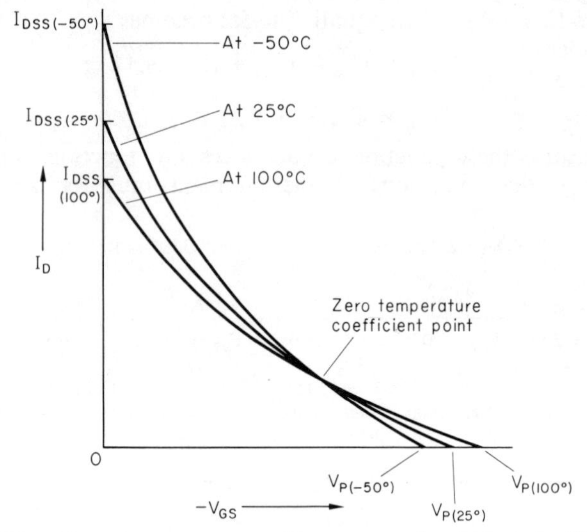

Fig. 5.9. Effect of temperature on transconductance curves for a high pinch-off FET

dealt with at length elsewhere[2, 3], the conclusions being that for a FET having a large value of pinch-off voltage, the expected cancellation is manifested by the cross-over of the transconductance characteristics, as shown in Fig. 5.9.

If a very low pinch-off device is involved, however, the characteristics may not cross over, and a zero temperature coefficient point will not be observed. That is, contact potential changes dominate, and I_D does not fall with rising temperature. If $V_P \simeq 0{\cdot}6$ volts, the curves coalesce near the Y-axis, and I_{DSS} varies but little with temperature.

It must be pointed out that, from a practical point of view, perfect temperature compensation is not possible by this method, but that with careful choice of I_D and V_{GS} for an individual FET, drifts as low as 50 μV/°C can be attained. The question of drift is of course critical when DC amplification is desired, and a comprehensive treatment is therefore presented in chapter 9.

Finally, the AC parameters will also vary with temperature. Because the slope of the transconductance curve is a function of temperature, g_{fs} will vary accordingly; and because the width of the depletion layer is in part a function of temperature, the various capacitances will vary also. Hence all these parameters must be defined for a given temperature, and the more important ones graphed.

Production spreads in FETs are (at the time of writing) considerably greater than those of bipolar devices. For example, Table 5.1 shows a 5:1 variation in I_{DSS} and a 4:1 variation in V_P for the 2N2607. Such variations are quite normal at present, and indeed, are much exceeded by the cheaper FETs currently available. This situation makes circuit design procedures somewhat more tedious than would be the case for smaller tolerances, and militates against optimum designs based on AC performance. Such procedures will be discussed later, and will be found to be assisted by the fact that I_{DSS} and V_P vary in the same sense, both being a function of the device geometry. Hence, a unit of high I_{DSS} will also have a high V_P and vice versa.

5.1.4 EQUIVALENT CIRCUITS FOR THE JUNCTION FET

The basic equivalent circuit has already been discussed and is depicted in Fig. 5.7. This equivalent can be reduced in complexity depending upon the frequencies which are to be considered in the design procedure. For example, if only very low audio-frequencies are involved, the capacitances may be omitted, as in Fig. 5.10. Here, the very high input resistances show that the FET is eminently suitable as an electrometer amplifier, for instance.

At higher frequencies (middle audio upwards), the leakage resistance can usually be ignored, while the capacitances become important. In the common-source configuration, the Miller capacitance $(1 - A)\,C_{GD}$ can be incorporated as part of the input capacitance according to equation 5.5 to give C_{in}. This simplification is shown in Fig. 5.11.

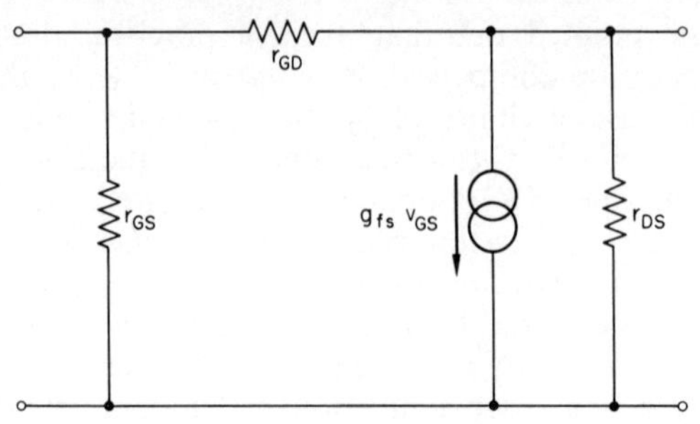

Fig. 5.10. Low frequency equivalent circuit

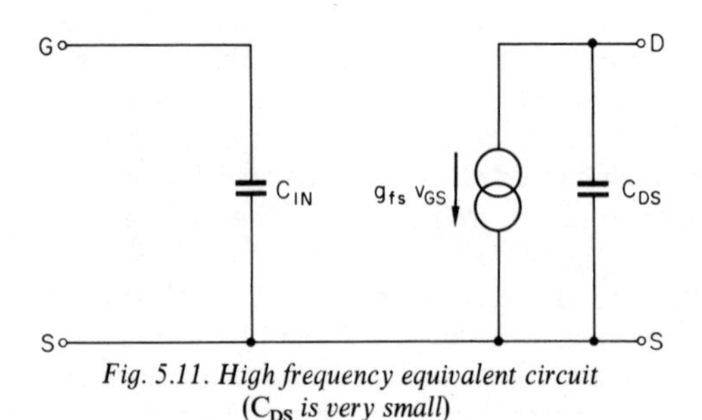

Fig. 5.11. High frequency equivalent circuit
(C_{DS} *is very small*)

In the preceding discussion, it has been assumed that the FET has been operating in the pinched-off region. For the triode region, the equivalent circuit will be somewhat different.

5.1.5 THE FET BELOW PINCH-OFF

Operated under suitable conditions the field effect transistor can exhibit the properties of an ohmic resistor so far as the circuit between source and drain terminals is concerned, but the value of the resistance is electrically controlled, being determined by the potential difference between the source and gate terminals. The exploitation of this variable resistance property in circuit applications has already been the subject of publications[4, 5], and there can be little doubt that it will play a significant role in the future development of electronic circuit technology, not least because it is the only variable linear passive element available for use in microcircuits.

In order that a FET shall behave like a ohmic resistor two conditions must be fulfilled:

(a) the potential difference between the gate region and the conducting channel must be such that no significant gate current flows (this condition does not, of course, apply to insulated gate devices) and
(b) the depletion region of the reverse-biased gate to channel *p–n* junction must at no point along the axis of the channel extend so far as to 'pinch-off' the channel.

If these two conditions are met then the channel will exhibit approximately ohmic properties, although there will be some departure from a linear voltage-current relationship as circuit conditions approach those at which pinch-off occurs. The value of resistance measured between drain and source will depend upon the voltage between the gate and source terminals of the device, increasing with the reverse bias applied to the gate junction due to the increasing penetration of the depletion region into the conducting channel, and hence the reduction of its effective cross-sectional area. The exact law of dependence of resistance on gate control voltage depends critically on the structural details of the device considered. Usually there is an approximately exponential increase, followed by a much sharper increase in resistance as pinch-off conditions are approached. The latter effect is rather undesirable, since it occurs at a gate voltage which varies widely from unit to unit. It also results in a marked increase in control sensitivity which may complicate design and set a limit to the resistance variation range which can be used. Development of transistor structures for variable resistance use will probably aim at 'remote pinch-off' types having an extended range of near exponential dependence of resistance on control voltage, and a structure which goes some way to meeting these requirements has been described[6]. Further developments in this sphere can be expected.

In approximate treatments, then, an exponential dependence of drain–source resistance on gate–source voltage may be assumed, that is

$$r_{DS} = r_0 \exp(\lambda V_{GS}) \tag{5.7}$$

where r_{DS} is the large signal (or chord) resistance of the channel, and r_0 and λ are constants characteristic of the FET concerned.

An exponential law makes for easy circuit analysis, in many cases, and it is likely that future remote pinch-off FET variable resistors will approximate quite closely to this law.

Where ordinary sharp pinch-off FETs (whether junction or IGFET) are used in variable resistance applications an alternative approximate resistance law, which holds good at gate–source voltages nearer the pinch-off values is a hyperbolic one of the form

$$r_{DS} = \frac{r_0}{1 - V_{GS}/V_P} \tag{5.8}$$

This also fails at $V_{GS} = V_P$ since it predicts infinite resistance, whereas in fact typically r_{DS} will be about $10^3\ r_0$ at this value of gate–source voltage. The hyperbolic law has some theoretical bases in that it can be derived from the Middlebrook charge controlled model of FET behaviour. Its failure at gate pinch-off may be related physically to neglect of the fact that the depletion regions are not in fact 'sharp edged', the charge carrier density varying continuously from zero to a maximum at the border of the depletion region. Thus, when depletion regions entering the channel from opposite sides meet, according to the simple 'sharp-edge' theory the channel conductance falls to zero: in fact a significant carrier density exists within the depletion region boundary, so that non-zero conductance is observed.

5.2. The insulated gate type (IGFET)

5.2.1 IGFET DC PARAMETERS

As has been explained in earlier chapters, an inversion layer, or *n*-channel, appears at an oxide-silicon interface due to bending of the Fermi levels. For this reason, the *p*-channel enhancement IGFET is the most common of the four possible devices, followed by the *n*-channel depletion IGFET. However, all four devices do exist, and Fig. 5.12 has been included to clarify their operation, and to relate their (I.E.E. recommended) symbols to their transconductance curves drawn in the correct quadrants.

Notice that in all cases, these curves have positive slopes, but that they cross the Y-axis only in the case of the depletion types. Hence, these types may actually operate both in the depletion *and* the enhancement regions; whereas the enhancement types operate *only* in the enhancement regions. (According to the JEDEC convention, a junction FET, which operates only in the depletion region, is a Type A FET; a 'depletion'-type IGFET, which operates in both regions, is a Type B FET; and an enhancement device is a Type C FET.)

Because the enhancement IGFETs are normally-OFF devices, their symbols incorporate broken lines to represent the normally-open-circuit channels; while the normally-ON depletion devices have channels symbolised by full lines. In both cases the polarity is denoted by an arrow representing the *p–n* direction of the channel-substrate junction. The substrate is therefore always marked, even though it may not be connected, or even brought out to a header lead.

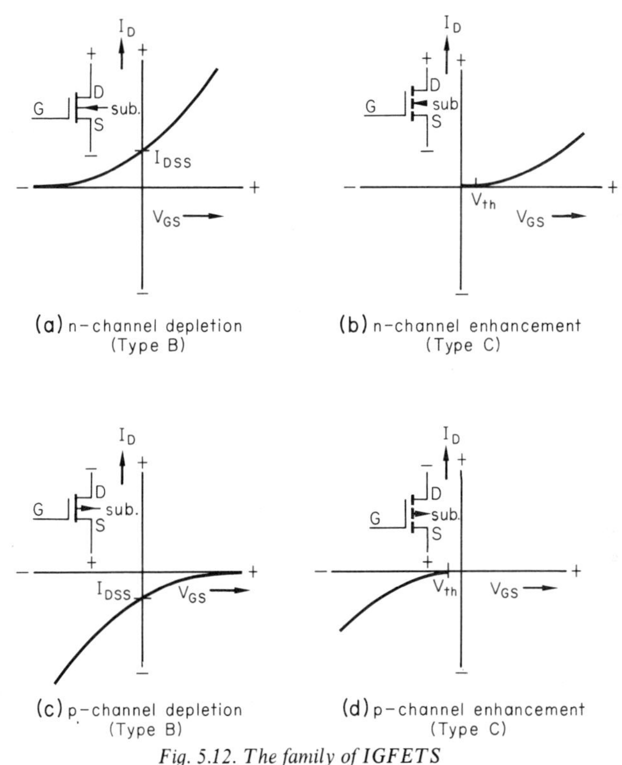

Fig. 5.12. The family of IGFETS

In both cases the gate is symbolised by a line separated from the channel line to denote the presence of insulation.

In the case of the depletion IGFETs, the transconductance curve is seen to cross the Y-axis at I_{DSS} (where $V_{GS} = 0$). However, by applying a 'forward' V_{GS}, currents greater than I_{DSS} may flow down the channel—a situation which is obviously impossible for junction FETs, wherein the junction would begin to conduct for a forward V_{GS} greater than a few hundred millivolts.

In the case of the enhancement types, I_{DSS} is extremely small, and is in fact quoted as a measure of the channel leakage in the $V_{GS} = 0$, or OFF condition. A parameter $I_{D(ON)}$ is also quoted, however, at specified values of V_{GS} and V_{DS}.

For most IGFETs, the drain current can be expressed as follows: For the triode, or below-pinch-off region, where $|V_{DS}| < |V_{GS} - V_{th}|$,

$$I_D \simeq -\beta\left[V_{DS}(V_{GS} - V_{th}) - \frac{V_{DS}^2}{2}\right] \tag{5.9}$$

Here, β is called the gain factor for the IGFET. This is actually both an inappropriate symbol and term, for not only may it be confused with the current gain of a bipolar transistor, but it obviously has the dimensions of mhos per volt. A better term for β is the 'device constant'. Richman[7] gives β as follows:

$$\beta = \frac{\varepsilon_{ox}\mu W}{T_{ox}L}$$

Here, a MOST structure is referred to, where

ε_{ox} = dielectric constant of the oxide layer
T_{ox} = thickness of the oxide layer
μ = carrier mobility
L = length of channel between source and drain diffusions
W = width of channel

For the saturation, or pinched-off region, where $|V_{DS}| \geqslant |V_{GS} - V_{th}|$,

$$I_D \simeq -\frac{\beta}{2}(V_{GS} - V_{th})^2 \tag{5.10}$$

In both equations, V_{th} is the threshold voltage, which is analogous to the pinch-off voltage for the junction FET. For a given value of V_{DS}, equation (5.10) may be re-written:

$$\sqrt{I_D} \simeq -\sqrt{\left(\frac{\beta}{2}\right)} \cdot V_{GS} + \sqrt{\left(\frac{\beta}{2}\right)} \cdot V_{th}$$

This is clearly a straight line equation whose slope is $\sqrt{(\beta/2)}$, and which cuts the X-axis at $\sqrt{(\beta/2)} \cdot V_{th}$. Actually this straight line must be extrapolated to cut the X-axis, because the square law relationship is not valid at low values of I_D.

In general, manufacturers do not define V_{th} in this way, but determine it by measuring V_{GS} when I_D is very low, often 10 μA. This simple procedure gives a V_{th} within a few per cent of the value obtained by extrapolation.

Whereas V_{th} is usually quoted for enhancement IGFETs, data sheets for depletion types often substitute the parameter $I_{D(\mathrm{OFF})}$. This is the maximum drain current which will flow when specific values of V_{GS} and V_{DS} are applied, and it is in fact a channel-to-substrate

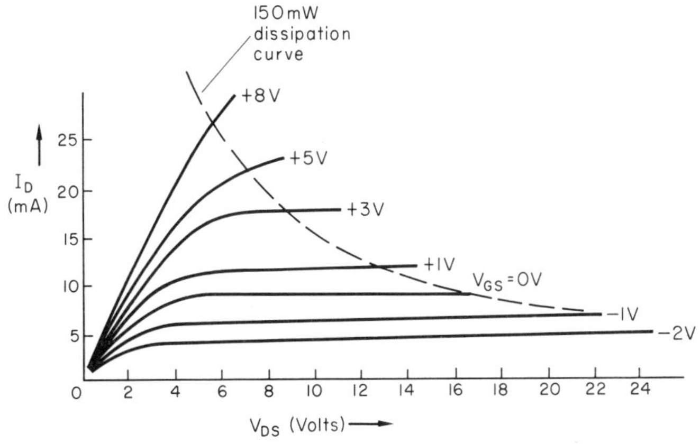

Fig. 5.13. Drain characteristic for the RCA 3N139 n*-channel depletion type* IGFET (Reproduced by permission)

leakage current. Table 5.3 gives $I_{D(\mathrm{OFF})}$ for a typical depletion IGFET, and also $I_{D(\mathrm{ON})}$ for a typical enhancement type.

Another useful parameter is the so-called forward leakage current, $I_{G(f)}$, which corresponds to I_{GSS} for the junction FET. This is the gate leakage current when the source and drain are short-circuited, but unlike I_{GSS}, it is entirely ohmic, for it represents a genuine leakage current through (and around) the insulating layer, and not a diode reverse current. It is therefore perfectly valid to

Table 5.3. SOME IGFET DC PARAMETERS

Fairchild BFX 78 *n*-Channel Depletion IGFET

I_{DSS}		$I_{D(\mathrm{OFF})}$	
$V_{GS} = 0$ min	$V_{DS} = 10$ V max	$V_{GS} = -10$ V min	$V_{DS} = 15$ V max
9 mA	26 mA		200 μA

Fairchild BSX 83 *p*-Channel Enhancement IGFET

I_{DSS}		$I_{D(\mathrm{ON})}$		V_{th}		$I_{G(f)}$	
$V_{GS} = 0$ min	$V_{DS} = -15$ V max	$V_{GS} = -15$ V min	$V_{DS} = -15$ V max	$V_{DS} = -15$ V min	$I_D = -10\,\mu$A max	$V_{GS} = -25$ V min	$V_{DS} = 0$ max
	−0·5 nA	−3 mA	−12 mA	−3 V	−6 V		−2·5 pA

assign a gate-channel resistance to an IGFET, and in the case of the BSX 83 of Table 5.3, this will be $25/(2.5 \times 10^{-12}) = 10^{13}\ \Omega$.

As already pointed out, both depletion and enhancement IGFETs may be either *n*-channel or *p*-channel, but for manufacturing reasons depletion types are usually *n*-channel and enhancement types are usually *p*-channel. It will be noticed that this applies to the two devices exemplified in Table 5.3, and it will be found instructive to compare the current and voltage polarities quoted with the transconductance curves of Fig. 5.12(*a*) and (*d*). If this is done, it will be clear that the numerical value of β must be negative for *n*-channel types and positive for *p*-channel types. This is so because the mobility μ appears in the expression for β, and μ_p is positive while μ_n is negative.

The breakdown voltages relevant to the IGFET are defined in terms of a short-circuited gate, or with the gate at some specific voltage. This is because, if the gate were open-circuited, quite a small charge accumulating on that gate could lead to a voltage capable of breaking down the insulating layer, since the gate resistance in the IGFET is too high to allow such a charge to leak away harmlessly.

5.2.2 IGFET AC PARAMETERS AND EQUIVALENT CIRCUIT

In the saturation region, the approximate drain current is given by equation (5.10):

$$I_D \simeq -\frac{\beta}{2}[V_{GS} - V_{th}]^2 \qquad (5.10)$$

where β is negative for *n*-channel and positive for *p*-channel types. For constant β (that is, constant V_{DS}), the transconductance will be:

$$g_{fs} = \frac{dI_D}{dV_{GS}} \simeq -\beta[V_{GS} - V_{th}] \qquad (5.11)$$

If this expression is evaluated using parameters relevant to any of the four transconductance curves shown in Fig. 5.12, it will be found that g_{fs} is always positive, as the slopes of these curves indicate.

Notice that equation (5.11) indicates that g_{fs} may be increased by making β large. Unfortunately this implies increasing the width of the gate, which of course, increases the gate-channel capacitance, and so degrades the high-frequency performance.

Equation (5.11) also indicates that g_{fs} may be improved by making the difference between V_{GS} and V_{th} large. In the case of operation in the depletion regions of diagrams (*a*) and (*c*), it implies making

V_{GS} approach zero. For operation in the enhancement regions of those diagrams, or in the case of diagrams (*b*) and (*d*), it implies making V_{GS} as high as is feasible.

Normally the transconductance curves provided by the manufacturer will be adequate for choosing a working point and initiating a bias design procedure. At the time of writing enhancement IGFETs are becoming prominent as switches because of the

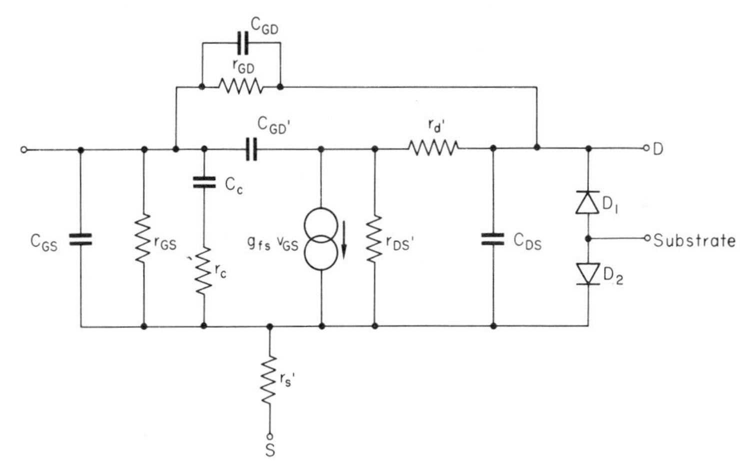

Fig. 5.14. Equivalent circuit for n-*channel* IGFET. *For* p-*channel type, reverse diode polarities* (After Griswold[8].)

comparative ease with which they can be fabricated in micro-miniature arrays, and here the limiting values of r_{DS} are of somewhat more importance than the values of g_{fs}.

Enhancement IGFETs can be very simply self-biased by virtue of their gate and drain voltages being in the same sense, which is sometimes a useful feature in amplifier circuits. Also, depletion IGFETs can inherently accept a gate input signal which may be both positive or negative-going with respect to the source, as is clear from the output characteristic reproduced in Fig. 5.13. These specialised advantages make IGFETs attractive for certain linear applications, especially in electrometer-type circuits where an exceptionally high input resistance is mandatory. Consequently, the small-signal equivalent circuit is of interest, and the pinched-off' version[8] is given in Fig. 5.14.

This diagram will be seen to be similar to that for the junction FET, (Fig. 5.7), but with the addition of several extra components. Notable among these are C_c and r_c. These are the approximate lumped equivalents of the distributed capacitance and resistance between the gate electrode and the channel thereunder. The more accurate network of capacitors and resistors which would be necessary to describe the true situation is prohibitively complex, so C_c and r_c are derived from the high-frequency performance of the IGFET, since they are primarily responsible for the limitation of this performance.

The incremental output resistance, r_{DS}, is very variable, even when only the 'pinched-off' values are being considered, as can be deduced from Fig. 5.13. However, it is much larger than the parasitic resistance r_d' under these conditions, though r_d' does become important when 'below pinch-off' applications are of interest. Both r_d' and r_s' represent those portions of the channel which are not directly under the influence of the gate voltage and the effect of r_s' in high-frequency work can be quite pronounced.

The diodes D_1 and D_2 represent the junctions between the substrate and the source and drain regions. These diodes do have an effect on the performance of the IGFET, particularly by virtue of their capacitances, which will tend to shunt high-frequency signals. More profound effects will occur when the substrate is connected, as will be explained later.

The other resistances and capacitances appear also in the junction FET equivalent circuit, the only significant difference being the markedly higher values of r_{GS} and r_{GD}. They will not therefore be treated further.

5.2.3 IGFETS OPERATED BELOW PINCH-OFF

Like junction gate devices, insulated gate FETs can be used as variable resistors. In this case, however, there is no requirement that the gate-channel voltage shall remain of one polarity since gate conduction cannot occur. The dependence of drain–source resistance on control voltage applied between gate and source is closely similar to that for a sharp cut-off junction device, and well approximated by a hyperbolic law.

From equation (5.9), the channel resistance in the triode region may be derived:

$$I_D = -\beta\left[V_{DS}(V_{GS} - V_{th}) - \frac{V_{DS}^2}{2}\right] \qquad (5.9)$$

from which

$$\frac{dI_D}{dV_{DS}} = \frac{1}{r_{DS}} = -\beta[V_{GS} - V_{th} - V_{DS}]$$

and when $V_{DS} \rightarrow 0$, which is the region where variable-resistance circuits normally operate, this becomes:

$$r_{DS} = \frac{1}{-\beta[V_{GS} - V_{th}]} \tag{5.12}$$

Remote pinch-off MOST structures have not yet been reported, although there seems in principle to be no reason why they should not be possible.

REFERENCES

1. Middlebrook, R. D. 'A simple derivation of field effect transistor characteristics', *Proc. I.E.E.E.* **51**, 1146 (August 1963).
2. Sevin, L. J. *Field-Effect Transistors,* Chapter 2. McGraw-Hill (1965).
3. Evans, L. L. 'Biasing FETs for zero DC drift, *Electrotechnology* No. 74, 93–96 (August 1964).
4. Gosling, W. 'Voltage controlled attenuators using field-effect transistors', *I.E.E.E. Trans. on Audio* **14**, 58–67 (September 1966).
5. Todd, C. D. *Junction Field-Effect Transistors,* Chapter 8. Wiley (1968).
6. Morgan, A. N. 'The FET as an electronically variable resistor' (letter), *Proc. I.E.E.E.* **54**, 892–893 (1966).
7. Richman, P. *MOS Field-Effect Devices,* p. 43. McGraw-Hill (1962).
8. Griswold, D. M. 'Understanding and using the MOSFET', *Electronics,* 66–70 (14th December 1967).

6

SINGLE-STAGE AMPLIFIERS

6.1. The three basic configurations

By virtue of its being a three-terminal device, the FET can be connected in three distinct ways, respectively the common-source (CS), common-drain (CD), and common-gate (CG) modes. By analogy with the bipolar emitter-follower, the CD mode is also referred to as the source-follower configuration.

The magnitudes of the associated components are very relevant to the small-signal performance of each of the three modes, and also to the problem of biasing the active devices in the first instance. However, the question of adequate biasing will be considered later in the chapter, and some general expressions for signal performance will first be derived.

Because the junction FET has a wider range of applications as a small-signal amplifier than does the IGFET (at the time of writing), it will be treated in more detail in the present chapter, while the specialised applications of the IGFET will be discussed elsewhere.

Throughout the discussion on small-signal amplifiers, the numerical value of g_{fs} will be taken as being positive, in accordance with the real slopes of the transconductance curves. In some publications, these curves are drawn in the first quadrant and g_{fs} is assigned a negative value. This is because an increase in the modulus of V_{GS} results in a fall in the modulus of I_D, so that it is natural, though mathematically incorrect, to think of g_{fs} as being negative. When IGFETs are being considered, this convention can lead to confusion, as is apparent from a study of Fig. 5.12.

When biasing procedures are treated, however, the first quadrant convention will be used in the interests of generality, and because the method of biasing dealt with will not apply to enhancement type IGFETs, as will become apparent.

6.1.1 THE COMMON-SOURCE STAGE

The basic form taken by this stage is illustrated in Fig. 6.1. Here, no sophisticated biasing circuit has been shown for reasons of clarity, but a battery symbol has been included to make the point that such a circuit must exist. The biasing components present a resistance at the input, and the combination of this and the generator resistance

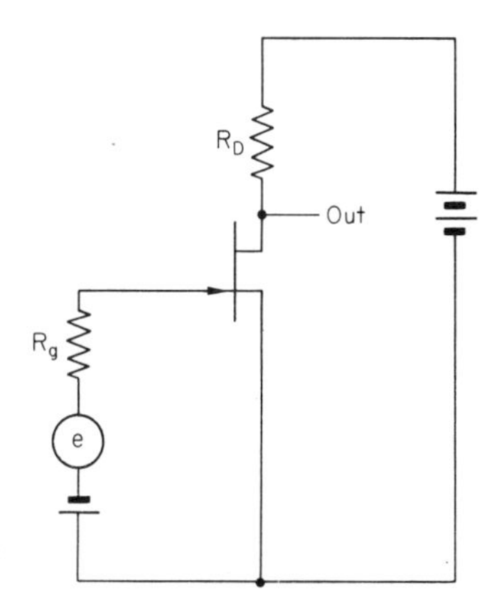

Fig. 6.1. Basic common source configuration

R_g will be termed R_G. Similarly, the drain resistor R_D will be combined with any parallel load resistance to give R_L.

It is now possible to draw an equivalent circuit for the stage, and in the first instance this has been done for low frequencies in Fig. 6.2. Here, the various interelectrode capacitances have been omitted.

From Fig. 6.2, the low frequency voltage gain may be determined:

$$v_{out} = -g_{fs} \cdot v_{GS}\left(\frac{r_{DS}R_L}{r_{DS}+R_L}\right)$$

and because $v_{GS} = v_{in}$

$$A_{v(l.f.)} = -\frac{g_{fs}r_{DS}R_L}{r_{DS}+R_L} = -\frac{g_{fs}R_L}{1+R_L/r_{DS}} \tag{6.1}$$

$$\text{If } r_{DS} \gg R_L, \text{ then } A_{v(l.f.)} \simeq -g_{fs}R_L \tag{6.2}$$

Notice that here, $A_{v(l.f.)}$ is in fact negative, because g_{fs} has a positive numerical value. This means that a phase reversal occurs from input to output.

The output resistance of the stage is obviously r_{DS}, because $r_{GD} \gg r_{DS}$; whilst the input resistance approximates to r_{GS} and r_{GD} in parallel, for the same reason.

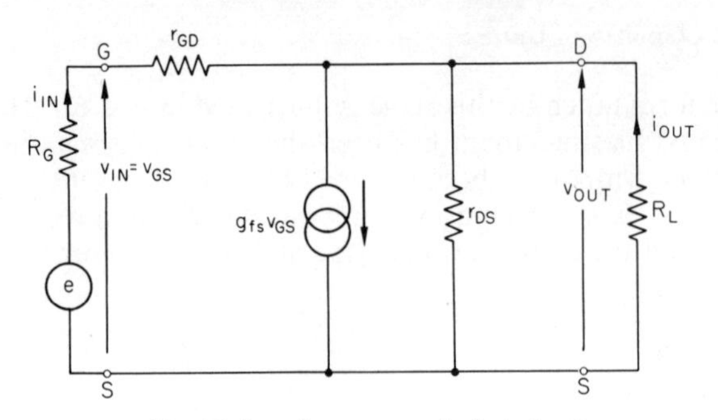

Fig. 6.2. Low-frequency equivalent circuit

At frequencies where C_{GS} and C_{GD} become important the equivalent circuit of Fig. 6.3 may be used to determine the circuit parameters. If reactances are to be discussed, then sinusoidal signals must be assumed, when the input impedance may be determined as follows:

$$Z_{\text{in}} = \frac{V_{\text{in}}}{I_{\text{in}}} = \frac{V_{\text{in}}}{V_{\text{in}} j\omega C_{GS} + (V_{\text{in}} - V_{\text{out}}) j\omega C_{GD}}$$

$$= \frac{1}{j\omega[C_{GS} + (1 - A_v)\, C_{GD}]} = \frac{1}{j\omega C_{\text{in}}} \qquad (6.3)$$

where $$C_{\text{in}} = C_{GS} + (1 - A_v)\, C_{GD}$$

The input capacitance is seen to be simply C_{GS} plus the Miller capacitance $(1 - A_v)\, C_{GD}$ discussed in section 5.1.2. The voltage gain A_v is, strictly speaking, that at the frequency under consideration, but at moderate frequencies, $A_{v(\text{l.f.})}$ can be taken as a close approximation, as will be shown.

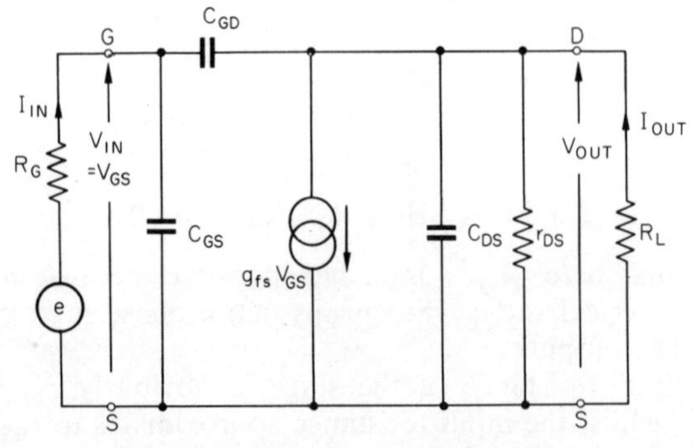

Fig. 6.3. High-frequency equivalent circuit

In the foregoing derivation, it was assumed that the load was purely resistive. However, had this load contained a reactive component, then Z_{in} would have had a real part. Further, had Z_L contained an inductive reactance, the real part of Z_{in} would have been negative. Unless the value of R_G were greater than this negative real part, instability would result.

To find A_v, it is convenient to convert the current generator of Fig. 6.3 into its equivalent voltage generator as shown in Fig. 6.4.

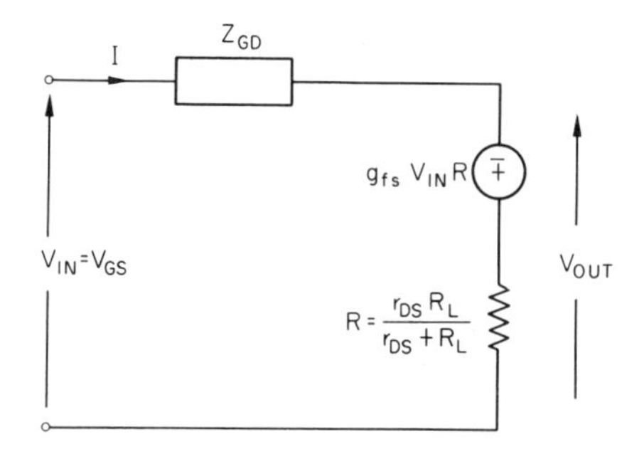

Fig. 6.4. Transformation of equivalent circuit for calculation of A

(Here, it has been assumed that $|X_{CDS}| \gg R$ so that C_{DS} may be omitted).

Ascribing a circulating current I to the series circuit,

$$V_{out} = -g_{fs}V_{GS}R + IR$$

$$= -V_{in}g_{fs}R + (V_{in} - V_{out}) \cdot \frac{R}{Z_{GD}}$$

$$= V_{in}A_{v(l.f.)} - (V_{in} - V_{out})\frac{A_{v(l.f.)}}{g_{fs}Z_{GD}}$$

$$\therefore A_v = \frac{V_{out}}{V_{in}} = A_{v(l.f.)} - (1 - A_v)\frac{A_{v(l.f.)}}{g_{fs} \cdot Z_{GD}}$$

or

$$A_v = A_{v(l.f.)}\left[\frac{g_{fs}Z_{GD} - 1}{g_{fs}Z_{GD} - A_{v(l.f.)}}\right] \tag{6.4}$$

At frequencies where $|X_{C_{gd}}| \ll r_{GD}$ this becomes:

$$A_v = A_{v(l.f.)}\left[\frac{\dfrac{g_{fs}}{j\omega C_{GD}} - 1}{\dfrac{g_{fs}}{j\omega C_{GD}} - A_{v(l.f.)}}\right] \tag{6.5}$$

From this equation, it is clear that if

$$\left|\frac{g_{fs}}{j\omega C_{GD}}\right| \gg 1, A_{v(\text{l.f.})} \quad \text{then } A_v \simeq A_{v(\text{l.f.})}$$

This observation is immediately applicable to the expression for input capacitance, which becomes:

$$C_{\text{in}} \simeq C_{GS} + (1 - A_{v(\text{l.f.})})\, C_{GD}$$

The very high resistance r_{GS} does of course appear in parallel with C_{in}, but normally, R_G is much smaller than r_{GS} so that the cut-off

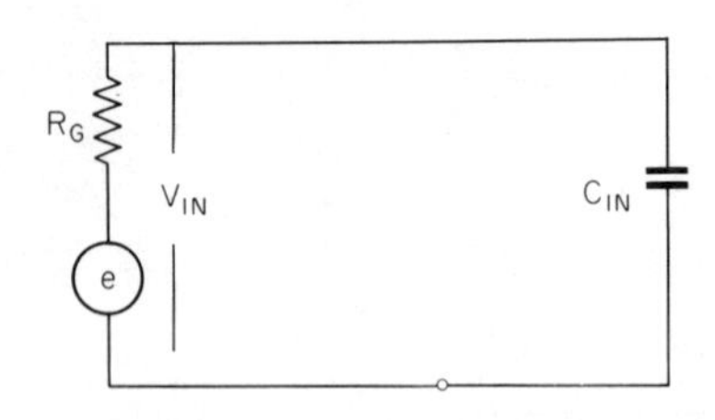

Fig. 6.5. Circuit for calculation of cut-off frequency

frequency for the input circuit is defined only by R_G and C_{in}. That is, where $R_G = 1$ MΩ and $R_L = 10$ kΩ. For the C95, $C_{GS} = 3$ pF, $C_{GD} = 3$ pF and $g_{fs} = 1$ mA/V under the relevant working conditions

Hence

$$f_h = \frac{1}{2\pi C_{\text{in}} R_G} \tag{6.6}$$

where $R_G = 1$ MΩ and $R_L = 10$ kΩ. For the C95, $C_{GS} = 3$ pF, $C_{GD} = 3$ pF and $g_{fs} = 1$ mA/V under the relevent working conditions.

$$A_{v(\text{l.f.})} \simeq -g_{fs} R_L = -(10^{-3})(10^4) = -10$$

and

$$C_{\text{in}} = C_{GS} + (1 - A_{v(\text{l.f.})})\, C_{GD}$$

$$= 3 + (1 + 10)\, 3 = 36 \text{ pF}$$

Hence,

$$f_h = \frac{10^6}{2\pi 36} = 4{\cdot}4 \text{ kHz.}$$

This is a very poor cut-off frequency, and it illustrates the fact that the simple CS stage is not suitable for exploiting the high input resistance offered by the FET unless only very low frequencies are of interest.

The CS circuit of Fig. 6.1 is shown with a gate bias battery in series with the input. In practice, however, it is usual to design for some form of automatic bias, and the most common circuit is the DC series feedback method depicted in Fig. 6.6. The operation of the circuit will be fully described later, but at present it is important to note that R_S must be bypassed by a large capacitor if the equations

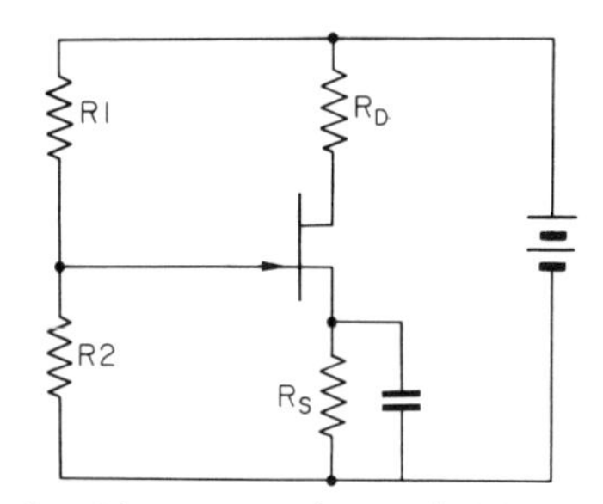

Fig. 6.6. Automatic bias method for CS stage

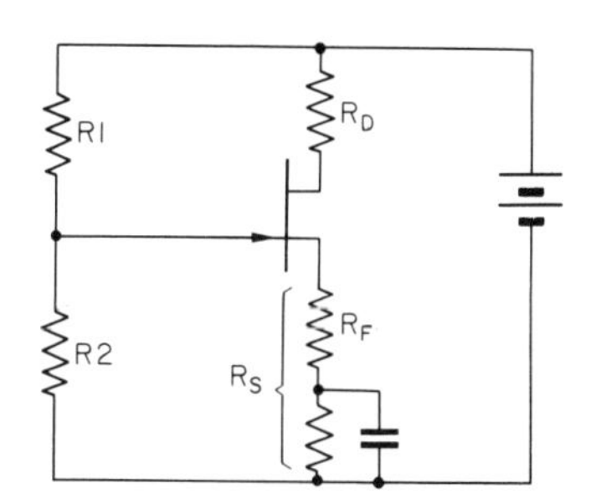

Fig. 6.7. Biased CS with partly bypassed source resistor

so far described are to be valid. However, if part of R_S is left unbypassed, as in Fig. 6.7, some signal feedback will occur which will tend to reduce the gain.

From the gate to the source, the voltage gain is shown in the next section (6.1.2) to be:

$$A_{v(CD)} = \frac{g_{fs}R_F}{1 + g_{fs}R_F} \tag{6.10}$$

where R_F is the unbypassed part of R_S.

Also, from the gate to the drain, the voltage gain is:

$$A_{v(CS)} = \frac{-i_d R_L}{v_{\text{in}}} = \frac{-i_d R_L R_F}{v_{\text{in}} R_F}$$

Here, $i_d R_F$ is the signal developed across R_F, which is by definition $v_{\text{in}} A_{v(CD)}$, so that inserting the expression for $A_{v(CD)}$,

$$A_{v(CS)} = \frac{-g_{fs}R_F R_L}{R_F(1 + g_{fs}R_F)} = -\frac{g_{fs}R_L}{1 + g_{fs}R_F} \tag{6.7}$$

and if $g_{fs}R_F \gg 1$ (which is *not* necessarily true), this expression reduces to $-R_L/R_F$.

6.1.2 THE COMMON-DRAIN OR SOURCE-FOLLOWER STAGE

The basic wiring diagram and equivalent circuit are shown in Figs. 6.8 and 6.6. From the latter, the voltage gain may be derived

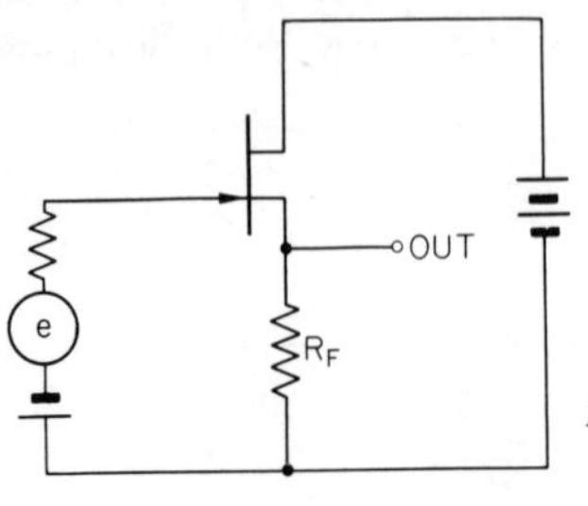

Fig. 6.8. Common drain stage (source-follower)

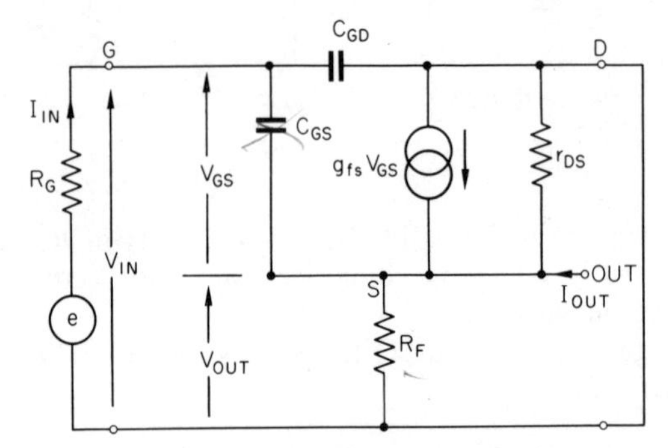

Fig. 6.9. Equivalent circuit for source follower

by summing currents at the junction of the source and load resistor R_F:

$$-g_{fs}V_{GS}+V_{\text{out}}\left(\frac{1}{R_F}+\frac{1}{r_{DS}}\right)+(V_{\text{out}}-V_{\text{in}})j\omega C_{GS}=0$$

and since $V_{GS}=-(V_{\text{out}}-V_{\text{in}})$ this becomes,

$$V_{\text{out}}\left(\frac{1}{R_F}+g_{DS}\right)+(V_{\text{out}}-V_{\text{in}})(j\omega C_{GS}+g_{fs})=0$$

or

$$V_{\text{out}}\left(\frac{1}{R_F}+g_{DS}+g_{fs}+j\omega C_{GS}\right)=V_{\text{in}}(j\omega C_{GS}+g_{fs})$$

giving:

$$A_v=\frac{V_{\text{out}}}{V_{\text{in}}}=\frac{g_{fs}+j\omega C_{GS}}{(1/R_F)+g_{DS}+g_{fs}+j\omega C_{GS}} \tag{6.8}$$

Notice that A_v is positive; that is, there is no phase reversal.
At very low frequencies, where $\omega C_{GS} \rightarrow 0$.

$$A_{v(\text{l.f.})} = \frac{g_{fs}}{(1/R_F) + g_{DS} + g_{fs}} \quad (6.9)$$

Because the FET is working in the pinched-off region, r_{DS} is likely to be much greater than R_F, in which case,

$$A_{v(\text{l.f.})} = \frac{g_{fs}R_F}{1 + g_{fs}R_F} \quad (6.10)$$

If $g_{fs}R_F \gg 1$, then $A_{v(\text{l.f.})} = 1$.

The CD stage is primarily useful as an impedance changer, and methods exist whereby the gain may be well defined over a wide dynamic range. The input capacitance is much lower than that of the CS stage as can be shown by summing currents at the gate in Fig. 6.9:

$$I_{\text{in}} = V_{\text{in}} \, . \, j\omega C_{GD} + (V_{\text{in}} - V_{\text{out}}) \, j\omega C_{GS}$$

giving

$$Y_{\text{in}} = \frac{I_{\text{in}}}{V_{\text{in}}} = j\omega[C_{GD} + (1 - A_v)\, C_{GS}]$$

That is

$$C_{\text{in}} = C_{GD} + (1 - A_v)\, C_{GS} \quad (6.11)$$

Here, A_v is a little less than unity and is positive, so that C_{in} is only slightly larger than C_{GD}.

Example

To illustrate the improved cut-off frequency implied by this small input capacitance, consider again the C95, and let it be connected as a source-follower with $R_G = 1\ \text{M}\Omega$ and $R_F = 10\ \text{k}\Omega$.

From equation (6.10),

$$A_{v(\text{l.f.})} = \frac{10^{-3} \times 10^4}{1 + 10^{-3} \, . \, 10^4} = \frac{10}{11}$$

so that

$$C_{\text{in}} = 3 + \left(1 - \frac{10}{11}\right) 3 = 3{\cdot}27\ \text{pF}$$

and

$$f_h = \frac{10^{12}}{2\pi 10^6 \times 3{\cdot}27} = 48\ \text{kHz}.$$

This represents a bandwidth some ten times better than that of the CS stage example.

The output impedance may be obtained by redrawing the equivalent circuit as in Fig. 6.10. Here, the input generator has been

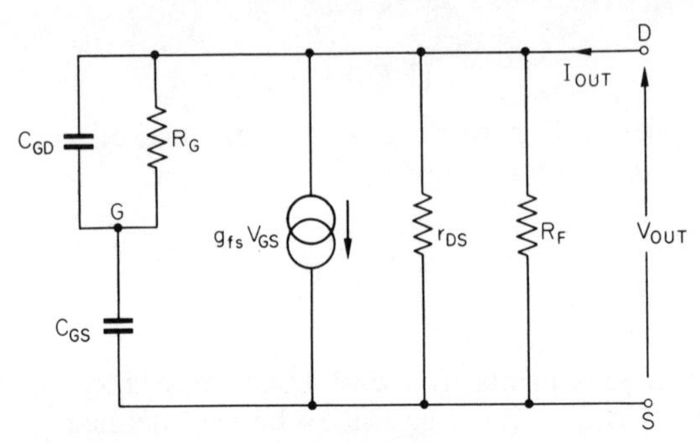

Fig. 6.10. Equivalent circuit for determination of Y_{out}

suppressed, and a voltage V_{out} applied externally at the output. Z_{out} is now given by V_{out}/I_{out}:

Letting the parallel combination of R_G and X_{CGD} be the complex impedance $\bar{Z}_{GD}$,

$$I_{out} = V_{out}\left[\frac{1}{R_F} + \frac{1}{r_{DS}} + \frac{1}{\bar{Z}_{GD} - jX_{C_{GS}}}\right] + g_{fs} \cdot V_{GS}$$

But

$$V_{GS} = V_{out}\left|\frac{-jX_{C_{GS}}}{\bar{Z}_{GD} - jX_{C_{GS}}}\right|$$

so that

$$I_{out} = V_{out}\left[\frac{1}{R_F} + g_{DS} + \frac{1 - jX_{C_{GS}} \cdot g_{fs}}{\bar{Z}_{GD} - jX_{C_{GS}}}\right]$$

$$\therefore \quad Y_{out} = \frac{I_{out}}{V_{out}} = \frac{1}{R_F} + g_{DS} + \frac{g_{fs} + j\dfrac{1}{X_{C_{GS}}}}{1 + j\dfrac{\bar{Z}_{GD}}{X_{C_{GS}}}} \tag{6.12}$$

At low frequencies, where X_{CGS} and X_{CGD} are large, this becomes:

$$G_{out} = \frac{1}{R_F} + g_{DS} + g_{fs} \tag{6.13}$$

or

$$R_{out} \simeq \frac{1}{g_{fs}} \quad \text{if} \quad g_{fs} \gg \frac{1}{R_F}, g_{DS}$$

For some calculations, the output admittance with the input short-circuited is useful. From equation (6.12), this is:

$$y_0 = \frac{1}{R_F} + g_{DS} + g_{fs} + j\omega C_{GS} \quad (6.14)$$

which again reduces to approximately g_{fs} at low frequencies.

Example

For the Semitron C95, $g_{fs} = 1$ mA/V so that when connected as a source-follower with $R_F \gg \frac{1}{g_{fs}}$,

$$G_{out} \simeq 10^{-3}\,\text{mhos}$$

or

$$R_{out} \simeq 1000\,\Omega$$

6.1.3 THE COMMON-GATE STAGE

This configuration, shown in Fig. 6.11, is the least useful of the three basic connections, because the predominant feature of the FET—high input resistance—is lost when the signal is injected into

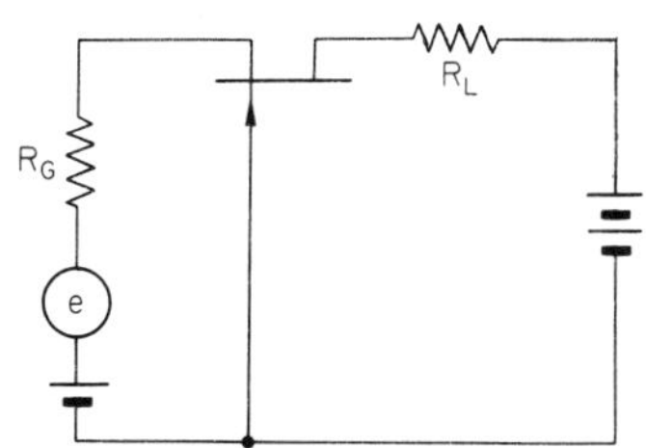

Fig. 6.11. Common gate stage

the source. However, it is still used in high-frequency circuits and its properties will therefore be examined.

The equivalent circuit is as shown in Fig. 6.12, and a conversion from current to voltage generator gives Fig. 6.13, from which the input impedance may be readily derived:

$$I_{in} = V_{in} j\omega C_{GS} + \frac{V_{in} - g_{fs} V_{GS} r_{DS}}{r_{DS} + R_L}$$

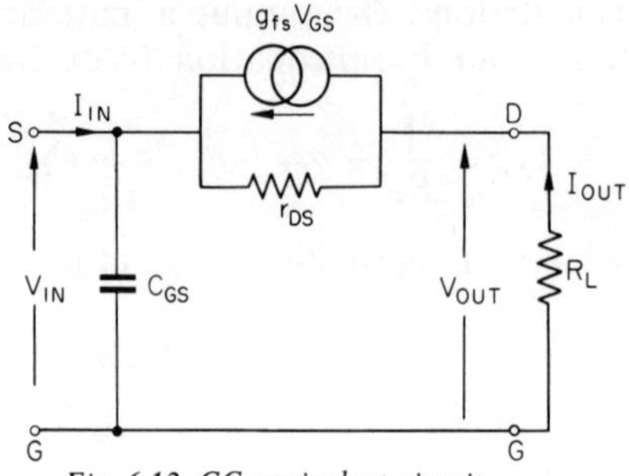

Fig. 6.12. CG equivalent circuit

and since $V_{GS} = -V_{in}$

$$Y_{in} = \frac{I_{in}}{V_{in}} = j\omega C_{GS} + \frac{1 + g_{fs}r_{DS}}{r_{DS} + R_L} \tag{6.15}$$

$$= \frac{1 + \mu}{r_{DS} + R_L} + j\omega C_{GS} \tag{6.16}$$

At low frequencies,

$$R_{in} = \frac{r_{DS} + R_L}{1 + \mu} \tag{6.17}$$

The voltage gain may be derived as follows:

$$V_{out} = -I_{out}R_L = \frac{(V_{in} - g_{fs}V_{GS}r_{DS} - V_{out})R_L}{r_{DS}}$$

whence

$$A_v = \frac{V_{out}}{V_{in}} = \frac{(1 + g_{fs}r_{DS} - A_v)R_L}{r_{DS}}$$

$$= \frac{(1 + g_{fs}r_{DS})R_L}{r_{DS} + R_L} = \frac{(1 + \mu)R_L}{r_{DS} + R_L} \tag{6.18}$$

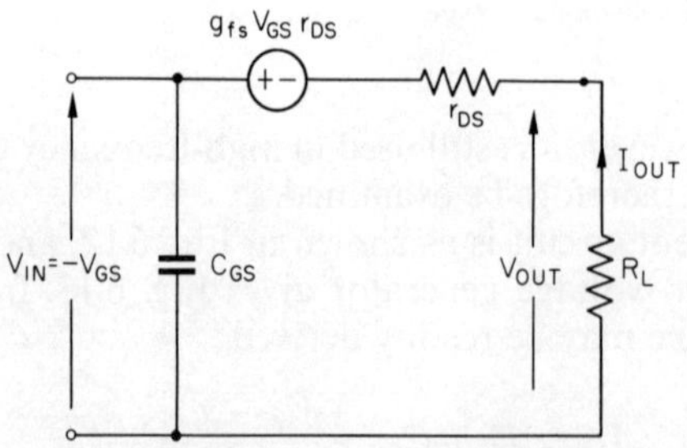

Fig. 6.13. CG equivalent circuit modified for calculation of Y_{in} *and* A_v

If typical parameters are inserted into these equations, it will be observed that the gain and input impedance are similar to those of a rather poor bipolar CE stage, except that there is no phase reversal. This again militates against the use of a CG FET stage.

6.1.4 THE BASIC CONFIGURATIONS COMPARED

The CS and CD stages have both been shown to present high input impedances. However, the bandwidth of the CS stage is less than that of the CD stage, and it has also been stated that the CS stage may become unstable if an inductive load is present. Further, under certain, rather rare conditions involving capacitive load, the CD stage may also become unstable. The voltage gain of the CS stage is not good compared with that for bipolar devices, and this leads to the concept of a combined CD and bipolar stage in the hope of exploiting the advantages of both.

The common-gate stage is not of interest in audio amplification because it presents a low input impedance and offers no advantage over a bipolar stage. It is, however, extremely useful for the amplification of high-frequencies, because the internal feedback capacitance C_{DS} is very small. This is particularly so if the FET is not encapsulated, and some high-frequency devices retain useful amplification into the gigahertz region when in chip form and using strip-line techniques. Were the FET used in the CS mode, then neutralisation would have to be employed, which would inevitably result in quite narrow bandwidths.

High-frequency amplification will be discussed in chapter 12, where devices other than simple junction FETs will be seen to have significant advantages.

6.2. Design of Bias Circuits

6.2.1 BIASING CONSIDERATIONS

At the time of writing, the published parameters for most FETs are subject to very wide tolerances, and this implies that the design of bias circuits must be quite stringent if these tolerances are to be accommodated. Fortunately, the square law transconductance curve is sufficiently accurate to allow such a biasing design to be undertaken.

The circuit of Fig. 6.14 shows the very common series DC feedback bias method. The operation of the circuit is simply that if the voltage V_B is held constant by R_1 and R_2 (which it is, there being no significant gate current), then any change in source current will produce a change in V_{GS} in such a direction as to bring I_S back to

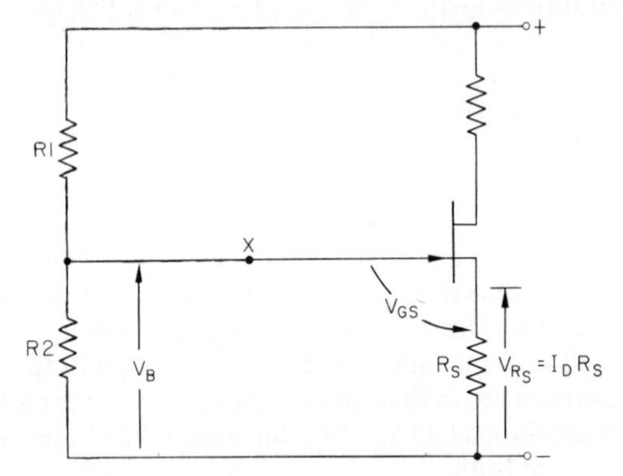

Fig. 6.14. The series DC feedback bias method

nearly its original value. This constitutes series negative DC feedback.

The basic equation of the system sums direct voltages around the input loop:

$$V_{GS} = R_S I_D - V_B \tag{6.19}$$

(because $I_D = I_S$)

This is a straight line equation, which when plotted on the relevant transconductance axes, gives a diagram like that of Fig.

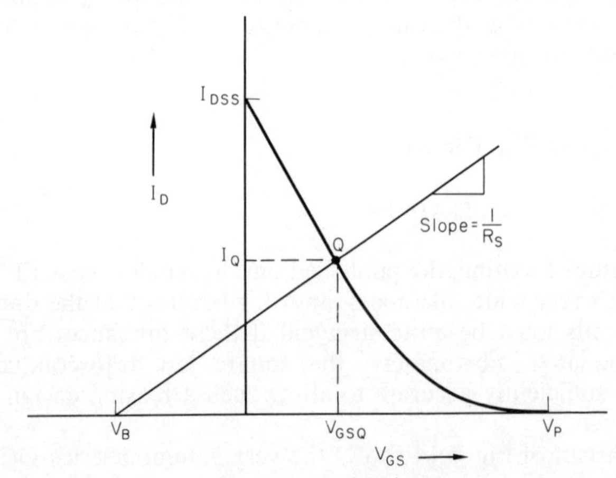

Fig. 6.15. The bias line

6.15. Clearly, the inverse slope of the line is R_S, while the crossing with the abscissa defines V_B. The quiescent working point Q occurs where the transconductance curve cuts the bias line.

It is now possible to extend this concept to cover the case where transconductance curves can be drawn for the two tolerance limits of I_{DSS} and V_P. Fortunately, there is a correlation between the spread of I_{DSS} and V_P because both vary with channel conductivity

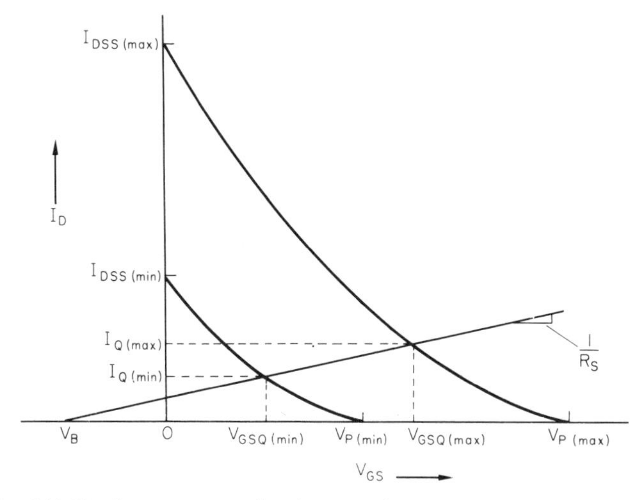

Fig. 6.16. Bias line superimposed on limiting tolerance transconductance curves

and dimensions in the same sense. Hence a FET of a given type having a high I_{DSS} will also have a high V_P, and vice versa.

Transconductance curves for a 'maximum' and a 'minimum' FET are shown in Fig. 6.16, and a bias line cutting the two is superimposed. If either the quiescent currents or the quiescent gate–source voltages are known as design parameters, the values for R_S and V_B may be obtained from this figure[1].

From equation (5.1), the quiescent gate–source voltages are given when $I_D = I_Q$:

$$V_{GSQ(\mathrm{min})} = V_{P(\mathrm{min})}\left[1 - \sqrt{\left(\frac{I_{Q(\mathrm{min})}}{I_{DSS(\mathrm{min})}}\right)}\right] \tag{6.20}$$

and

$$V_{GSQ(\mathrm{max})} = V_{P(\mathrm{max})}\left[1 - \sqrt{\left(\frac{I_{Q(\mathrm{max})}}{I_{DSS(\mathrm{max})}}\right)}\right] \tag{6.21}$$

The inverse slope of the bias line gives R_S:

$$R_S = \frac{V_{GSQ(\max)} - V_{GSQ(\min)}}{I_{Q(\max)} - I_{Q(\min)}} = \frac{\Delta V_{GSQ}}{\Delta I_Q} \tag{6.22}$$

The point at which the bias line crosses the abscissa defines V_B, and from simple co-ordinate geometry, this is:

$$V_B = \frac{V_{GSQ(\min)} I_{Q(\max)} - V_{GSQ(\max)} I_{Q(\min)}}{I_{Q(\max} - I_{Q(\min)}} \tag{6.23}$$

Example

To illustrate the use of equations (6.20) through (6.23), consider a circuit employing the Siliconix 2N4119 *n*-channel FET, which has the following parameters:.

$$I_{DSS} = 0{\cdot}2 \text{ mA min.,} \quad 0{\cdot}6 \text{ mA max.}$$

$$|V_P| = 2 \text{ V min.,} \quad 6 \text{ V max.}$$

Defining $I_{Q(\min)} = 0{\cdot}1$ mA and $I_{Q(\max)} = 0{\cdot}15$ mA, equations (6.20) and (6.21) give:

$$V_{GSQ(\min)} = 2\left[1 - \sqrt{\left(\frac{0{\cdot}1}{0{\cdot}2}\right)}\right] = 0{\cdot}58 \text{ V}$$

and

$$V_{GSQ(\max)} = 6\left[1 - \sqrt{\left(\frac{0{\cdot}15}{0{\cdot}6}\right)}\right] = 3{\cdot}0 \text{ V}$$

Equation 6.22 gives

$$R_S = \frac{3{\cdot}0 - 0{\cdot}58}{0{\cdot}15 - 0{\cdot}1} \simeq 47 \text{ k}\Omega$$

(to nearest preferred value)

Finally, equation (6.23) gives V_B:

$$V_B = \frac{0{\cdot}58 \times 0{\cdot}15 - 3{\cdot}0 \times 0{\cdot}1}{0{\cdot}15 - 0{\cdot}1} = 4{\cdot}3 \text{ V}$$

This is a simple procedure providing that realistic values of I_Q have been assigned, otherwise impractical values for R_S and V_B can result—a situation which is alleviated by the use of nomograms if many such calculations have to be made. (Suitable nomograms are included in the Appendix.)

V_B is defined by R_1 and R_2 in Fig. 6.14, and because it is convenient to use fairly low values here, a high value resistor R_3 is often inserted at point X. (The signal is, of course, taken to the gate itself.) If

$R_3 \gg R_1$, R_2 then for the small-signal calculations described previously, R_G is given simply by R_3 combined with any generator resistance R_g.

If R_3 is large, it is possible that the gate current may produce a significant voltage drop $I_G R_3$. This voltage would be additive with V_B, resulting in a rise in I_Q; a situation which can be represented by moving the bias line bodily to the left by an amount $I_G R_3$.

The magnitude of I_G is often taken as approximately $\frac{1}{2} I_{GSS}$ under operating conditions, but this is not very accurate[2], particularly for n-channel devices at high values of V_{GD}. Also, the temperature

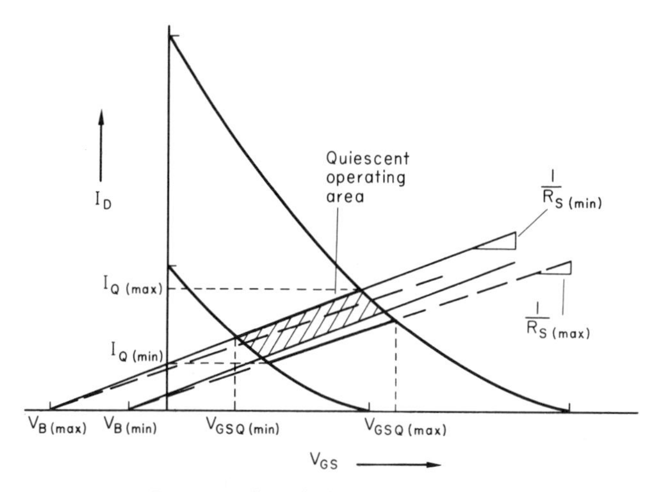

Fig. 6.17. Effect of tolerances in R_S *and* V_B

dependence of I_G must be taken into account, and whereas this is normally assumed to double for every 10°C temperature rise, some caution is again required when the operating value of I_G is involved, particularly at high values of V_{GD}.

The tolerancing of R_3 (and/or R_1 and R_2) also means that V_B will be subject to a spread of values, and this must also be taken into account when determining the position of the bias line.

The slope of the bias line is given by $1/R_S$, and because R_S is subject to tolerancing, there must also be a spread in this slope.

These various effects can be plotted on the transconductance curves as shown in Fig. 6.17, where it will be seen that a quiescent operating area is defined which encloses all possible operating points. Here it has been assumed that the temperature variations in I_{DSS} and V_P have also been taken into account so that the relevant maximum and minimum values represent the true situation for the temperature extremes called for in the design.

This method of biasing is applicable to both the CS and CD configurations, and results in a closely defined value for R_S. However, this value may not be optimum from the point of view of the small signal analysis, particularly in the CS case. Here, it may be necessary to shunt all, or part of R_S with a bypass capacitor as shown in Fig. 6.6. An optimum compromise between biasing and signal requirements is the goal of the design procedure, and at this point further consideration of the relevant signal requirements becomes necessary.

6.2.2 THE DESIGN PROCEDURE (CS STAGE)

For a simple single-stage CS amplifier, the usual requirement is that of highest possible gain for a given bandwidth. Equation (6.2) implies that for high gain, a FET of high g_{fs} should be chosen:

$$A_{v(\mathrm{l.f.})} \simeq -g_{fs} \cdot R_D \tag{6.2}$$

Unfortunately, the geometry called for in the realisation of high g_{fs} also results in a high interelectrode capacitance, and this is much augmented by the Miller capacitance, which itself is a function of $A_{v(\mathrm{l.f.})}$. However, a compromise can be made, and according to equation (6.2), the next choice is that of a high R_D.

The peak output signal required will dictate the choice of power supply voltage, and the proportion of it which will be dropped across the drain resistor, V_{R_D}. Having established V_{R_D}, it is clear that the absolute value of R_D will be in inverse proportion to I_Q, the quiescent drain current. That is,

$$R_D = V_{R_D}/I_Q \tag{6.24}$$

For a high voltage gain, therefore, I_Q must be as low as possible. Recalling that from the transconductance curve, I_D falls according to a square law, whilst g_{fs} falls according to a linear law, it is reasonable to suppose that if I_D is low, the consequent reduction in g_{fs} is more than compensated by the increase in R_D. To show this, consider equation (5.1):

$$I_D = I_{DSS}\left(1 - \frac{V_{GS}}{V_P}\right)^2 \tag{5.1}$$

giving

$$g_{fs} = \frac{\mathrm{d}I_D}{\mathrm{d}V_{GS}} = -\frac{2I_{DSS}}{V_P}\left(1 - \frac{V_{GS}}{V_P}\right) \tag{6.25}$$

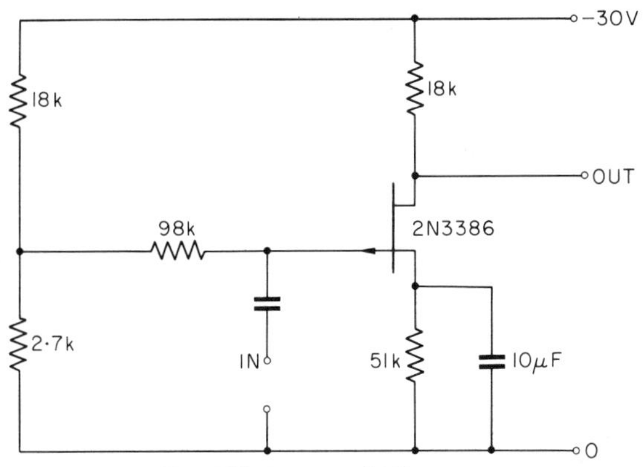

Fig. 6.18. A practical CS stage

Putting (5.1) and (6.25) into (6.2) and (6.24) gives:

$$A_{v(\text{l.f.})} \simeq -g_{fs} \cdot \frac{V_{R_D}}{I_Q} = \frac{\dfrac{2I_{DSS}}{V_P}\left(1 - \dfrac{V_{GS}}{V_P}\right)V_{R_D}}{I_{DSS}\left(1 - \dfrac{V_{GS}}{V_P}\right)^2}$$

or

$$A_{v(\text{l.f.})} \simeq \frac{2V_{R_D}}{V_P - V_{GS}} \tag{6.26}$$

Equation (6.26) is very approximate, because the square-law relationship of equation (5.1) does not hold near $V_{GS} = V_P$. Nevertheless, it does indicate that maximum gain will result if the FET is biased near pinch-off.

In practice, this means that I_Q should be low; but it should of course remain much higher than I_{GDO}. Typically, I_Q may be made about 1% of I_{DSS}, providing that it is also at least two orders of magnitude greater than I_{GDO}.

The power supply voltage and the value of R_D will depend upon the peak value of the output signal required, $V_{\text{out(pk)}}$. At $I_{Q(\text{min})}$ the voltage drop across R_D will be minimal, but must still be greater than $V_{\text{out(pk)}}$:

$$|I_{Q(\text{min})}| \cdot R_D > V_{\text{out(pk)}} \tag{6.27}$$

At $I_{Q(\text{max})}$, the voltage drop across R_D will be maximal, and under these conditions, the source–drain voltage must still be larger than $V_{\text{out(pk)}}$ *plus* $|V_{P(\text{max})}|$. This is so that the FET remains in the pinch-off region, and $|V_{P(\text{max})}|$ is that value of $|V_P|$ associated with the 'maximum' FET which would give rise to $I_{Q(max)}$ in the first place. That is,

$$|V_{DD}| - |V_{P(max)}| - |I_{Q(max)}|(R_S + R_D) > V_{out(pk)} \qquad (6.28)$$

Combining equations (6.27) and (6.28) gives:

$$\frac{V_{out(pk)}}{|I_{Q(min)}|} < R_D < \frac{|V_{DD}| - V_{out(pk)} - |V_{P(max)}| - |I_{Q(max)}|R_S}{|I_{Q(max)}|} \qquad (6.29)$$

Here, V_{DD} is the supply voltage. Because R_S is known from biasing calculations, trial values of V_{DD} and R_S can be inserted into the inequality until a suitable combination is determined.

To obtain a high voltage gain, it is clear that a FET having both high I_{DSS} and low V_P should be chosen, otherwise, inequality (6.29) will be difficult to satisfy for realistic values of V_{DD}, R_S and R_D. An example using a FET whose V_P is fairly marginal in this respect will illustrate the procedure.

Example

The following parameters apply to the Siliconix 2N3386 at 25°C:

$$I_{DSS(min)} = -15\text{ mA} \qquad I_{DSS(max)} = -50\text{ mA}$$

$$V_{P(min)} = 4\text{ V} \qquad V_{P(max)} = 9{\cdot}5\text{ V}$$

Let $I_{Q(min)} = I_{DSS(min)}/100 = -0{\cdot}15$ mA and let $I_{Q(max)} = -0{\cdot}25$ mA. From equations (6.20) and (6.21) (or nomogram A),

$$V_{GSQ(min)} = 4\left[1 - \sqrt{\left(\frac{0{\cdot}15}{15}\right)}\right] = 3{\cdot}6\text{ V}$$

$$V_{GSQ(max)} = 9{\cdot}5\left[1 - \sqrt{\left(\frac{0{\cdot}25}{50}\right)}\right] = 8{\cdot}8\text{ V}$$

From equation (6.22) (or nomogram B),

$$R_S = \frac{8{\cdot}8 - 3{\cdot}6}{0{\cdot}25 - 0{\cdot}15} = \underline{52\text{ k}\Omega}$$

From equation (6.23) (or nomogram C),

$$V_B = 9{\cdot}0 - 13{\cdot}2 = \underline{-4{\cdot}2\ V}$$

If a maximum peak-to-peak output voltage of 5 V is required, then letting $V_{DD} = -30$ V, then inequality (6.29) gives:

$$\frac{2{\cdot}5}{0{\cdot}15} < R_D < \frac{30 - 2{\cdot}5 - 9{\cdot}5 - (0{\cdot}25 \times 52)}{0{\cdot}25}$$

or $16{\cdot}7 < R_D < 20$

Here, a suitable value for R_D would be $\underline{18\text{ k}\Omega}$

The voltage gain of the stage may now be calculated from equation (6.2):

$$A_{v(\mathrm{l.f.})} = -g_{fs}R_D \tag{6.2}$$

Inserting a value for g_{fs} from equation (6.25) gives:

$$A_{v(\mathrm{l.f.})} = \frac{2I_{DSS} \cdot R_D}{V_P}\left(1 - \frac{V_{GS}}{V_P}\right)$$

For 'minimum' parameters, this is:

$$A_{v(\mathrm{l.f.})(\mathrm{min})} = -\frac{2 \times 15 \times 18}{4}\left(1 - \frac{3{\cdot}6}{4}\right) = \underline{-13{\cdot}2}$$

$$A_{v(\mathrm{l.f.})(\mathrm{max})} = -\frac{2 \times 50 \times 18}{9{\cdot}5}\left(1 - \frac{8{\cdot}8}{9{\cdot}5}\right) = \underline{-14}$$

Observing that if R_1 R_2 are small, R_3 appears in parallel with C_{in}, then the high-frequency cut-off point can be found. If the input generator is represented by its Norton equivalent, that is, a perfect current source in parallel with an internal resistance R_g, then f_h will be given when

$$\frac{R_g \cdot R_3}{R_g + R_3} = |X_{C_{in}}|$$

or

$$f_h = \frac{1}{2\pi R C_{in}}$$

where

$$R = \frac{R_g \cdot R_3}{R_g + R_3}$$

In this case, the 'typical' values of C_{gs} and C_{ds} are each 4·8 pF, so taking the average value of A_v as $-13{\cdot}6$,

$$C_{in} = 4{\cdot}8 + (1 + 13{\cdot}6)\,4{\cdot}8 \simeq 75\ \mathrm{pF}$$

By way of example, a very high resistance was connected in series with a signal generator as that $R \simeq R_3$, and R_3 was made 100 kΩ, giving

$$f_h = \frac{10^{12}}{2\pi 10^5 \times 75} \simeq \underline{21\ \mathrm{kHz}}$$

The low-frequency cut-off point will be defined by either the input time constant (if a generator series capacitor exists); or the output time constant formed by C_S in parallel with the output resistance at the source. This output resistance is given by the expression relevant to the source-follower, equation (6.13) which in this case is:

$$R_{\text{out}(CD)} \simeq 1/g_{fs}$$

Here, g_{fs} is given by equation (6.25) as ranging from 0·73 to 0·78 mA/V. Hence,

$$R_{\text{out}} \simeq 1{\cdot}36 \text{ to } 1{\cdot}27 \text{ k}\Omega$$

Hence, if $C_S = 10$ microfarads, the low-frequency cut-off point will be:

$$f_1 = \frac{10}{2\pi 1{\cdot}36 \times 10} = 11{\cdot}7 \text{ Hz.}$$

The circuit relevant to this example is shown in Fig. 6.18, and using an arbitrary 2N3386, some measurements were made:

$$I_Q = -0{\cdot}16 \text{ mA}$$

$$A_{v(1 \text{ kHz})} = -13{\cdot}3$$

$$f_h = 21 \text{ kHz}$$

$$f_1 \simeq 11{\cdot}4 \text{ Hz}$$

Although the input resistance is high, this stage does not have either a high gain or a good bandwidth. This is typical of currently available FETs and means that it is usually a better design philosophy to specify a CD connected FET followed by a bipolar transistor where both high input impedance, good bandwidth and good voltage gain are required simultaneously. Such pairs will be treated later; for the present, the design of the CD stage will be briefly considered.

6.2.3 THE DESIGN PROCEDURE (CD STAGE)

The design of a bias network for the source-follower is obviously identical to that for the CS stage, and will lead to the simple circuit of Fig. 6.19. For this stage, the gain will always be positive and less than unity, so that there is no reason to reduce the quiescent current to a very small value as was the case for the CS stage. Instead, I_Q may be chosen to give convenient resistor and power supply voltages having regard to the nature of the envisaged load, and the output voltage swing required.

If the peak output voltage is to be $V_{\text{out(pk)}}$, then,

$$|I_{Q(\text{min})}|R_F > V_{\text{out(pk)}} \tag{6.30}$$

Here, R_F is the unbypassed source resistor.

$$\text{Also,}\quad |V_{DD}| - |V_{P(\max)}| - |I_{Q(\max)}|R_F > V_{\text{out(pk)}} \tag{6.31}$$

Combining equations (6.30) and (6.31) gives:

$$\frac{V_{\text{out(pk)}}}{|I_{Q(\min)}|} < R_F < \frac{|V_{DD}| - |V_{P(\max)}| - V_{\text{out(pk)}}}{|I_{Q(\max)}|} \tag{6.32}$$

From this equation, R_F may be chosen to suit the desired values of I_Q by trial and error, and the bias design completed by using

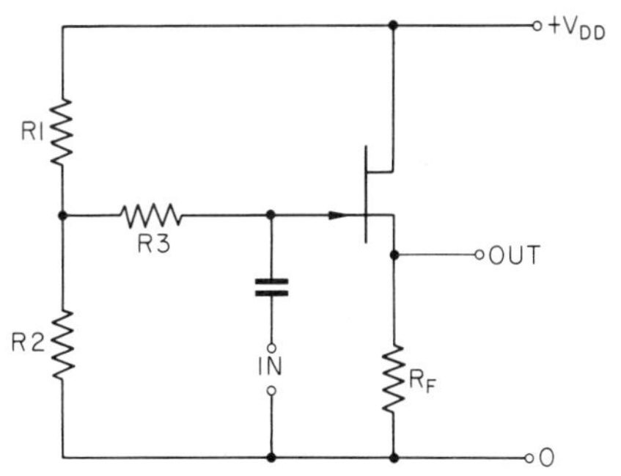

Fig. 6.19. Simple source follower

equations (6.20) through (6.23) or the nomograms. Again, R_1 and R_2 can be made small, and the input resistance defined by a series gate resistor R_3.

A simple, and very useful circuit which achieves an improved input resistance, and is fully self-biasing, is the *bootstrap-follower*, shown in Fig. 6.20. Bootstrapping refers to the technique of applying positive feedback to increase the apparent value of the input resistor, and this will be demonstrated. First, however, consider the bias conditions relevant to Fig. 6.20. Clearly, $V_{GSQ} = I_Q R_{F1}$, so that,

$$I_Q = I_{DSS}\left[1 - \frac{I_Q R_{F1}}{V_P}\right]^2$$

which, if a value for I_Q is postulated, leads to a design equation for R_{F1},

$$R_{F1} = \frac{V_P}{I_{DSS}}\left[1 - \sqrt{\left(\frac{I_Q}{I_{DSS}}\right)}\right] \tag{6.33}$$

The tolerances and temperature variations in V_P I_{DSS} and R_{F1} may, of course, be inserted to give a range of quiescent currents $I_{Q(\min)}$ to $I_{Q(\max)}$.

R_{F2} may then be found simply by establishing that V_{DD} is sufficiently high to provide all the voltage drops required plus the signal swing, as detailed in the foregoing discussions.

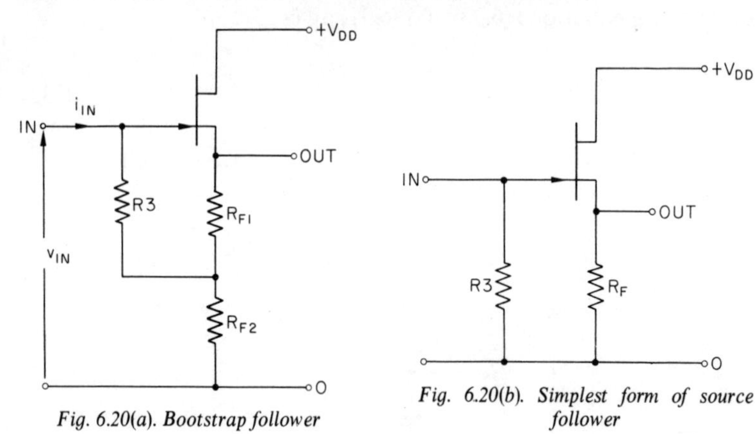

Fig. 6.20(a). Bootstrap follower

Fig. 6.20(b). Simplest form of source follower

Fig. 6.20(c). Fully bootstrapped AC source follower

The input resistance may be determined by combining the expressions for i_{in} and v_{gs}, which are, by inspection:

$$i_{\text{in}} = \frac{v_{\text{in}} - i_d R_{F2}}{R_3} = \frac{v_{\text{in}} - g_{fs} v_{gs} R_{F2}}{R_3}$$

and $v_{gs} = v_{\text{in}} - i_d(R_{F1} + R_{F2}) = v_{\text{in}} - g_{fs} v_{gs}(R_{F1} + R_{F2})$

A little manipulation shows that,

$$R_{\text{in}} = \frac{v_{\text{in}}}{i_{\text{in}}} = R_3 \left[\frac{1 + g_{fs}(R_{F1} + R_{F2})}{1 + g_{fs} R_{F1}} \right] \tag{6.34}$$

Notice that if R_3 is taken to the common-line, the circuit becomes the simplest form of source-follower. That is, $R_{F2} = 0$ so that $R_{in} = R_3$. The circuit is now as shown in Fig. 6.20(*b*). Notice also that the difference between the input and the output voltages—that is, the offset voltage—is simply V_{GS}. If the stage is intended to accept DC signals, this offset can be reduced to zero by tapping off the output signal part-way down R_F, though part of the gain will thereby be lost. Inevitably, drift will occur with temperature, but this may be minimized by working the FET near the zero-temperature-coefficient point.

Returning to Fig. 6.20(*a*), if R_{F1} were made zero, so that R_3 was returned directly to the source, then from equation (6.34):

$$R_{in} = R_3(1 + g_{fs}R_{F2})$$

Recalling that $A_{v(l.f.)}$ for a source-follower is given by

$$A_{v(l.f.)(CD)} = \frac{g_{fs}R_F}{1 + g_{fs}R_F} \tag{6.10}$$

this becomes,

$$R_{in} = \frac{R_3}{1 - A_{v(l.f.)(CD)}} \tag{6.11}$$

For a FET having a high value of g_{fs}, the voltage gain can approach $+1$, so making R_{in} very high indeed.

From a biasing point-of-view, however, the absence of R_{F1} means that, apart from the very small volt drop due to the bridge of silicon between the source lead and the active part of the channel, the value of $I_D R_S$, and hence of V_{GSQ}, would be zero. In other words, the drain current would be I_{DSS}, and the stage would not operate for signals of more than a few mV. However, the circuit can be realized for AC signals by retaining R_{F1}, and shunting it with a capacitor as shown in Fig. 6.20 (*c*). Here, R_{F1} continues to set the bias conditions, but the full AC output signal of the stage is applied to R_3, so rendering equation (6.11) valid.

6.3. The IGFET as a small-signal amplifier

As a voltage amplifier, the IGFET has a somewhat poorer performance than does the junction FET, because in general it exhibits an even lower value of g_{fs}. Also, its noise performance is worse than that of the junction FET. However, its phenomenally high input resistance makes the IGFET suitable for applications in electrometry, and the depletion type has applications where input voltages of both senses must be accepted.

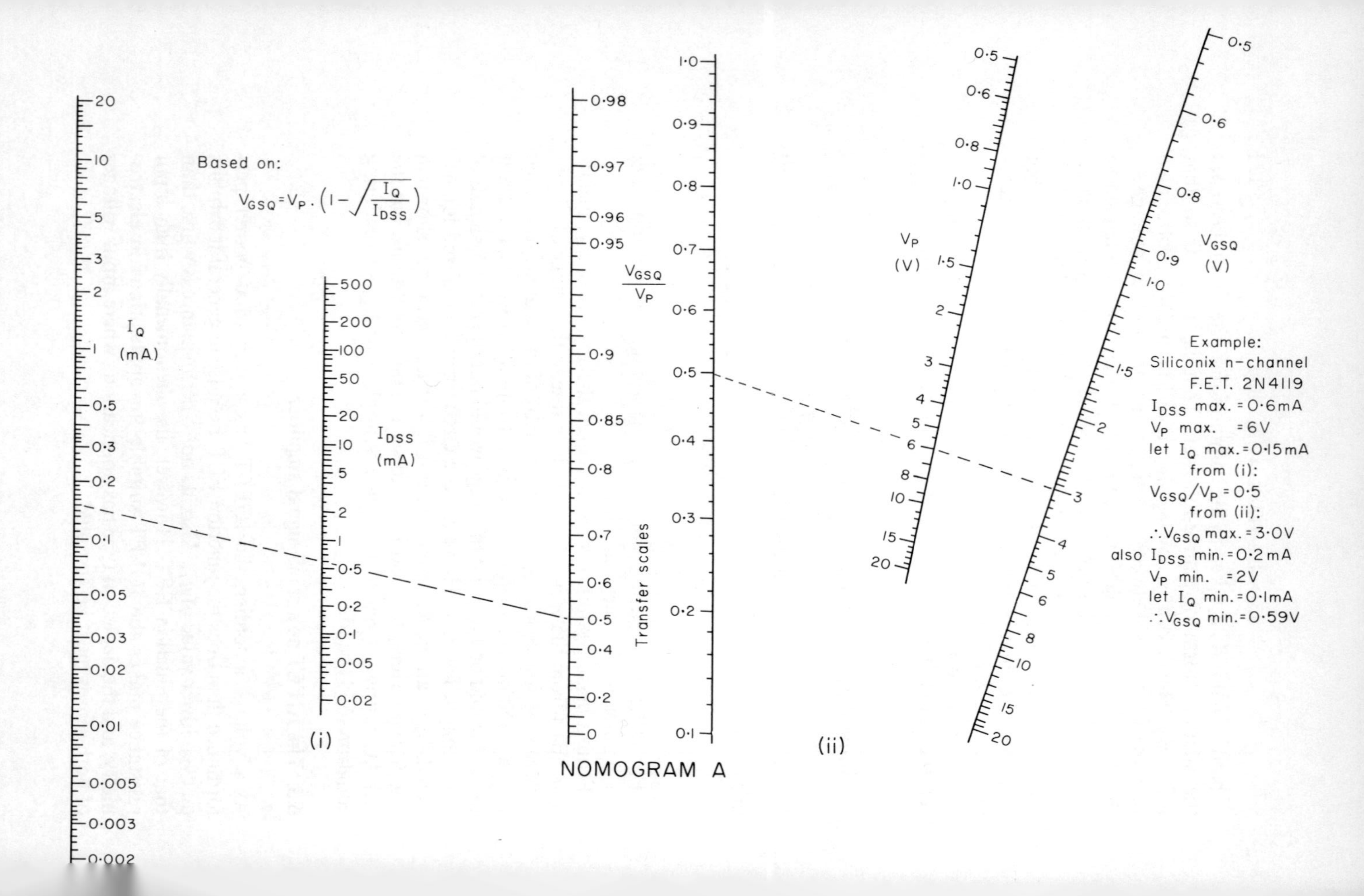

Based on:
$V_{GSQ} = V_P \cdot \left(1 - \sqrt{\frac{I_Q}{I_{DSS}}}\right)$
I_Q (mA)
I_{DSS} (mA)
$\frac{V_{GSQ}}{V_P}$
Transfer scales
V_P (V)
V_{GSQ} (V)
(i)
(ii)
Example:
Siliconix n-channel
F.E.T. 2N4119
I_{DSS} max. = 0·6 mA
V_P max. = 6V
let I_Q max. = 0·15 mA
from (i):
V_{GSQ}/V_P = 0·5
from (ii):
∴ V_{GSQ} max. = 3·0V
also I_{DSS} min. = 0·2 mA
V_P min. = 2V
let I_Q min. = 0·1 mA
∴ V_{GSQ} min. = 0·59V

NOMOGRAM A

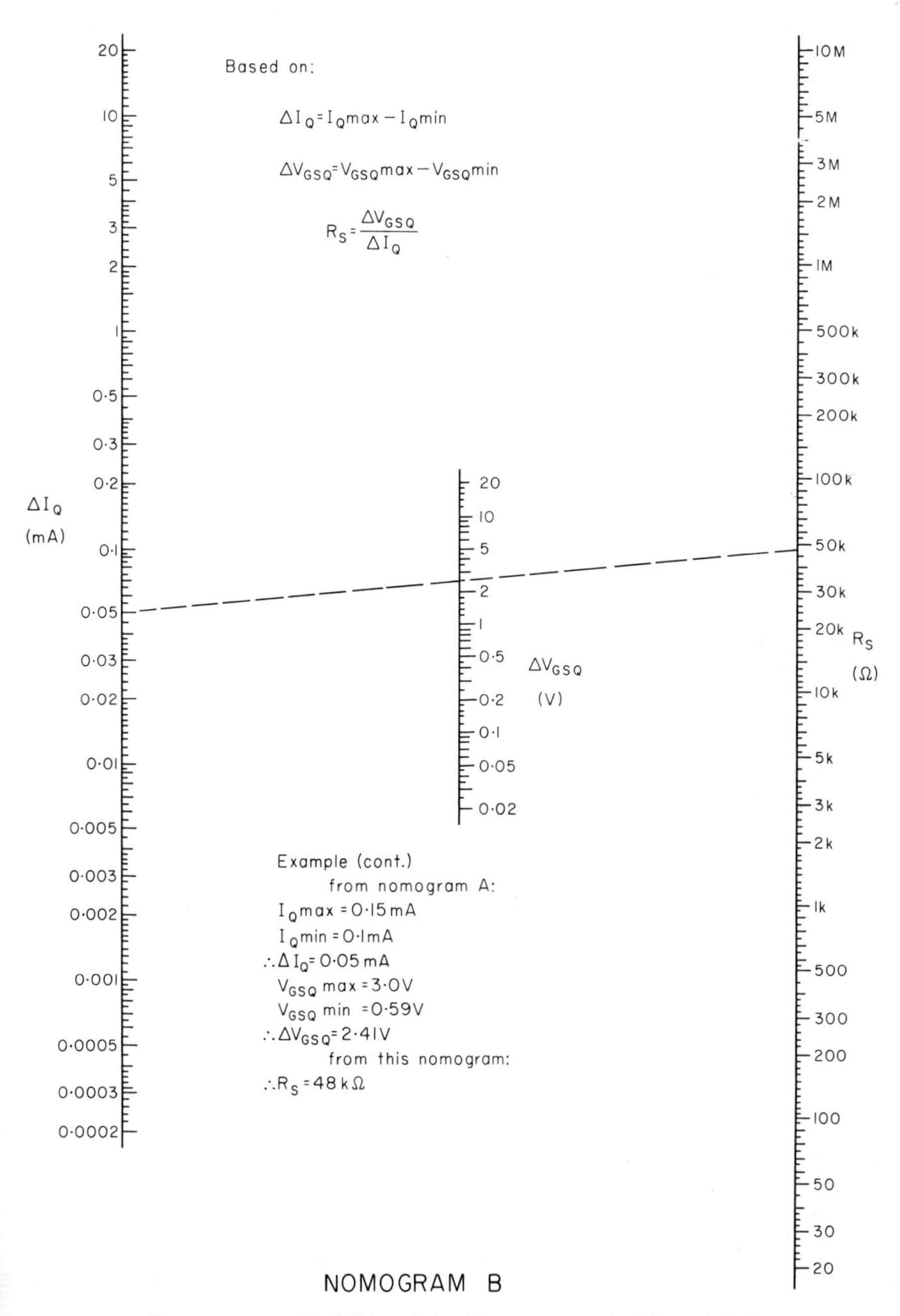

(Nomograms prepared by W. E. Eder (Univ. of Calgary) and reproduced by permission)

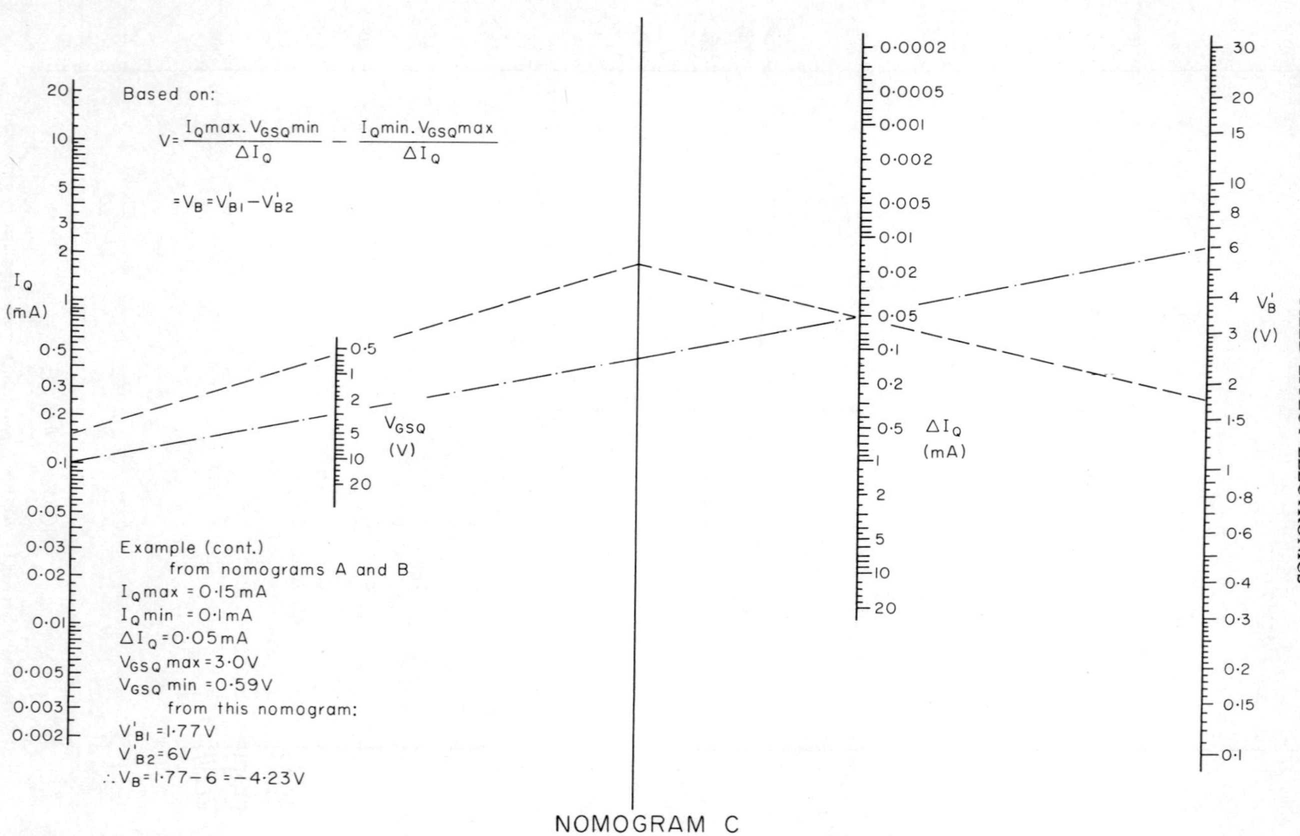

Based on:
$V = \frac{I_Q max. V_{GSQ} min}{\Delta I_Q} - \frac{I_Q min. V_{GSQ} max}{\Delta I_Q}$
$= V_B = V'_{B1} - V'_{B2}$
I_Q (mA)
20 10 5 3 2 1 0·5 0·3 0·2 0·1 0·05 0·03 0·02 0·01 0·005 0·003 0·002
V_{GSQ} (V)
0·5 1 2 5 10 20
ΔI_Q (mA)
0·0002 0·0005 0·001 0·002 0·005 0·01 0·02 0·05 0·1 0·2 0·5 1 2 5 10 20
V'_B (V)
30 20 15 10 8 6 4 3 2 1·5 1 0·8 0·6 0·4 0·3 0·2 0·15 0·1
Example (cont.)
from nomograms A and B
$I_Q max = 0·15 mA$
$I_Q min = 0·1 mA$
$\Delta I_Q = 0·05 mA$
$V_{GSQ} max = 3·0 V$
$V_{GSQ} min = 0·59 V$
from this nomogram:
$V'_{B1} = 1·77 V$
$V'_{B2} = 6 V$
$\therefore V_B = 1·77 - 6 = -4·23 V$

NOMOGRAM C

The discrete IGFET is rarely used as a small-signal amplifier, but because of its comparative ease of manufacture in large numbers, it has become the predominant active device in integrated circuits.

Should it be desired to use the discrete IGFET as an amplifier, however, the design procedures developed for the junction FET can usually be used. Fig. 5.14 shows the equivalent circuit for the IGFET, and though it differs in detail from that of the junction FET, it is sufficiently similar for the junction FET equations to remain valid. These equations will therefore not be treated further.

6.3.1 BIASING PROCEDURE

The biasing procedure outlined for the junction FET is also applicable to an IGFET operating in the depletion mode. For the enhancement mode, however, the much simpler self-biasing method of Fig. 6.21 is available. Here, a biasing resistor R_B connects the

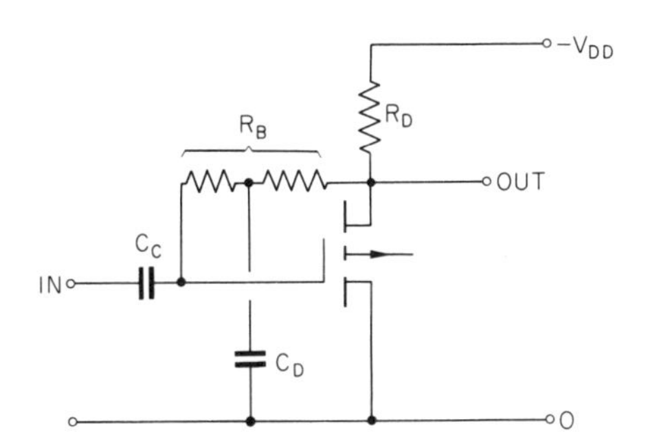

Fig. 6.21. Self-biased enhancement IGFET

gate to the drain, and because the gate leakage current is essentially zero, this means that the gate is at the same potential as the drain. The locus of points where $V_{GS} = V_{DS}$ may be superimposed on the output characteristics as shown in Fig. 6.22, and the intersection of the load line with this locus defines the quiescent working point for the IGFET.

This working point is defined entirely by R_D, but the presence of R_B implies that signal feedback can take place between drain and gate. If required, therefore, a decoupling capacitor may be inserted half way along R_B as shown in Fig. 6.21. Normally $R_B \gg R_D$, so that the input impedance for frequencies where X_{C_D} and X_{C_C} are very small will be simply $R_B/2$.

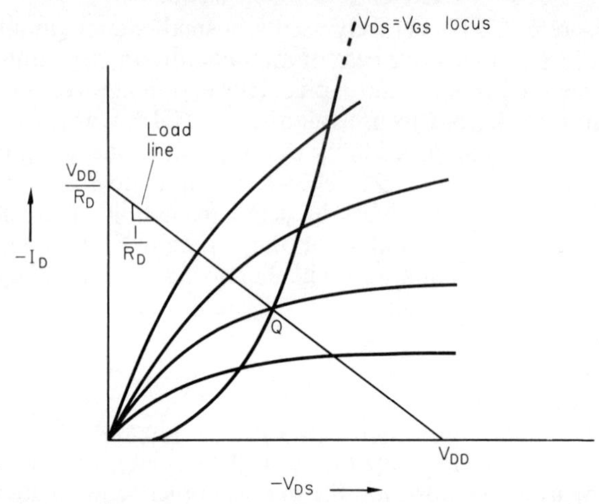

Fig. 6.22. Output characteristic showing working point for self-biased enhancement IGFET

This method of biasing is analogous to the shunt feedback biasing technique often used with bipolar transistors.

REFERENCES

1. Watson, J. 'Biasing considerations in FET Amplifier Stages'. *Electronic Engineering,* **40**, 489 (Nov. 1968).
2. MacDonald, C. L. 'Behaviour of FET Gate Current'. *Siliconix Inc. Application Tip*. (April 1969).

7

UNIPOLAR/BIPOLAR AMPLIFIERS

Field-effect transistors have extremely high current gain, since virtually no current flows in the input circuit in the usual CS or CD amplifier configurations. The voltage gain obtainable, however, is rather modest, typically between $\times 10$ and $\times 50$. By contrast bipolar transistors have a lower current gain but can have higher voltage gain. A combination of the two might thus be expected to give high power gain and very substantial amplification of both voltage and current.

The field effect device has the desirable property of high input impedance, whilst the bipolar transistor can have fairly low output impedance and also bottoms to a lower voltage than that typical of a FET and so has a higher possible power efficiency as an output stage. It is thus reasonable to direct attention to two stage amplifiers using a FET in the first stage and a bipolar device in the second[1].

To attain high input impedance the FET stage should be in the CS or CD configuration whilst the second, bipolar, stage may be CE, CC or CB, although the last will give a rather high output impedance, in many cases. Thus there are six circuit configurations of intrinsic interest: CS–CE, CS–CC, CS–CB, CD–CE, CD–CC, CD–CB. The properties of the amplifiers are very sharply dependent on the configuration of the second stage, and it will therefore be convenient to discuss them in three groups divided on this basis.

In the first instance it will be assumed that the FET is in all cases operated with a sufficient drain-source voltage to ensure that the channel is pinched-off. Interesting circuit properties can, in fact, be obtained when this is not the case, but consideration of circuits of this kind will be delayed until the properties of those with more conventional operating conditions have been discussed.

7.1. Amplifiers with CE output stages: introduction

The CE amplifier has high current gain with reasonably large voltage gain. Its output impedance is intermediate between that of CC and CB amplifiers. These characteristics largely determine the properties of two stage amplifiers using CE output stages.

Considering first the CD–CE amplifier; a typical circuit is shown in Fig. 7.1(*a*). In the junction gate form the gate will normally (in the

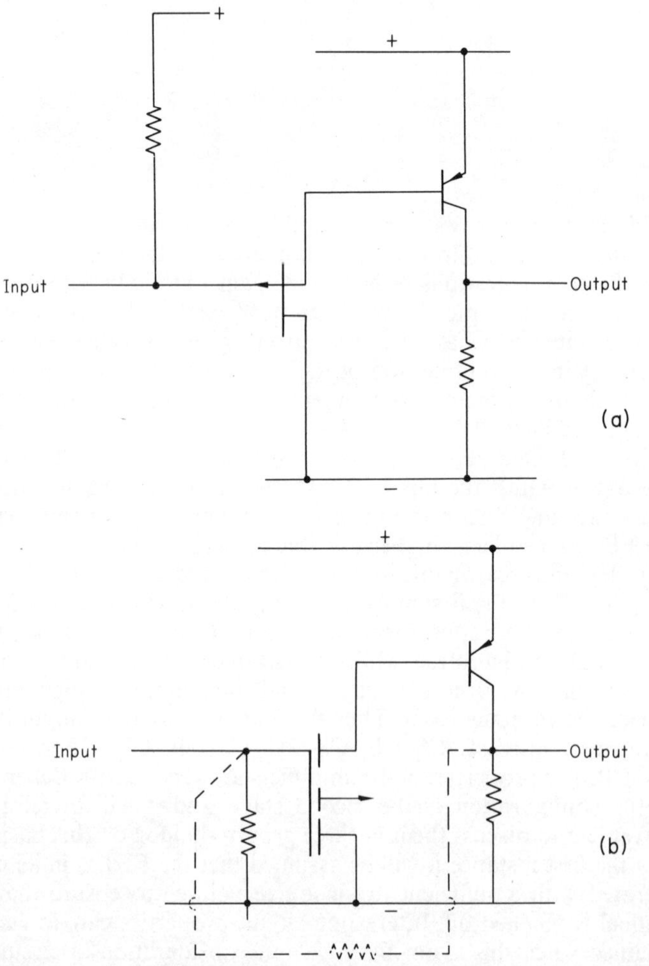

Fig. 7.1(a). A CD–CE amplifier using a junction FET(a) and enhancement MOST(b) The resistor shown in broken line (b) introduces a shunt negative voltage feedback

p-channel case) be at a voltage more positive than the positive supply rail. This is inconvenient, since it implies that a separate bias supply will be required. With an enhancement mode MOST, Fig. 7.1(*b*), this difficulty does not arise. The gate potential falls between the two supply rails and since the amplifier is phase inverting, excellent DC stabilisation can be achieved by overall DC feedback as shown, decoupled for signal frequencies.

The gate circuit resistors need to be large in value to achieve best possible low frequency response with minimum capacitor values. For example, on integrated circuits where use of high resistor values is forbidden, these components may be replaced by pinch resistors on reverse biased *p–n* junctions of small cross-sectional area. With this circuit configuration shunt voltage negative feedback as shown in Fig. 7.1(*b*), is relatively easily applied, as in series current feedback (by inclusion of a resistor in the bipolar transistor emitter lead) but shunt current feedback and series voltage feedback cannot easily be arranged. This last is particularly unfortunate, since it is negative feedback of this type which raises input impedance and lowers output impedance, both of which changes are usually design objectives.

Provided that the rise in output impedance resulting from the use of current feedback can be tolerated, however, the introduction of a resistor in series with the bipolar transistor emitter will give improved gain stabilisation and tend to reduce the apparent input capacitance of the amplifier. It will also modify the DC conditions, and by making the emitter of the bipolar device, and hence the FET source, negative relative to the positive supply rail, it becomes possible to operate a junction or depletion mode *n*-channel FET with the gate at the potential of the positive rail.

In calculating the gain (without feedback) of the CD–CE amplifier it is most useful to begin by calculating the overall transfer conductance $g_{f,1}$. The voltage gain of the CD stage is just

$$a_1 = \frac{g_{fs}\, r_{\text{in}}}{1 + g_{fs}\, r_{\text{in}}}$$

provided that $r_{\text{in}} \ll r_{DS}$, where r_{in} is the input resistance of the CE amplifier, g_{fs} is the transfer conductance of the FET, and r_{DS} is the incremental drain–source impedance of the FET.

Hence it follows at once that

$$g_{f,1} = \frac{h_{fe}\, g_{fs}}{1 + g_{fs}\, r_{\text{in}}} \tag{7.1}$$

The value of r_{in} may, of course, be obtained quite readily from the hybrid-pi parameters of the bipolar device as

$$r_{in} = h_{ie} - \frac{h_{fe}h_{re}}{(1 + R_L h_{oe})} R_L \tag{7.2}$$

The exact significance of equation (7.1) depends on the relative magnitudes of g_{fs} and r_{in}. Two limiting cases are of particular interest. For FETs of low g_{fs}, particularly when operating into bipolar transistors having large values of load resistor, the product $g_{fs}r_{in}$ may be small compared with unity. In this case

$$g_{f,1} \rightarrow h_{fe}g_{fs}$$

This, as will be shown subsequently, is just the value that would be obtained with the CS–CE configuration, and in fact the properties of the CD–CE configuration are, in this case, closely similar in respect of gain to those of the variant with the FET connected in common source.

Modern FETs, however, have values of g_{fs} large enough to make $g_{fs}r_{in}$ of the order of, or even larger than, unity. In this case, from equation (7.1)

$$g_{f,1} \rightarrow \frac{h_{fe}}{r_{in}}$$

This represents, in effect, a limiting value of transfer conductance, beyond which it is impossible to go by increasing g_{fs}. Taking r_{in} as of the order of a few hundred ohms and h_{fe} as of the order of one hundred, it will be seen that the maximum transfer conductance for the amplifier as a whole ranges from a few tens of millimhos up to a few mhos, depending on the transistors chosen. This seems to suggest that high voltage gains should be attained, particularly as the output impedance of the amplifier is high. Since the output impedance of the CD stage is approximately $1/g_{fs}$ the output impedance of the CE bipolar device is

$$r_{out,1} = \frac{h_{ie} + 1/g_{fs}}{\Delta h + h_{oe}/g_{fs}} \tag{7.3}$$

which tends to $1/h_{oe}$ for small g_{fs}.

However, where, as is usually the case, the CD–CE amplifier works into a resistive load, high voltage gains are not easily attained. Although the forward transfer conductance has been greatly increased relative to g_{fs} by the addition of the CE stage (in fact, has been increased by a factor of up to h_{fe}) the standing DC current has also been increased from I_D to I_C which, with the simple direct coupled circuits described, will be of the order of $h_{FE}I_D$. For any amplifier with a resistive load, the factor determining gain is the ratio of forward conductance to standing DC current, assuming high output impedance, and in the present case this has only improved in the ratio h_{fe}/h_{FE}. This ratio is typically about unity or only a little larger. Thus, the simple direct-coupled configuration

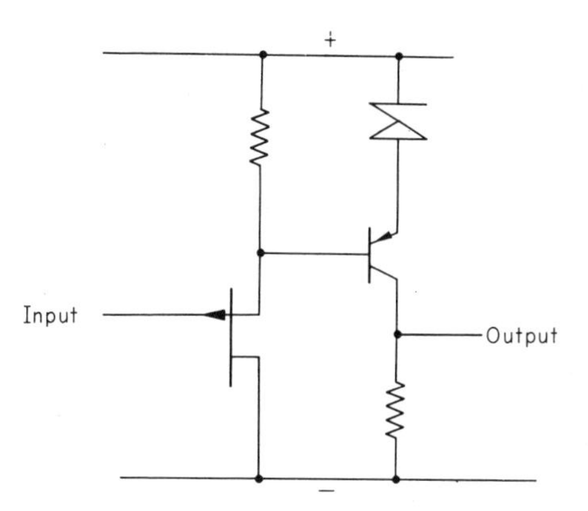

Fig. 7.2. A CD–CE amplifier with Zener diode coupling network

gives little voltage gain advantage (compared with the same FET in CS configuration) into a resistive load, although it might well be worth considering when operated into a transformer coupled load.

Simple circuit modifications can effect a substantial improvement. For example, the zener diode coupling network of Fig. 7.2 will substantially reduce I_C with only a minimal reduction of $g_{f,1}$. In this way very large voltage gains can be obtained into a resistive load. However, the drawback with this circuit is that variations in drain current of the FET due to temperature changes, or other causes, appear undiminished at the base of bipolar device, and thus correspond to a much larger fractional change of collector current in circuits of this kind than in the simple circuits of Fig. 7.1.

Where DC response is not required some kind of overall DC stabilising feedback circuit provides a solution to the problem which is all the more effective because of the relatively large forward-path gain of the amplifier. Better stabilisation can be achieved in this way than by using an AC coupling between the CD and CE stage and stabilising the stages independently, also an AC coupling at this point is not desirable because the low input impedance of the bipolar device will involve the use of a relatively high value capacitor for a given lower cut-off frequency.

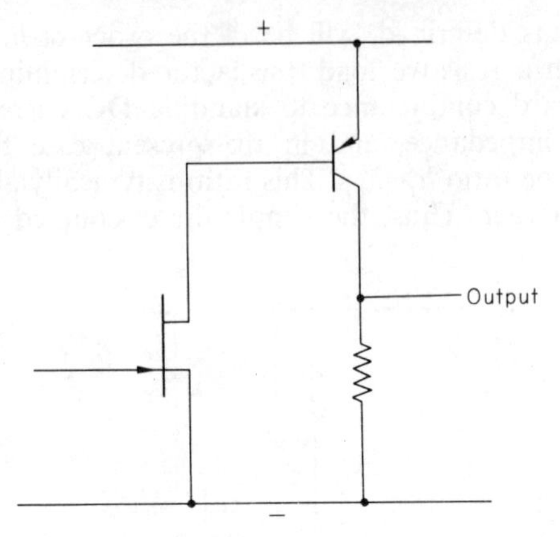

Fig. 7.3. A CS–CE amplifier, using a n*-channel FET*

The CS–CE amplifier (Fig. 7.3) presents very similar features, except that in this case the forward transfer conductance is

$$g_{f,2} = g_{fs}h_{fe} \tag{7.4}$$

and does not reach a limiting value for large g_{fs}, as in the CD–CE case. Thus for transistors of large g_{fs} the CS configuration for the first stage will often result in a higher transfer conductance than would have been obtained with a CD stage. The output impedance must now be calculated in terms of a source impedance to the bipolar

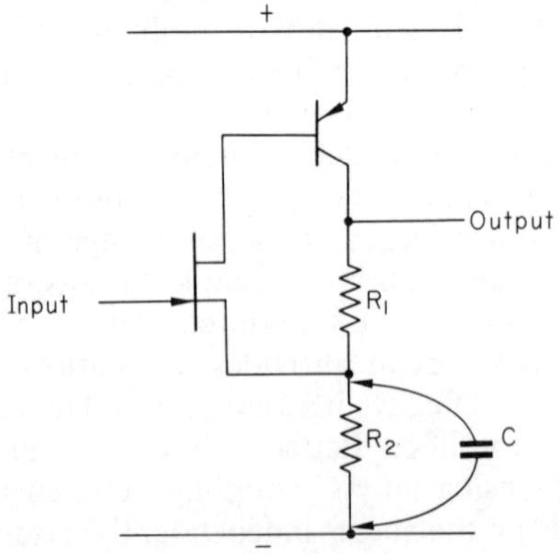

Fig. 7.4. A CS–CE amplifier with overall negative series voltage feedback. The voltage gain will be close to $(R_1 + R_2)R_2$. *Feedback may be made inoperative at* AC *by connecting a suitable capacitor C as shown*

transistor of r_{DS} (instead of $1/g_{fs}$ as in the CD case) and is, therefore, very close to $1/h_{oe}$, since r_{DS} is usually large.

The same problem of high collector current arises as for the CD–CE form, and it is treated in the same way. The amplifier is non-phase-inverting so that negative feedback is best provided from the bipolar device collector to the source terminal of the FET. In this way, series voltage feedback is most easily attained (Fig. 7.4), a strongly advantageous feature of the circuit configuration, since it results in the high input impedance and low output impedance which are very common design goals.

Alternatively, if high AC gain is required the feedback path can be bypassed for signal frequencies by a capacitor connected in shunt with the resistor R_2, and in this way DC stabilisation is retained.

By omitting the resistor R_1 a unity gain amplifier of very low output impedance results.

7.1.1 FREQUENCY RESPONSE

Two stage amplifiers of the type described, with CE output stage, can have a low frequency response extending down to zero frequency, when the usual DC amplifier drift problems will be encountered. Consideration of the techniques of circuit design used to reduce drift will be found in chapter 9. Since, like other semiconductor devices, the FET has a $1/f$ noise spectral density function, the low-frequency response should not be extended beyond the limits of the signal spectrum or needless degradation of the amplifier noise factor will result.

So far as high frequency limitations are concerned, three factors are important in amplifiers of this kind:

(a) capacitative loading of the amplifier output,
(b) the combined effect of amplifier input capacitance and non-zero signal source impedance, and
(c) a fall in forward transfer conductance at high frequencies due to the large equivalent input capacitance of the bipolar device.

The first of these is not peculiar to amplifiers of the present type, its effect can be calculated in an entirely conventional manner, and it therefore need not be further discussed here. So far as the second is concerned, the frequency at which gain would fall by 3 dB due to this cause alone is just

$$\omega_1 = \frac{1}{R_G C_{in}} \qquad (7.5)$$

where R_S is the signal source impedance, and C_{in} is the input capacitance of the amplifier.

The CD and CS input stages differ slightly in their input capacitance. In the case of the CD amplifier the gate–source capacitance is 'bootstrapped' and the input capacitance is thus

$$C_{in,1} = C_{GD} + C_{GS}(1 - a_1)$$

$$= C_{GD} + C_{GS} \cdot \frac{1}{1 + g_{fs} r_{in}} \tag{7.6}$$

By contrast, in the CS case the component of input capacitance derived from C_{GD} is increased by Miller effect, so that in this case the input capacitance is

$$C_{in,2} = C_{GS} + C_{GD}(1 + g_{fs} r_{in} \tag{7.7}$$

It will be seen that $C_{in,2}$ is invariably larger than $C_{in,1}$ so that the CS amplifier will fare worse than the CD in respect of high frequency response, unless the signal source impedance is negligibly small.

So far as the high frequency characteristics of the bipolar transistor are concerned, it is convenient for the purposes of discussion to

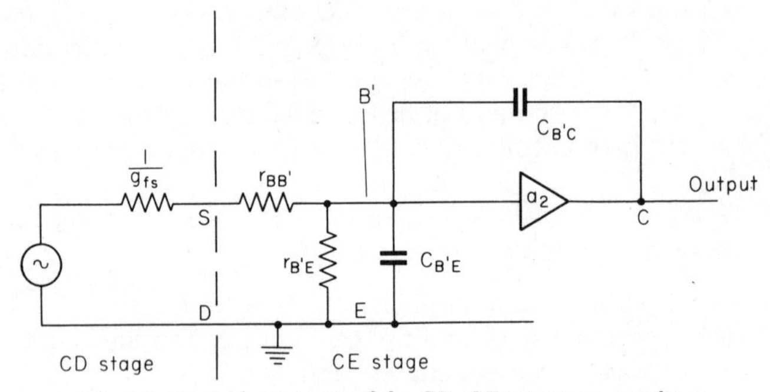

Fig. 7.5. Equivalent circuit of the CD–CE interstage coupling

represent this device by its hybrid-pi equivalent circuit. Then the interstage coupling between the two transistors will look, in the CD–CE case, like the equivalent circuit shown in Fig. 7.5, assuming that the output impedance of the CD stage may be approximated as $|1/g_{fs}|$. By inspection the cut-off frequency for the interstage coupling is

$$\omega_2 = \left\{ \frac{r_{B'E}(1 + g_{fs} r_{BB'})}{1 + (r_{B'E} + r_{BB'}) g_{fs}} \cdot \left[C_{B'E} + (1 - a_2) C_{B'C} \right] \right\}^{-1} \tag{7.8}$$

or, if, as usual for high g_{fs} FETs,

$$\frac{1}{g_{fs}} + r_{BB'} \ll r_{B'E}$$

the cut-off frequency is well approximated by the simpler expression

$$\omega_2 = \frac{g_{fs}}{(1 + g_{fs}\, r_{BB'})\, C'_{\text{in}}} \tag{7.9}$$

where $C'_{\text{in}} = C_{B'E} + (1 - a_2)\, C_{B'C}$

For the CS–CE amplifier an expression similar to equation (7.8) is valid, except that the output impedance of the first stage is now not $|1/g_{fs}|$ but r_{DS}. Since r_{DS} is much larger than $r_{B'E}$ an approximation comparable with equation (7.9) is not possible, but instead, provided that

$$r_{DS} \gg r_{B'E}$$

it follows that in this case

$$\omega'_2 = \frac{1}{r_{B'E} C'_{\text{in}}} \tag{7.10}$$

Comparing (7.9) and (7.10)

$$\frac{\omega^2}{\omega'_2} = \frac{r_{B'E}\, g_{fs}}{1 + r_{BB'}\, g_{fs}}$$

This ratio is invariably greater than unity and tends to $r_{B'E}/r_{BB'}$ for large g_{fs}. Thus the CD–CE amplifier has an advantage in high frequency response which becomes more marked for larger values of g_{fs}.

7.1.2 EFFECT OF A BASE-EMITTER RESISTOR

An interesting variant of the CD–CE and CS–CE amplifiers results when the voltage of the zener diode in the coupling circuit of Fig. 7.3 is reduced to zero, so that the circuit becomes the simple direct-coupled configuration of Figs. 7.1 and 7.2, but with a resistor connected between base and emitter of the bipolar device. Since some of the FET current flows through this resistor, the collector current will be reduced. Other effects depend on the particular configuration chosen.

In the CD–CE circuit the output impedance of the FET is so low (assuming reasonably large g_{fs}) that substantial diversion of direct current from the base of the bipolar device is possible without very much loss of signal. If the base-emitter voltage of the bipolar device is V_{BE} and the resistance is of value R then

$$R = \frac{V_{BE}}{I_D - I_B} \gg \frac{1}{g_{fs}}$$

for negligible signal loss, but

$$g_{fs} = \frac{2I_{DSS}}{V_P}\left\{1 - \frac{V_{GS}}{V_P}\right\}$$

This expression applies to a junction FET, but a very similar one is also quite a good approximation in the case of a MOST. Substituting the value of I_D to eliminate V_{GS} from the expression

$$g_{fs} = \frac{2}{V_P}\left\{I_D I_{DSS}\right\}^{\frac{1}{2}}$$

Hence the condition for negligible signal loss is

$$\frac{I_D V_{BE}^2}{(I_D - I_B)^2} \ll \frac{V_P^2}{I_{DSS}}$$

If I_B is small compared with I_D, this expression may be approximated to

$$\frac{I_D}{I_{DSS}} \gg \left\{\frac{V_{BE}}{V_P}\right\}^2 \tag{7.11}$$

This condition sets a lower limit on I_D, relative to I_{DSS}, which is, however, easily met since V_{BE} (at about $\frac{1}{2}$ volt for silicon device) is usually at least half an order of magnitude less than V_P. Thus subject to this condition, the inclusion of a resistor between base and emitter of the bipolar device in the CD–CE amplifier reduces the collector current to any required degree, whilst producing no significant reduction in overall signal power gain.

In the case of the CS–CE amplifier the effect of the resistor is quite different. Since the output impedance of the CS amplifier, r_{DS}, is very high, a substantial reduction in gain (as well as collector current) will result. If the base-emitter resistor is R_B then the forward transfer conductance is reduced to $g'_{f,2}$ where

$$g'_{f,2} = \frac{R_B}{r_{in} + R_B} g_{f,2}$$

$$= \frac{R_B g_{fs} h_{fe}}{r_{in} + R_B} \tag{7.12}$$

But

$$R_B = \frac{V_{BE}}{I_D - I_B}$$

or if $I_B \ll I_D$, to reasonable approximation

$$R_B = \frac{V_{BE}}{I_D}$$

Hence

$$g'_{f,2} = \frac{V_{BE} g_{fs} h_{fe}}{V_{BE} + r_{in} I_D} \tag{7.13}$$

This interesting result shows that, provided that $r_{in} I_D$ can be kept no larger than V_{BE}, the loss in gain will not be large. Since V_{BE} is virtually a constant, this implies an upper limit on I_D for a given r_{in}. However, r_{in} decreases with increasing I_C, due to the working point dependence of the transistor parameters. Thus a limit is implied on the extent to which I_C can be reduced for given I_D. For real transistors the dependence of r_{in} on I_C is not particularly simple, hence an explicit relationship indicating the limit on the value of the ratio of I_D to I_C is not easily derived. In practical situations, however, once I_C has been chosen r_{in} can easily be calculated and then a value of I_D chosen so that $I_D r_{in}$ is not too large compared with V_{BE}, thus minimising gain loss.

It will be noted that when the resistor R_B is included, the signal source impedance at the input of the CE stage falls from r_{DS} to a little less than R_B. This fall in impedance results in an improvement in the frequency at which $g'_{f,2}$ falls due to input capacitance effects.

In fact the introduction of the base-emitter resistor tends to reduce the differences between the characteristics of the CS–CE and CD–CE amplifiers. Without the resistor, the CS circuit shows higher gain but poorer frequency response than the CD circuit. With R_B present the gain of the CS–CE amplifier falls but its frequency response improves insofar as it had previously been limited by the properties of the interstage coupling.

7.1.3 TEMPERATURE EFFECTS

The effects of temperature on unipolar transistor parameters have been discussed in chapters 2 and 3, and for bipolar transistors excellent discussions of temperature effects may be found in the literature[2]. When transistors are cascaded the individual temperature dependent effects combine in a perfectly straightforward way. Only two special points need to be made.

Circuits in which the standing current in the second stage is reduced without significant reduction in small signal gain present, as has already been indicated, considerable stabilisation problems, as the changes of FET drain current constitute very much larger fractional variations in collector current. The effect can be minimised by using overall DC negative feedback or, particularly where amplifier response must extend down to zero frequency, by the use of a compensation technique in the first stage, such as so-called zero drift biasing (see chapter 9).

The second point of interest is that, in some cases, thermal compensation for the overall amplifier can be achieved by balancing the thermal characteristics of the first stage against those of the second. For example[3], in the CS–CE amplifier, having a base-emitter resistor R_B, if the FET is operated at or near zero gate-source voltage the main temperature dependent effect is a fall in the drain-current with increasing temperature due to a decline in charge carrier mobility. The effect of gate-channel contact potential variations will be negligible provided that V_P is of the order of at least a few volts, and effects due to gate current may also be disregarded provided that the resistance in the gate circuit is not excessive. If a negative coefficient of drain current is assumed, the base-emitter voltage of the bipolar device will fall with rising temperature, thus tending to offset the fall in internal contact potential of the base-emitter junction and reducing the dependence of collector current on temperature

Provided that the base current I_B is negligibly small compared with I_D

$$V_{BE} = I_D R_B$$

so that writing

$$\alpha = \frac{1}{I_D} \cdot \frac{\mathrm{d}I_D}{\mathrm{d}T}$$

$$\frac{\mathrm{d}V_{BE}}{\mathrm{d}T} = \alpha I_D R_B = \alpha V_{BE}$$

But if the internal base-emitter contact potential is ϕ, it is known that ϕ decreases almost exactly linearly with temperature for silicon junctions in the room temperature range, and thus

$$\frac{\mathrm{d}\phi}{\mathrm{d}T} = \lambda$$

where λ is a constant equal to between -2 and -3 mV per degree C for a silicon device, depending on the working point of the transistor.

The condition for zero drift is that the rate of change of V_{BE} shall equal that of the contact potential, hence

$$\frac{\mathrm{d}V_{BE}}{\mathrm{d}T} = \alpha V_{BE} = \frac{\mathrm{d}\phi}{\mathrm{d}T} = \lambda$$

or

$$V_{BE} = \frac{\lambda}{\alpha}$$

under which conditions

$$R_B = \frac{\lambda}{\alpha I_D} \tag{7.14}$$

This circuit is not very useful as it stands, since λ/α is typically 300 mV, which is too low a value of base-emitter voltage to give a

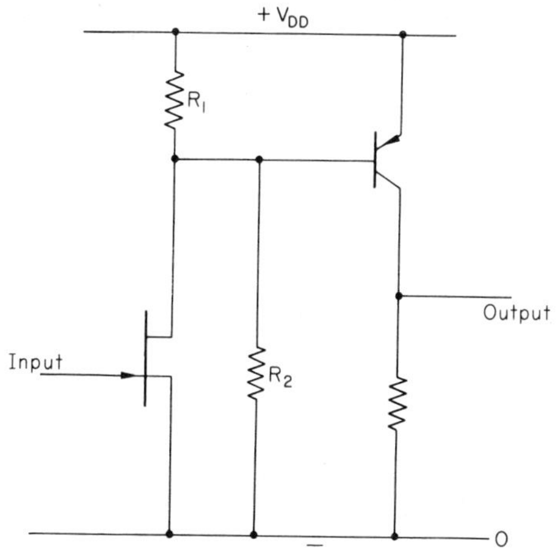

Fig. 7.6. A 'zero-drift' CS–CE amplifier

useful collector current with a silicon transistor. By adding a second resistor, as in Fig. 7.6, the zero drift condition becomes attainable at different values of V_{BE}. It can readily be shown that for given values of α, γ, I_D, V_{DD} and V_{BE}, the values of R_1 and R_2 for zero drift are

$$R_2 = \frac{\lambda V_{DD}}{I_D(\alpha V_{BE} - \lambda)} \tag{7.15}$$

and

$$R_1 = \frac{\lambda V_{DD}}{\{\alpha(V_{DD} - V_{BE}) + \lambda\}\, I_D} \tag{7.16}$$

provided that $V_{BE} > \lambda/\alpha$ and $I_D \gg I_B$.

This treatment neglects the effect of I_{CE0} and h_{FE} variations in the bipolar device. The latter assumption may not be justified, in which case the values of R_2 and R_1 may be very slightly modified, but compensation is still possible.

7.1.4 AMPLIFIERS WITH CE OUTPUT STAGES

From all the foregoing it will be seen that the CD–CE amplifiers are distinguished primarily in the respects indicated in the following table:

Table 7.1. COMPARISON OF THE CD–CE AND CS–CE AMPLIFIERS

	CD–CE	CS–CE
Gain	–	Best
High frequency response	Best	–
Phase inversion	Yes	No
Negative feedback:		
Series current	Yes	Not easily
Shunt current	Not easily	Yes
Series voltage	Not easily	Yes
Shunt voltage	Yes	Not easily

To achieve maximum voltage gain provision must be made to reduce the standing collector current of the bipolar device. If this is done simply by connecting a resistor between its base and emitter the effect on AC characteristics of the amplifier is rather small in the case of a CD first stage but more marked in the CS case, tending to modify the performance of the CS–CE amplifier in such a way as to diminish the differences between it and the CD–CE form.

7.2. Amplifiers with CB output stages

When the bipolar device is operated in the common base configuration its current gain is less than unity. Thus the forward transfer conductance of a CD–CB or CS–CB amplifier is very substantially less than that of comparable amplifiers having the output transistor in the CE configuration. In fact the main function of the CB transistor is to give a degree of isolation between input and output circuits and an impedance transformation. Since the input impedance of the common base amplifier is small, even compared with the output impedance of a CD FET, the forward transfer conductance of the two amplifiers is not very different, being in both cases different from that of the FET alone by the factor h_{fb}, which is only slightly less than unity. The output impedance of both amplifiers is high, approximating closely for the CS–CB and (provided g_{fs} for the FET is not too large for the CD–CB to $1/h_{ob}$). This is typically of the order of megohms for a small transistor.

An amplifier of this type, in which the forward transfer conductance is rather modest but the output impedance high, can only give high voltage gain when working into very high values of load. It is thus not well suited to operation with resistive loads and is likely to give poor high frequency response with even modest values of capacitance shunting the load. The amplifiers having CB output stages, therefore, seem best suited to use as tuned narrow band amplifiers. The main problem with amplifiers of this kind is to produce designs which are stable and not significantly regenerative, preferably without the use of neutralisation (chapter 12).

Early FETs had relatively low values of g_{fs} and were fairly easy to stabilise in either of the configurations considered, but more modern units with much higher transfer conductance may present substantial problems. The CS configuration presents particular difficulties. With available units the drain-gate feedback capacitance, C_{GD}, is of the order of picofarads, and thus regeneration and even oscillation is readily provoked by an inductive component in drain load impedance. This will be present when the parallel resonant circuit in the collector of the bipolar device is tuned slightly above the signal frequency.

By contrast the CD amplifier can become unstable on a capacitative load, but due to the low output impedance a markedly capacitative load phase angle is essential before significant regeneration occurs. As a consequence it is easier to design stable CD–CB amplifiers using high g_{fs} FETs than the CS variant.

The detailed design procedure for tuned CD–CB amplifiers is too extensive to present here. The approach is quite conventional, but to ensure stability capacitative loading of the CD stage must be minimised.

7.3. Amplifiers with CC output stages

The two amplifiers using CC output stages have very different properties. Where the CS input stage is adopted the voltage gain is substantially that of the CS stage alone, but the output impedance is much reduced. When the input transistor is in the common drain configuration the output impedance is even lower, but the overall voltage gain of the two stage amplifier is less than unity.

7.3.1 THE CS–CC AMPLIFIER

The CS–CC amplifier (Fig. 7.7) presents no very unexpected features. The CC stage acts merely as an impedance transformer, reducing the output impedance by a factor very near to $1/(1 + h_{fe})$ relative

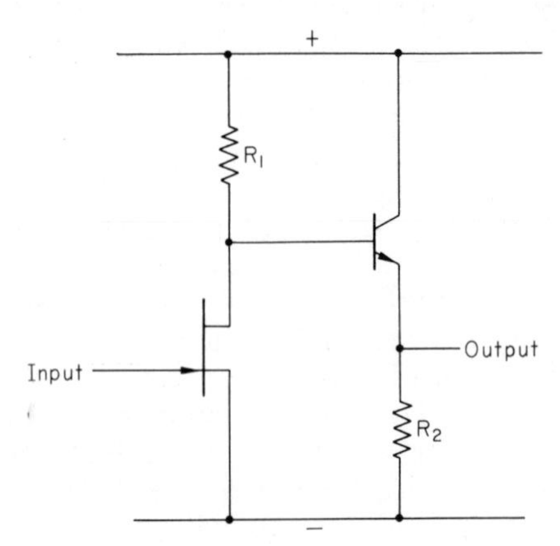

Fig. 7.7. A CS–CC amplifier

to that of the CS stage alone. Provided that R_2 is of the same order as R_1 the output impedance is well approximated by

$$r_{out} = \frac{R_1 r_{DS}}{(R_1 + r_{DS})(1 + h_{fe})} \tag{7.17}$$

whilst the voltage gain is very close to A_V, where

$$A_V = \frac{-g_{fs}}{\dfrac{1}{r_{DS}} + \dfrac{1}{R_1} + \dfrac{1}{(1 + h_{fe})R_2}} \tag{7.18}$$

In many cases $R_1 \ll r_{DS}$ and $R_1 \ll (1 + h_{fe}) R_2$ in which event

$$r_{out} = \frac{R_1}{1 + h_{fe}} \tag{7.19}$$

and

$$A_V = -g_{fs}R_1 \tag{7.20}$$

to good approximation.

Due to the low output impedance, CS–CC amplifiers give a large gain-bandwidth product when cascaded. Otherwise, however, the large input capacitance of amplifiers of this kind (due to the substantial voltage gain, and hence Miller effect, of the CS stage which results from its operation into the relatively large input impedance of the CC bipolar amplifier) militates against high frequency operation from signal sources of high internal resistance. However, where a double gate tetrode MOST is used as the first stage, input capacitance of as little as a few picofarads is possible.

The stability of the DC level at the output terminal of the amplifier depends on the combined effects of change of drain current and fall in the base-emitter potential with increasing temperature.

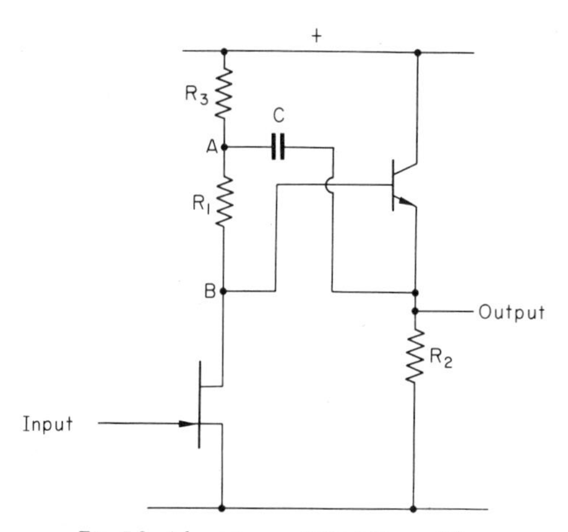

Fig. 7.8. A bootstrapped CS–CC amplifier

A degree of compensation can be achieved by operating the FET with a gate-source voltage slightly less than the zero drift value.

Even higher voltage gain can be achieved with the CS–CC amplifier by 'bootstrapping' the load resistor (Fig. 7.8). If the capacitor C is of negligible impedance at signal frequencies the resistors R_3 and R_2 are effectively connected in parallel as the load resistors of the CC stage. The point A is thus at a signal voltage only a little less than, and in phase with, the point B. If the voltage gain of the CC stage is a_2, the load seen by the CS stage is thus R_L where, to good approximation (assuming r_{DS} large and $1/h_{ob} > r_{DS}$)

$$\frac{1}{R_L} = \frac{(1 - a_2)}{R_1} + \frac{R_2 + R_3}{(h_{fe} + 1) R_2 R_3} \tag{7.21}$$

the first term on the RHS being the effective conductance (to the common line) due to R_1 and the second deriving from the input conductance of the CC stage, suitably approximated. Usually a value of R_1 is chosen large enough to make the first of these two terms negligible and also R_2 can be made large compared with R_3, thus to fair approximation.

$$R_L = (h_{fe} + 1) R_3$$

If C were disconnected the load resistance would be $(R_1 + R_3)$, thus the improvement of gain due to bootstrapping (assuming g_{fs} unchanged) is just

$$\frac{R_L}{R_1 + R_3} = (h_{fe} + 1)\frac{R_3}{R_1 + R_3} \tag{7.22}$$

Subject to these conditions, increase of voltage gain by a factor of 10 or more is relatively easily achieved. The inevitable disadvantage

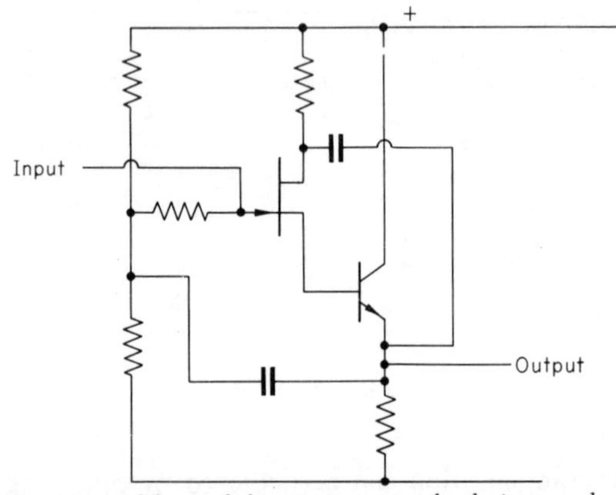

Fig. 7.9. A CD–CC amplifier with bootstrapping to the drain, to reduce the effective input capacitance and the low potential end of the gate continuity resistor, to reduce input conductance. Drain bootstrapping can be omitted if double gate tetrode FETs are used in the first stage

of this circuit is that the output impedance of the amplifier rises substantially, up to a limiting value of $r_{DS}/(1 + h_{fe})$. Use of this circuit will be limited to low frequency applications where the relatively high input capacitance (which, due to Miller effect, may amount to hundreds of picofarads) is not too embarrassing. Again, this difficulty is largely overcome when the first stage is a double gate tetrode.

7.3.2 THE CD–CC AMPLIFIER

Since both the CD and CC amplifiers have a voltage gain less than unity that of the CD–CC combination is also less than one. It is used as an impedance transformer when the reduction in impedance level afforded by the CD stage alone is insufficient.

Analysis of the CD–CC amplifier is entirely straightforward and will not be treated here. The main design point of interest is the use of bootstrapping to increase the effective input impedance. A typical arrangement showing bootstrapping to the low potential end of the gate continuity resistor, to reduce input conductance, and to the drain, to reduce the component of input capacitance due to C_{GD}, is shown in Fig. 7.9. Since C_{GS} is already bootstrapped the input capacitance and conductance can be very small, typically down to a fraction of picofarad and a few hundred picomhos.

7.4. Two stage amplifiers in which the FET is not pinched-off

In all the amplifiers previously considered, the field effect transistor has been operated with a sufficient drain–source voltage to ensure that the conducting channel is pinched-off. If this is not the case the output impedance, and to a lesser extent, the forward transfer conductance of the FET (whether junction or insulated gate type) will fall. However, if the transistor is operating into a relatively low impedance load, such as the input of a CB or CE bipolar transistor amplifier, the fall in output impedance does not produce any significant loss of power gain.

Operation with low drain–source voltage has certain marked advantages. Power dissipated in the transistor is reduced, with a consequent increase in input impedance, or alternatively very high current FETs, which have correspondingly high transfer conductance, can be used without overdissipation. If the value of drain–source voltage is reduced to the base-emitter voltage of the following stage a particularly simple circuit arrangement results. Finally, in junction gate FETs, but possibly not in enhancement mode insulated gate devices, low voltage operation seems to reduce noise levels.

The detailed analysis of the operation of amplifiers of this kind has been described elsewhere[4]. Of the four possible configurations, CD–CE, CS–CB, CD–CB and CS–CE, the first two are of greatest interest, since they are phase inverting and thus lend themselves to DC stabilisation by means of overall feedback. As in the case of the amplifiers already described, the CD–CE gives the better gain when operating into a moderate value of resistive load, and it is therefore upon this configuration that attention has primarily concentrated. The gain which can be obtained is slightly less than that possible in the comparable amplifier with the channel pinched-off, but the frequency response, so far as signal transfer from the FET to the bipolar device is concerned, is likely to be marginally better since the FET presents a lower output impedance.

The CD–CE amplifier has the important further property that the gate-channel and base-emitter contact potentials in the unipolar and bipolar devices respectively operate in opposite sense so far as the effect of their temperature dependence on the CE stage collector current is concerned. Since these two temperature coefficients are approximately equal, that of the unipolar device being slightly greater, by choosing the (less than unity) voltage gain of the CD stage correctly almost exact cancellation of temperature dependent characteristics of the amplifier as a whole can be achieved. Although cancellation cannot, at least in macro-circuits, be close enough to be useful for any but the least critical DC amplifier applications, it certainly can greatly simplify the design of AC amplifiers for satisfactory operation over a range of temperatures.

The choice between pinched-off and non-pinched-off operating conditions for the FET in a given amplifier design is not always straightforward. The small gain loss which results from non-pinched-off operation can usually be countered by changing the FET for a type of slightly larger channel cross section, whilst the difference in frequency response is rarely significant. Where the CD–CE or CS–CB configurations are acceptable, in respect, for example, of the operating input and output impedance levels and types of feedback which can be applied, the circuit simplicity of the non-pinched version will commend it, particularly in integrated applications. Other configurations are best considered only in forms with substantial drain–source voltages, since the circuits will inevitably be more complex due to the need to introduce DC stabilisation in an amplifier which does not have input–output phase inversion.

7.5. Experimental results

Some typical two-stage amplifier circuits are described in reference (1). The results obtained are not necessarily particularly representative, and certainly do not indicate the maximum performance which can be attained, but may be regarded as general evidence for the correctness of the conclusions which have been drawn.

7.5.1 A CS–CE AMPLIFIER

A CS–CE amplifier was constructed as in Fig. 7.10. Performance characteristics were as follows:

(a) without the bypass capacitor *C*
Voltage gain (at 1 kHz) 3(+9·5 dB)
Input impedance (at 1 MHz) 2·3 pfd in parallel with 10 MΩ
−3 dB frequencies 0 and 14 MHz
(a peak of +6 dB occurred at 8·4 MHz)
Signal handling capacity (at output) 1·5 V p–p
Output impedance **56Ω**

(b) with bypass capacitor *C*
Voltage gain (at 1 kHz) 39 (+32 dB)
Input impedance (at 1 MHz) 4 pfd in parallel with 10 MΩ
−3 dB frequencies 35 Hz and 1·2 MHz
Signal handling capacity (at output) 2 V p–p
Output impedance 850 Ω.

It will be noted that a large gain peak occurred at 8·4 MHz when the amplifier was operated with maximum signal negative feedback by omitting *C*. This is, of course, due to phase shift within the feedback loop, and can be eliminated by connecting a capacitor of

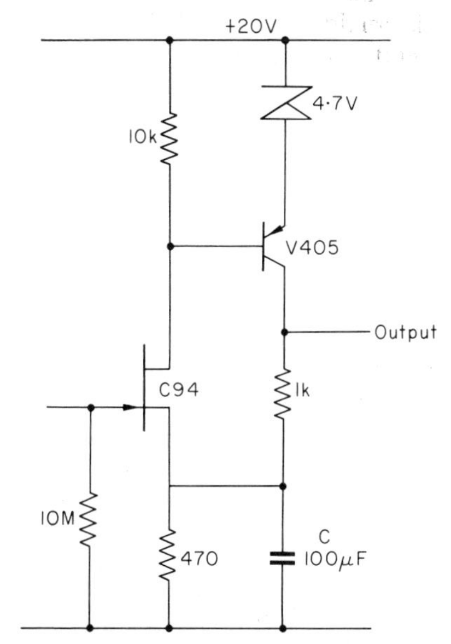

Fig. 7.10. An experimental CS–CE amplifier

15 pfd between the collector of the CE stage and the source of the CD stage, in which case the upper −3 dB frequencv becomes 9·5 MHz.

7.5.2 A CD–CB AMPLIFIER

A 465 kHz IF amplifier shown in Fig. 7.11. This was constructed using valve-type double tuned IFTs. A stage gain of 660 (+56 dB) was obtained without any evidence of regeneration or instability,

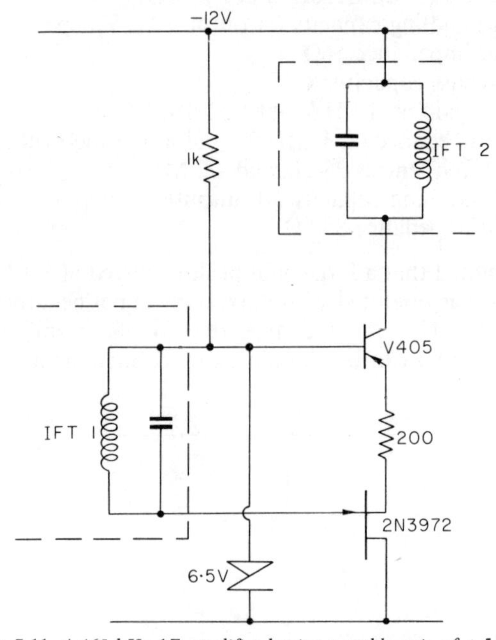

Fig. 7.11. A 465 kHz 1F amplifier having a stable gain of +56 dB

regardless of signal level. The stability was maintained despite ±30% changes in supply voltage. A bandpass characteristic 20 dB down at ±10 kHz was observed.

Note that the resistor of 200 Ω between the source of the CD stage and the CB emitter stabilises the circuit by modifying the phase angle of the input impedance of the second stage.

7.5.3 A CD–CE AMPLIFIER OPERATING BELOW PINCH-OFF

The circuit of Fig. 7.12 shows a CD–CE amplifier in which the CD stage is operated with insufficient drain–source voltage to pinch-off

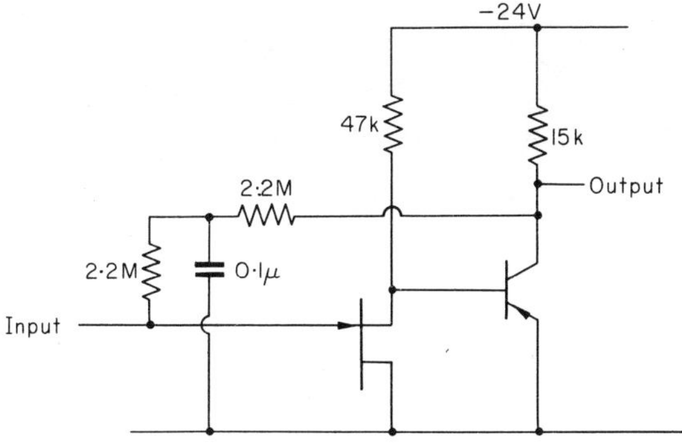

Fig. 7.12. A CD–CE amplifier with very low FET drain–source voltage

the conducting channel. The voltage gain was 232 (+47 dB), falling by 3 dB at 2·5 Hz and 90 kHz when the amplifier was driven from a constant voltage signal source. The input impedance was 2·2 MΩ shunted by 30 pfd, and the output impedance was 15 kΩ.

REFERENCES

1. Gosling, W. 'Amplifiers combining bipolar and field-effect transistors', *Electronic Engineering*, **39**, 478–482 (August 1967).
2. Searle, C. L., *et al. Elementary circuit properties of transistors*, Wiley (1964).
3. Gosling, W. 'A drift-compensated FET-bipolar hybrid amplifier', *Proc. I.E.E.E.*, **53**, No. 3, 323–324 (April 1965).
4. Gosling, W. 'Amplifiers using bipolar and unipolar transistor with limited drain–source voltage' *Proc. I.E.E.*, **113**, No. 10, 1580–1586 (October 1966).

8

AF AMPLIFIER TOPICS

8.1. AF amplifiers

The attributes of the FET are basically those of low noise and high input impedance, which make it ideal as the input stage for many AF amplifiers. It is, however, capable of only modest voltage gains (as was pointed out in chapter 6) which means that it is pointless to design a discrete amplifier consisting of a series of cascaded stages as shown in Fig. 8.1. This configuration appears to confer no advantage whatsoever, and in fact, Meindl[1] has pointed out that it consumes more power than would a bipolar amplifier of the same gain-bandwidth performance by one to two orders of magnitude.

It has therefore become customary to precede a bipolar transistor amplifier with an FET stage, which means that a discussion of the frequency and phase response of such a combination reduces to a treatment of the bipolar transistor amplifier itself, which is out of place within the context of the present book. An exception to this is where a closely coupled FET-bipolar pair is concerned, and these have been considered in chapter 7.

At the time of writing, sundry discrete component, encapsulated operational amplifiers are available which utilise FET input stages, and as manufacturing technology progresses, monolithic equivalents of these will become more common. However, for certain comparatively simple amplifiers, such as those accepting AC signals from transducers, including tape heads, gramophone pick-up cartridges, and inductive or capacitive position sensitive elements, it is still economical to use wired discrete amplifiers. In particular, where special-purpose feedback is required, such as is provided by record or tape equalising networks, it is convenient to use a high voltage-gain FET-bipolar pair as the input stage.

Where highly linear pre-amplification at a high input impedance is required, the distortion characteristics of the FET are of

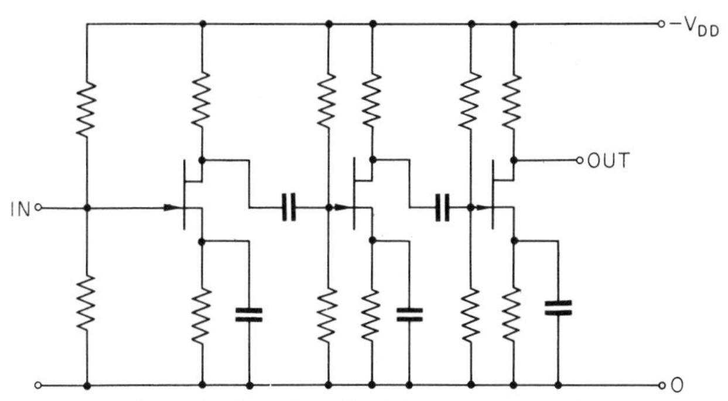

Fig. 8.1. Cascade of FETs (not recommended)

considerable importance, because although it is possible to reduce distortion by the application of negative feedback, gain will also be reduced proportionally.

8.2. Distortion

8.2.1 DISTORTION IN FET STAGES

A complete analysis of distortion in FET stages is quite complex if based upon a comprehensive equivalent circuit[2]. However, a comparatively simple analysis, based on the 'minimal' equivalent

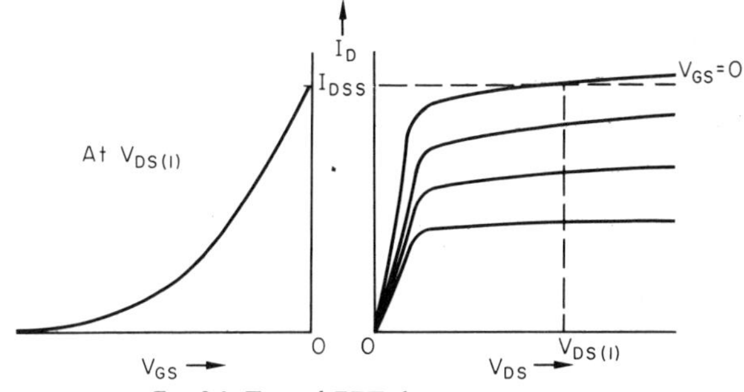

Fig. 8.2. Typical FET characteristics

circuit of Fig. 5.7, and assuming the validity of the square-law transconductance curve, is as follows[3,4]:

An examination of typical transconductance and output characteristics (exemplified by Fig. 8.2) indicates the presence of two predominant non-linearities. The first is the square law curvature of the transconductance characteristic; and the second is the variation in output resistance r_{DS} made evident by the fact that the slopes of the output characteristics are functions of both V_{DS} and V_{GS}.

Equation (6.1) can be rewritten to give v_{out}:

$$v_{out} = -\frac{v_{in} g_{fs} R_L}{1 + R_L/r_{ds}} \tag{8.1}$$

This equation shows that the distortion is directly proportional to changes in g_{fs}; and is a more complex function of changes in r_{DS} Further, it is clear that the effect of changes in r_{DS} may be minimized by making R_L large, which implies a high value of V_{DD}. A more careful examination of the equation, along with Fig. 8.2. will show that the two effects are combative; that is, they act in opposition, so that at some value of V_{DS}, minimum distortion may be expected. Experimental plots of such minima have been made by Sherwin[3], who shows that they occur, unfortunately, very close to the drain–source pinch-off voltage. Immediately below this voltage the FET enters the triode region, r_{DS}, decreases, and hence the distortion rises rapidly.

The major part of the distortion is due to the curvature of the transconductance characteristic, and because this curvature is minimal at $V_{GS} = 0$, then for the lowest possible distortion, the FET must have its quiescent working point in this region. This, however, has two important disadvantages. Firstly, it is here that the voltage gain will be at its lowest, for section 6.2.2 showed that for maximum voltage gain, V_{GSQ} should approach V_P; and secondly, the input signal must have a very small excursion, otherwise the junction may be driven into forward conduction.

This latter point is in fact a special case of a generality also illustrated by equation (8.1); that distortion is also proportional to the magnitude of v_{in}. This is another self-evident point, because as v_{in} increases, it clearly encompasses progressively more of the transconductance curve.

The following sections will treat the aforementioned points more rigorously, and will demonstrate that the transconductance curve gives rise to both harmonic and intermodulation distortion.

8.2.2 HARMONIC DISTORTION

If an FET is operated as a small-signal amplifier, then the total instantaneous value of drain current i_D' may be represented as the usual square law function of the total instantaneous gate–source voltage v'_{GS}. Note that $i'_D = (I_Q + i_D)$ and $v'_{GS} = (V_{GSQ} + v_{GS})$.

$$i_D = I_{DSS}\left(1 - \frac{v_{GS}}{V_P}\right)^2$$

If $v_{GS} = V \sin \omega t$, this becomes:

$$i'_D = I_{DSS}\left(1 - \frac{V_{GSQ} + V \sin \omega t}{V_P}\right)^2$$

Multiplying out, this becomes:

$$i'_D = \frac{I_{DSS}}{V_P^2}[(V_{GSQ} - V_P)^2 + V^2 \sin^2 \omega t + 2V(V_{GSQ} - V_P) \sin \omega t] \quad (8.2)$$

In this expression, the first term gives I_Q because

$$I_Q = I_{DSS}\left(1 - \frac{V_{GSQ}}{V_P}\right)^2 = \frac{I_{DSS}}{V_P^2}(V_P - V_{GSQ})^2;$$

the second term can be written $\frac{1}{2} V^2(1 - \cos 2\omega t)$; and because

$$g_{fs} = \frac{-2I_{DSS}}{V_P}\left(1 - \frac{V_{GS}}{V_P}\right),$$

the third term becomes $g_{fs} V \sin \omega t$. Equation (8.2) now reduces to:

$$i_D = \left[I_Q + \tfrac{1}{2} I_{DSS}\left(\frac{V}{V_P}\right)^2\right] - \tfrac{1}{2} I_{DSS}\left(\frac{V}{V_P}\right)^2 \cos 2\omega t + g_{fs} V \sin \omega t \quad (8.3)$$

This expression demonstrates two important points:

(a) that the quiescent drain current is increased by a very small amount which is proportional to the square of the signal amplitude, and
(b) that only a second harmonic exists, and that this is also proportional to the square of the signal amplitude.

Note that the effect of the DC distortion component is to move the quiescent operating point further towards $V_{GS} = 0$, which means that if the FET is already biased too near this axis, then one extreme of the input signal excursion may force the gate junction into conduction.

Because the transconductance curve almost follows a square law, it gives rise to only a second harmonic. (This is in contrast to the bipolar transistor, whose transconductance curve can be expressed as a power series.) If the harmonic distortion percentage, D_h, is expressed as the ratio of the r.m.s. value of the distortion component to that of the fundamental component, then from equation (8.3),

$$D_h = \frac{I_{DSS}V^2 100}{2V_P^2 g_{fs} V} = \frac{I_{DSS}V}{2V_P^2} \cdot \frac{(-V_P^2 \, . \, 100)}{2I_{DSS}(V_P - V_{GSQ})} \%$$

or

$$D_h = \frac{25V}{V_{GSQ} - V_P} \% \tag{8.4}$$

These equations formalise two qualitative statements already made: that distortion is proportional to the magnitude of the input signal, and is minimal near $V_{GSQ} = 0$. It is also smaller when an FET of high V_P is chosen.

8.2.3 INTERMODULATION DISTORTION

The amplification of a pure sinusoid is a comparatively rare requirement; more often a complex speech or music wave, for example, forms the input signal to an FET preamplifier. For this reason it is necessary to determine the magnitude of the intermodulation distortion, for any complex wave may be represented by a sine wave series, and any two components of this series will give rise to sum and difference frequencies.

Using equation (8.1), and letting the input signal consist of two sinusoids, then if

$$v'_{GS} = V_{GSQ} + V_A \sin \omega_A t + V_B \sin \omega_B t$$

$$i'_D = I_{DSS} \left[1 - \frac{V_{GSQ} + V_A \sin \omega_A t + V_B \sin \omega_B t}{V_P} \right]^2$$

Multiplying out and rearranging, this becomes,

$$\begin{aligned} i'_D = \frac{I_{DSS}}{V_P^2} [V_P^2 - 2V_P V_{GSQ} + V_{GSQ}^2 \\ + V_A^2 \sin^2 \omega_A t + V_B^2 \sin^2 \omega_B t + 2V_A V_B \sin \omega_A t \, . \, \sin \omega_B t \\ + 2(V_{GSQ} - V_P)(V_A \sin \omega_A t + V_B \sin \omega_B t)] \end{aligned} \tag{8.5}$$

The first line of this equation is obviously $I_{DSS}(1 - V_{GSQ}/V_P)^2$, that is, I_Q. The DC components derived from the square terms in the second line augment this value:

$$I_D = I_Q + \frac{I_{DSS}}{2V_P^2}(V_A^2 + V_B^2) \tag{8.6}$$

Equation (8.6) is similar to the first term of equation (8.3), and if generalised, would simply give an indication of the amount by which the quiescent point is shifted when a complex wave is applied.

The final term in the second line of equation (8.5) may be re-written:

$$2V_A V_B \sin \omega_A t \,.\, \sin \omega_B t = V_A V_B[\cos(\omega_A - \omega_B)t - \cos(\omega_A + \omega_B)t] \quad (8.7)$$

This is the intermodulation component of the distortion.

The third line of equation (8.5) gives the fundamental components of the output current. Knowing that the r.m.s. value of a series of sinusoids is the root of the sum of the squares of these sinusoids, then the r.m.s. value of the fundamentals is:

$$V_{\mathrm{r.m.s.}(f)} = \frac{2I_{DSS}}{V_P^2}(V_{GSQ} - V_P)\sqrt{(V_A^2 + V_B^2)} \quad (8.8)$$

Also, the r.m.s. value of the intermodulation terms is:

$$V_{\mathrm{r.m.s.}(m)} = \frac{(\sqrt{2})\,I_{DSS}}{V_P^2}(V_A V_B) \quad (8.9)$$

The percentage intermodulation distortion factor may be defined as the ratio of the two r.m.s. values:

$$D_m = \frac{100\,V_{\mathrm{r.m.s.}(m)}}{V_{\mathrm{r.m.s.}(f)}} = \frac{100\,V_A V_B}{(V_{GSQ} - V_P)\sqrt{[2(V_A^2 + V_B^2)]}}\,\% \quad (8.10)$$

Here again it is seen that the distortion may be reduced by letting V_{GSQ} approach zero, and choosing an FET having a high pinch-off voltage.

Finally, because equation (8.3) is obviously a special case of equation (8.5) where V_A or $V_B = 0$, the second harmonic distortion components in the second line of equation (8.5) will not be considered further.

8.2.4 OUTPUT IMPEDANCE, DISTORTION AND LOAD CONSIDERATIONS

Mentioned in section 8.2.1 was the fact that if R_L is made much greater than r_{DS}, the effect of changes in r_{DS} would be reduced. If, however, I_Q is designed to be high, in the interests of harmonic and intermodulation distortion minimisation, then R_L cannot be significantly increased because V_{DD} would become inconveniently large. A solution to this problem is shown in Fig. 8.3(*a*). Here, R_D has been replaced by an FET operating as a constant current source, which means that the incremental load of T_f1 is now simply r_{DS}

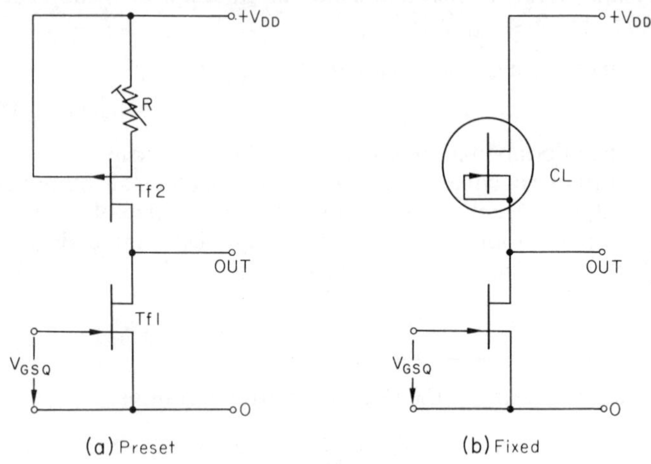

Fig. 8.3. FET stages with current-limiter loads

for T_f2, which will have a high value because T_f2 is working in the pinched-off region. The value of R will define the quiescent current.

An alternative to this three terminal load is an FET used as a two-terminal device, the gate being internally connected to the source as shown in Fig. 8.3(*b*). Such a unit is called a current limiter (CL) and the specified current is simply I_{DSS} for the device. This means that

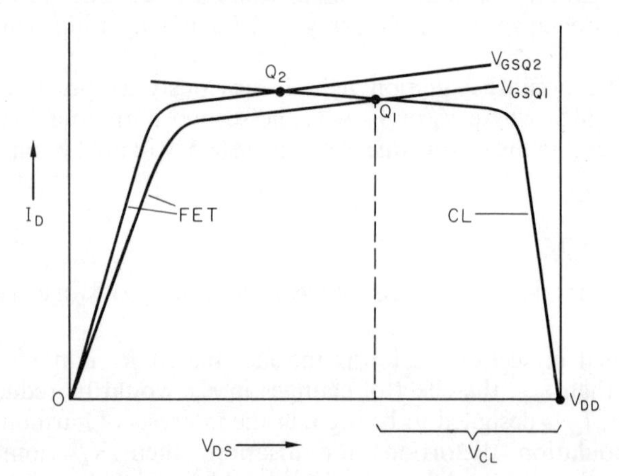

Fig. 8.4. Division of voltage across FET and CL

the amplifier FET must have a larger I_{DSS} than the current limiter. This is not normally a problem, for commercial units cover wide ranges; for example, the Siliconix CL series covers 0·22 to 4·7 mA with incremental resistances from 6 to 0·25 MΩ. Notice, however, that the high gains promised by these high resistances will not be realised if the load impedance is comparatively small, which is usual.

A CL may also be used as a source resistor, when the gain of the stage will very closely approach unity according to equation (6.10).

The outstanding problem associated with the use of the current limiter is that of voltage division across (*a*) the FET and (*b*) the CL itself. Fig. 8.4 shows that the supply voltage is divided at the point of intersection of the characteristics of the FET and the CL. Clearly, a very small shift in gate bias will move this operating point to one extreme or the other, as will a small change in temperature if the FET and the CL do not track very closely.

8.3. Operating Conditions

8.3.1 CHOICE OF QUIESCENT OPERATING POINT

The effects of operating an FET at different points on the transconductance curve are considered in various sections of the book, and may be conveniently summarised here. Referring to Fig. 8.5, for a given quiescent voltage drop across the drain resistor, R_D, the following conclusions apply for the various regions of the characteristic,

(i) Near $V_{GSQ} = V_P$ (Low I_Q)

As shown in section 6.2.2, this is the region where the voltage gain is maximal. The condition arises because, whereas g_{fs} falls according to $\sqrt{I_D}$, the drain resistor has a value proportional to $1/I_D$. Hence,

$$A_{v(\text{l.f.})} \simeq g_{fs}R_D \propto 1/\sqrt{(I_D)}$$

The disadvantages of operating near this point are (*a*) that it is in a region of high distortion and (*b*) that I_D is extremely small, and could be incapable of driving a following bipolar amplifier.

(ii) At the 'zero-drift point'.

This has been treated in section 5.1.3 (see chapter 5, Ref. 5). The disadvantage here is that although the drift can indeed be made very small, the quiescent point is unique to each FET, so that a general design procedure cannot be formulated.

(iii) Near $V_{GS} = 0$.
Although this represents the operating point for minimal distortion, the disadvantages include (a) restricted signal amplitude, so that forward gate bias is avoided and (b) low voltage gain.

Although the foregoing discussion appears to lead to a compromise between the three basic choices, depending upon which is the predominant factor in the design requirement, there are other

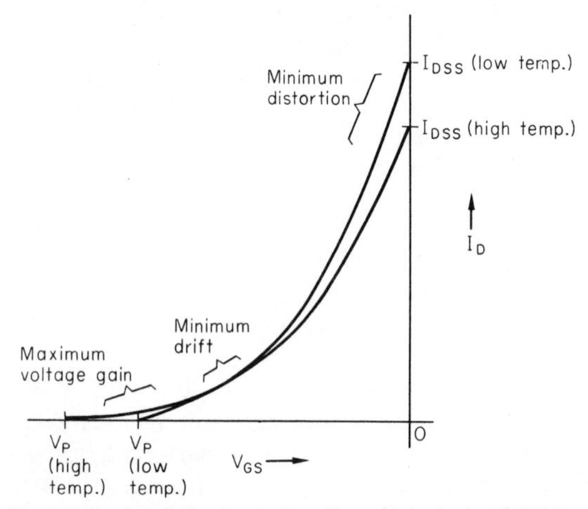

Fig. 8.5. Preferred biasing regions (for a high pinch-off FET)

restrictions which must be taken into account. Firstly, chapter 6 showed that because the parameters for present-day FETs have wide manufacturing tolerances, the operating area on the transconductance graph is usually large (see Fig. 6.17), and this alone militates against accurately defined biasing. Secondly, the requirements of the following stage must be taken into account. In particular, if a bipolar transistor is to be used, it should be established that there is sufficient current to drive it. If one of the pairs discussed in chapter 7 is utilised, then the operating point of the FET is fully defined by the feedback loop itself and is open only to modest variation. This is particularly true if the FET is operated in its 'below-pinch-off' mode.

8.3.2 FEEDBACK IN AUDIO AMPLIFIERS

The generalised effect of the application of negative feedback is that gain is reduced, and so is internally generated noise and distortion.

In the case of FET amplifiers, it is usually better to treat each individual case separately, rather than to apply general feedback theory. This is because it is often difficult to determine whether a

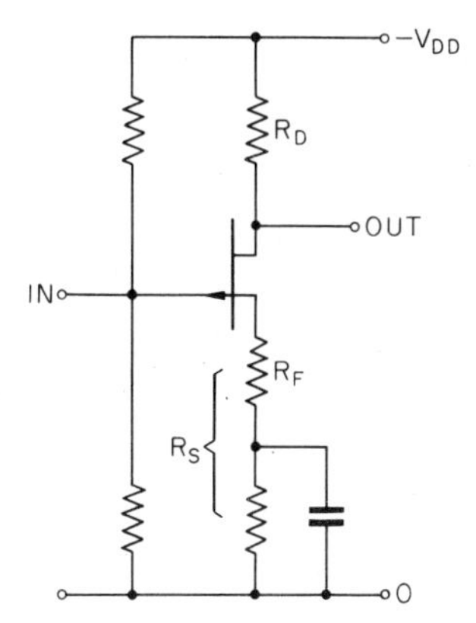

Fig. 8.6. CS stage with source degeneration

voltage or current feedback signal is being derived, and whether it is being applied as series or shunt feedback.

However, a simple case which clearly demonstrates the effect of feedback is provided by the CS stage in which part of the source resistor is unbypassed, as shown in Fig. 8.6. This is clearly a case of current derived series feedback, because the feedback signal $i_s R_F$ is proportional to the output current, and it is applied in series with the input signal.

By definition, the voltage gain is:

$$A_v = -\frac{v_{out}}{v_{in}} = -\frac{i_d R_D}{v_{in}} = -\frac{i_s R_D}{v_{in}} \qquad (8.11)$$

But the gain from the gate to the source is that of a source-follower, which, from equation (6.10), results in,

$$i_s R_F = v_{in} A_{v(CD)} = \frac{v_{in} g_{fs} R_F}{1 + g_{fs} R_F}$$

Equation (8.11) can be rewritten to utilise this expression:

$$A_v = \frac{-i_s R_D}{v_{in}} = \frac{-i_s R_F R_D}{v_{in} R_F} = -\frac{g_{fs} R_D}{1 + g_{fs} R_F} \qquad (8.12)$$

and if $g_{fs} R_F \gg 1$ (which is not always so), this reduces to

$$A_v \simeq -\frac{R_D}{R_F} \qquad (8.13)$$

Equation (8.12) may be compared with the ideal feedback expression,

$$A_{v(FB)} = \frac{A_v}{1 - A_v B}$$

Here, the denominator is the feedback factor, and thus corresponds to $(1 + g_{fs}R_F)$ in equation (8.12). Hence, the internally generated noise and distortion will be reduced by this factor. This means that if a portion R_F of a source resistor remains unbypassed then the harmonic and intermodulation factors become:

$$D_{h(FB)} = \frac{25V}{(V_{GSQ} - V_P)(1 + g_{fs}R_F)} \% \qquad (8.14)$$

and

$$D_{m(FB)} = \frac{100\, V_A V_B}{(V_{GSQ} - V_P)\sqrt{[2(V_A^2 + V_B^2)]}\,(1 + g_{fs}R_F)} \% \qquad (8.15)$$

8.4. Narrow-band Amplifiers

8.4.1 FREQUENCY SELECTIVE AMPLIFIERS USING RC NETWORKS

Numerous frequency and phase sensitive networks exist which, when combined with active elements in the proper manner, result in selective amplifiers. Fig. 8.7(*a*) shows how a frequency selective network may be connected between two wideband amplifiers to produce such a result. This is the simplest form taken by the selective amplifier, and a more satisfactory system from the point of view of selectivity and component economy is shown in Fig. 8.7(*b*). Here, the passive frequency sensitive network (f.s.n.) is connected as a feedback element across a single amplifier.

This system can operate in two distinct ways, depending upon whether the forward gain of the amplifier is negative or positive; that is, whether or not a phase reversal exists. Conclusions regarding the transfer function of the f.s.n. can be drawn as follows:

(a) When the forward gain is negative, the transfer characteristic of the f.s.n. must appear as in Fig. 8.8(*a*). When such a network is used as the feedback element, the overall gain of the selective amplifier will be greatest at the frequency f_0, where minimal negative feedback occurs. For frequencies above and below this, feedback will increase and the overall gain will fall.

(b) When the forward gain is positive, the f.s.n. characteristic must appear as in Fig. 8.8(*b*). In this case, maximum positive

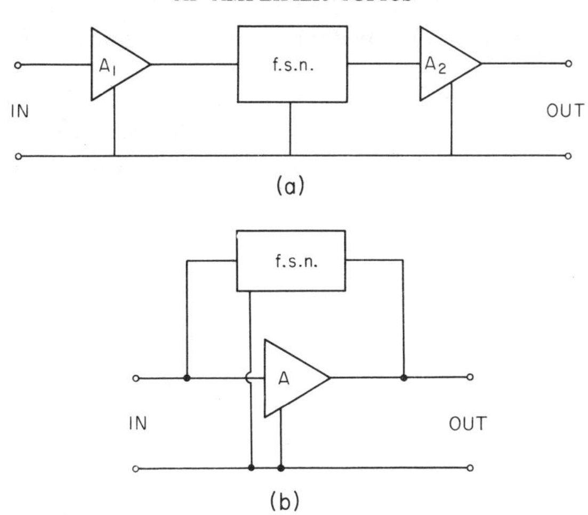

Fig. 8.7. Application of frequency sensitive networks

feedback occurs at f_0, which means that at this frequency, the loop gain must be carefully restricted lest the complete system become unstable. The general feedback equation illustrates this point:

$$A_{v(FB)} = \frac{A_v}{1 - A_v B}$$

Here, $A_v B$ must obviously be less than unity if the system is to remain stable.

Two very common frequency sensitive networks may be used to exemplify the two types of selective amplifier system, namely the twin-T and the Wein bridge. In both cases, the analyses relevant to these networks assume that they are to be driven from perfect voltage sources, and feed into open circuits. It is here where FET amplifiers have a distinct advantage, for although it is easy to produce a bipolar transistor amplifier with a very low output impedance, the

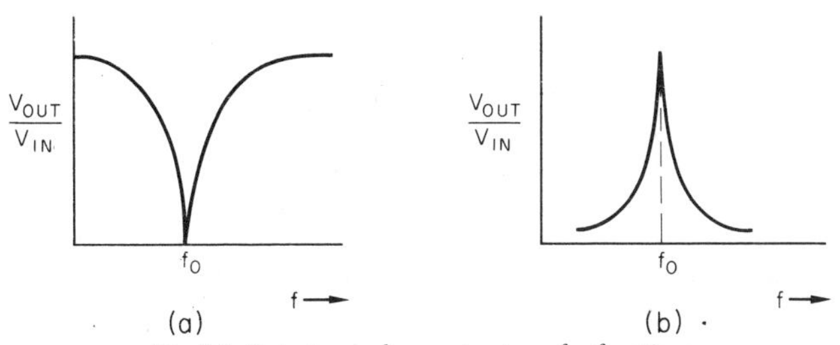

Fig. 8.8. Rejector and acceptor transfer functions

provision of a high input impedance is difficult. As an alternative, it is of course possible to design a dual network for any given f.s.n. which depends on feeding into a short-circuit for optimum operation[6].

The following two sections will illustrate the use of FET amplifiers with the twin-T and the Wein bridge frequency-sensitive networks.

8.4.2 TWIN-T FREQUENCY SELECTIVE AMPLIFIER

The circuit of the twin-T f.s.n. is shown in Fig. 8.9. If the capacitors and resistors are well-matched, so that $C_1 = C_2 = \frac{1}{2}C_3$ and $R_1 = R_2 = 2R_3$, then the rejection frequency is given by:

$$f_0 = \frac{1}{4\pi C_1 R_3} \tag{8.16}$$

Noting that a DC path exists through the arms of the twin-T, it is reasonable to connect the network in such a manner that bias

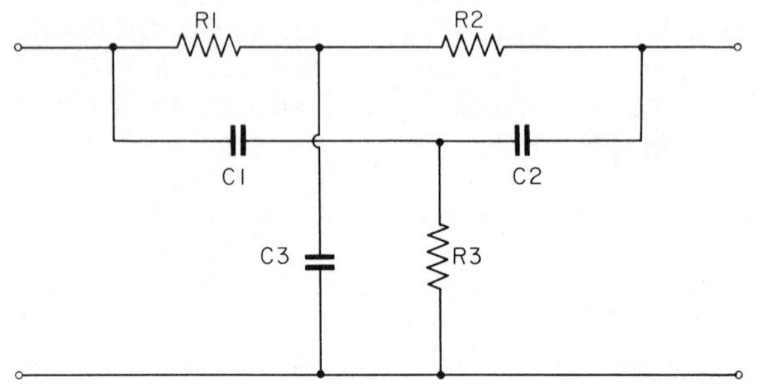

Fig. 8.9. Twin-T rejector f.s.n.

stabilisation is effected concurrently with signal feedback. An amplifier which utilises this principle is shown in Fig. 8.10, and it will be seen that, because the first stage is a differential amplifier (or long-tailed pair), the actual input point is isolated from the twin-T network itself. The drain of $T_f1(a)$ feeds the amplified incoming signal to the base of the bipolar transistor Tr2, where it is amplified further and injected into the twin-T f.s.n. Thus, the feedback signal is applied to the gate of the alternate half of the dual FET, $T_f1(b)$. The negative feedback loop also biases $T_f1(b)$ so that no further components are needed at the relevant gate.

As a design example, the currents and voltages for the circuit were arbitrarily defined, and led to the resistor values shown. Drain currents of 0·5 mA were chosen because $I_{DSS(min)}$ for the Siliconix 2N5199 is quoted as 0·7 mA. Almost any small-signal *pnp* silicon

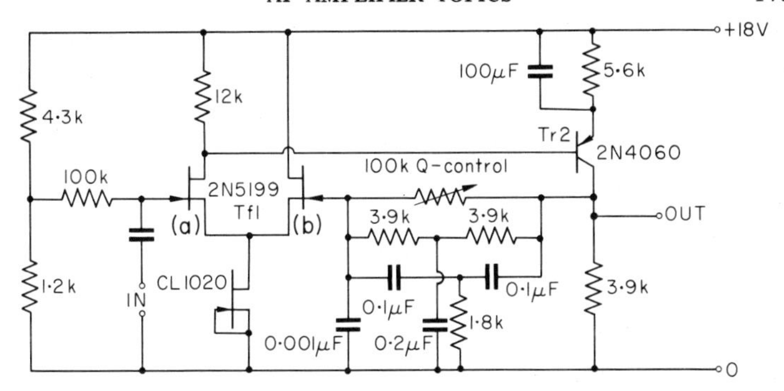

Fig. 8.10. Twin-T frequency-selective amplifier (f_0 = *400H2*)

bipolar transistor will be suitable as Tr2, and the collector current of 1 mA was arbitrarily chosen.

The twin-T f.s.n. itself was designed for 400 Hz by letting $C_1 = C_2 = \frac{1}{2}C_3 = 0{\cdot}1$ μF; and $R_1 = R_2 = 3{\cdot}9$ kΩ while $R_3 = 1{\cdot}8$ kΩ, these being convenient preferred values.

The transfer functions of both the twin-T f.s.n. alone and the

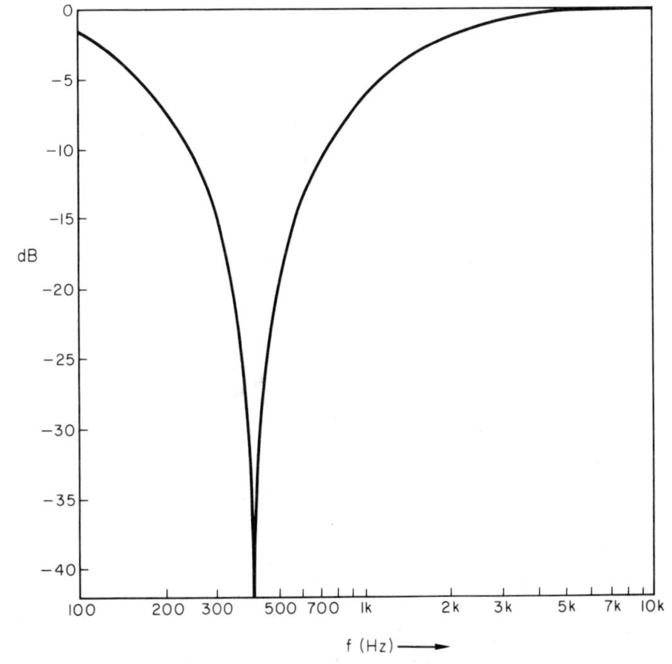

Fig. 8.11. Measured transfer function of twin-T filter

complete selective amplifier are given in Figs. 8.11 and 8.12. The Q-value for the amplifier, defined as the centre-frequency divided by the −3 dB bandwidth was approximately 130, which is extremely

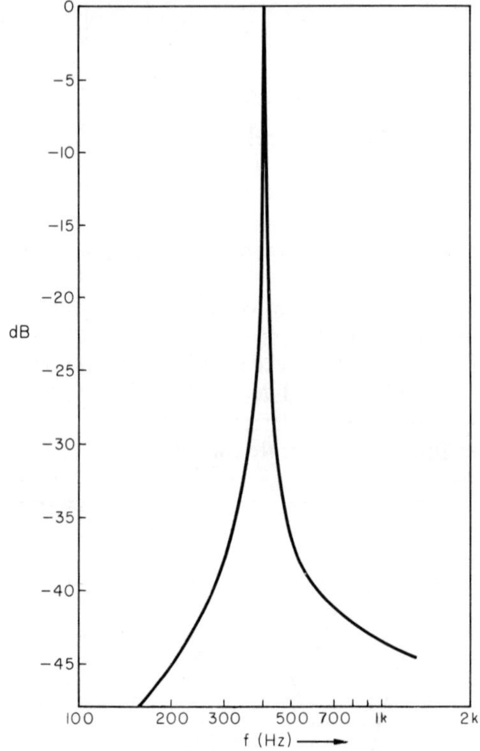

Fig. 8.12. Measured transfer function of selective amplifier of Fig. 8.10. (dB below max. gain of approx × 550)

good for a circuit of this simplicity. The calculation of selectivity will not be pursued, because the aim of this exercise is to demonstrate the applicability of the FET to filter circuits, not to analyse the filters themselves.

8.4.3 THE WEIN BRIDGE FREQUENCY SELECTIVE AMPLIFIER

Two of the arms of the original Wein four-arm bridge can be used as a f.s.n., illustrated in Fig. 8.13. The transfer function can be written[6]:

$$\frac{v_{out}}{v_{in}} = \frac{1}{\left(\dfrac{X_{C_1}}{X_{C_2}} + \dfrac{R_1}{R_2} + 1\right) + j\left(\dfrac{R_1}{X_{C_2}} - \dfrac{X_{C_1}}{R_2}\right)} \tag{8.17}$$

This is of the form $1/a + jb$, and for v_{out} to be in phase with v_{in}, $b = 0$. That is, $R_1/X_{C_2} = X_{C_1}/R_2$.

Therefore,

$$R_1R_2 = \frac{1}{\omega^2 C_1 C_2}$$

or

$$f_0^2 = \frac{1}{4\pi^2 R_1 R_2 C_1 C_2}$$

If $R_1 = R_2 = R$ and $C_1 = C_2 = C$, then

$$f_0 = \frac{1}{2\pi RC} \tag{8.18}$$

This is the centre frequency, and at this point, the attenuation of the circuit may be found by substituting R and C into equation (8.17), giving

$$\left(\frac{v_{out}}{v_{in}}\right)_{f=f_0} = \frac{1}{3}$$

This shows that if the Wein network is used as a positive feedback element, then the internal gain of the amplifier must be less than 3, otherwise instability will result.

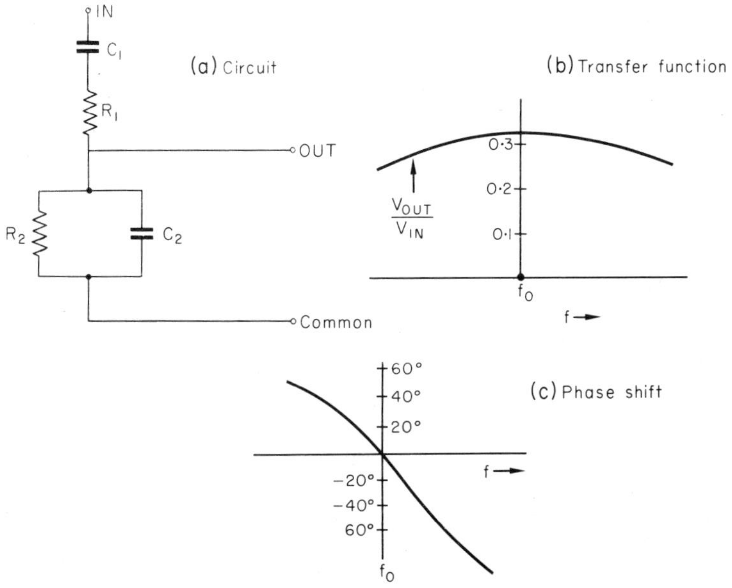

Fig. 8.13. The Wein Bridge f.s.n. and its characteristics

If equation (8.17) is plotted, it will be found that the acceptor characteristic is not at all sharp, but appears as in Fig. 8.13(*b*). However, the phase shift, shown in Fig. 8.13(*c*), is very marked, which accounts for the success of the network as a feedback element.

The analysis of the Wein f.s.n. again depends on the input to the network being a voltage generator (i.e. low impedance) and the output feeding an open circuit. These requirements, plus that of an

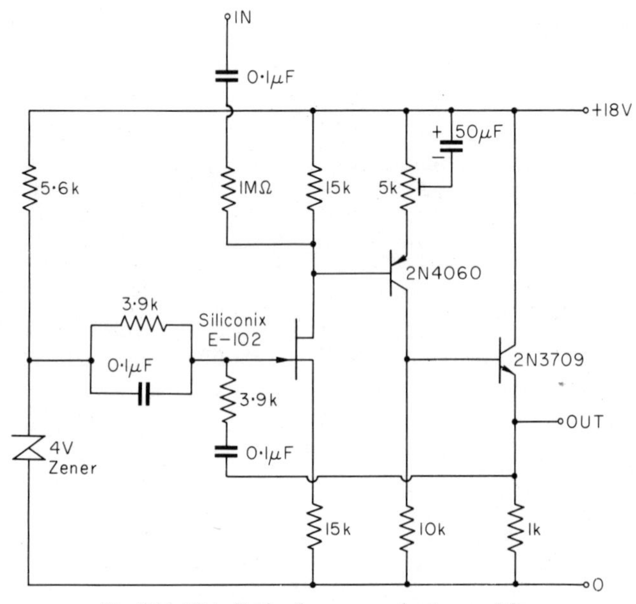

Fig. 8.14. Wein Bridge frequency-selective amplifier

internal gain of less than 3, can be realised by the simple FET circuit of Fig. 8.14. Here, the gain of the FET is approximately unity and that of the bipolar transistor is given by R_C/R_F, where R_F is the unbypassed part of the emitter potentiometer, RV1. The emitter-follower stage provides the low impedance drive for the Wein bridge f.s.n., and the output point provides feedback to the gate of the FET. The forward gain of the amplifier may be adjusted by means of RV1, and as it approaches 3, the selectivity increases rapidly until self-oscillation begins. A response curve near $A_v = 3$ is given in Fig. 8.15.

Unfortunately, this type of frequency selective amplifier can be temperature unstable, because if the forward gain is set close to 3, a

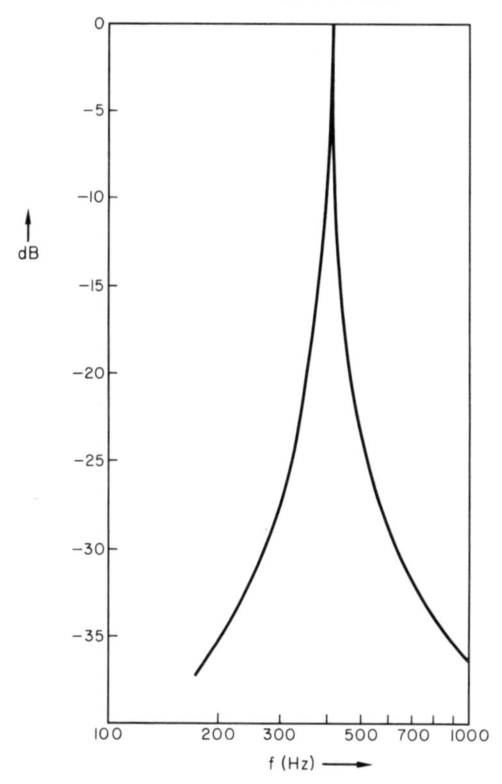

Fig. 8.15. Frequency response of Wein Bridge selective amplifier of Fig. 8.14

temperature rise may increase it far enough to allow undesirable transient instability or even oscillation.

8.5. Noise considerations

In chapter 4 the ways in which noise arises in FETs were discussed. For the junction FET, it will be recalled that the predominant mechanisms are thermal, or Johnson noise generated by the channel and excess or flicker noise at low frequencies. A less important component is the shot noise due to the gate leakage current.

It will also be recalled that the noise characteristics can be conveniently represented by the two-generator convention of Fig. 8.16. Here, the thermal and flicker noise is represented largely by $\overline{e_n}$, while the gate-current shot noise contributes most of $\overline{i_n}$. The correlation at medium and low frequencies is practically zero, because of the clear separation of the two noise functions. At high frequencies, however, the correlation increases significantly due to noise breakthrough via the interelectrode capacitances.

The noise factor is given by

$$NF = 1 + \frac{1}{4kT}\left[\frac{\overline{e_n^2}}{R_G} + \overline{i_n^2}R_G\right] \tag{8.19}$$

from equation (4.11).

Because the gate current is very small, its noise contribution is also very small, so that for most values of R_G.

$$\frac{\overline{e_n^2}}{R_G} \gg \overline{i_n^2}R_G$$

Hence,

$$NF \simeq 1 + \frac{\overline{e_n^2}}{4kTR_G} \tag{8.20}$$

Returning to equation (8.19), the value of R_G which results in a minimum noise factor may be obtained by equating the derivative of NF to zero, giving,

$$R_{G(\text{opt})} = \frac{\overline{e_n}}{\overline{i_n}} \tag{8.21}$$

For a FET, $R_{G(\text{opt})}$ is invariably large, which means that if a low noise factor is desired, then R_G should also be large. This implies,

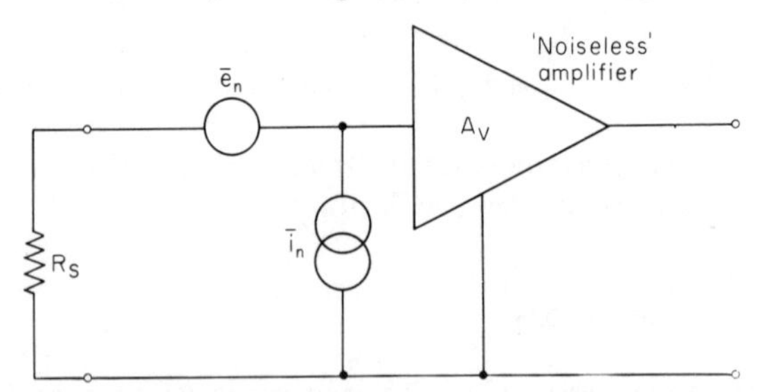

Fig. 8.16. Two-noise-generator equivalent circuit

however, that R_G will itself generate considerable Johnson noise, which will be amplified along with the signal, and will appear at the output. This is simply an expression of the fact that the noise factor is a measure of the *degradation* in signal-to-noise ratio from the input to the output of an amplifier, and is in no way a measure of the absolute value of the noise. Consequently, if R_G happens to be small, it is quite likely that a bipolar transistor stage will produce less noise than a FET stage.

A very convenient way of looking at the relationship between the noise factor and R_G is to convert $\overline{e_n}$ and $\overline{i_n}$ to equivalent noise resistances, as recommended by Faulkner[7]. This has also been done in chapter 4, the results being:

$$R_{nv} = \frac{\overline{e_n^2}}{4kT} \simeq \frac{\overline{e_n^2}}{16} \text{ at } 25\,^\circ\text{C if } \overline{e_n} \text{ is in nV}/\sqrt{\text{Hz}} \qquad (8.22\text{(a)})$$

$$R_{ni} = \frac{4kT}{\overline{i_n^2}} \simeq \frac{16}{\overline{i_n^2}} \text{ at } 25\,^\circ\text{C if } \overline{i_n} \text{ is in pA}/\sqrt{\text{Hz}} \qquad (8.22\text{(b)})$$

Equation (8.19) becomes:

$$NF = 1 + \frac{R_{nv}}{R_G} + \frac{R_G}{R_{ni}} \qquad (8.23)$$

from which it is obvious that the noise factor decreases as the ratio R_{ni}/R_{nv} increases. Fig. 8.17 gives a plot of equation (8.23) for several values of this ratio, and it will be seen that when R_{ni}/R_{nv} is

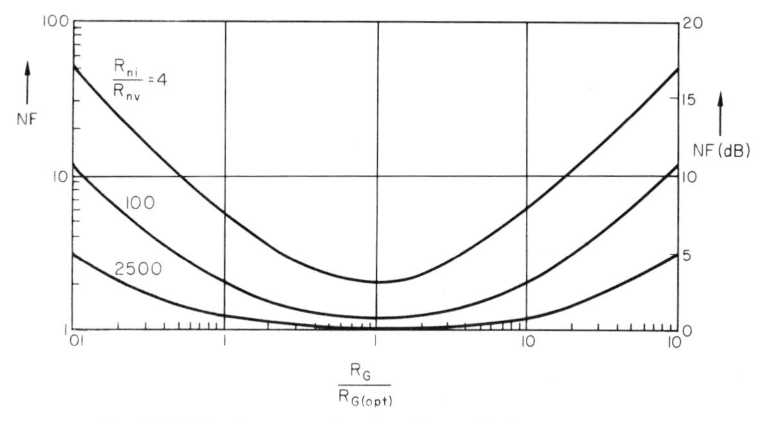

Fig. 8.17. Noise figure as a function of normalised source resistance

large, not only is the noise factor low, but it remains so over a wide range of R_G. Note that the noise factor has also been expressed in the more usual logarithmic form, which is termed the noise figure. That is,

$$NF_{(\text{dB})} = 10 \log_{10}(NF)$$

R_{ni}/R_{nv} is in fact high for all FETs, which means that a wide range of (high) input generator resistances can be accepted without a significant reduction of the noise figure.

An example will illustrate the foregoing points. For the Siliconix

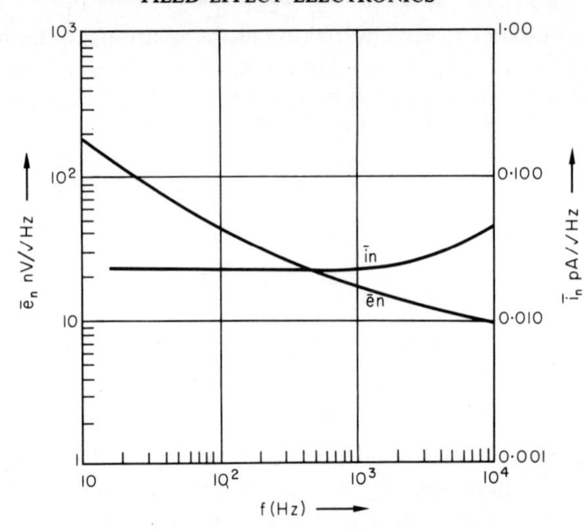

Fig. 8.18. Noise generator characteristics for Siliconix U168 (Reproduced by permission)

U168, the data sheet curves of $\bar{e}_n$ and $\bar{i}_n$ versus frequency have been reproduced in Fig. 8.18, from which,

$$\left.\begin{aligned}\bar{e}_n &\simeq 45\ \text{nV}/\sqrt{\text{Hz}}\\ \bar{i}_n &\simeq 0{\cdot}022\ \text{pA}/\sqrt{\text{Hz}}\end{aligned}\right\}\quad \text{at 100 Hz}$$

Hence, from equation (8.21),

$$R_{G(\text{opt})} = \frac{45 \times 10^{-9}}{0{\cdot}022 \times 10^{-12}} \simeq 2\ \text{M}\Omega$$

From equations (8.22),

$$R_{nv} \simeq \frac{45^2}{16} \simeq 126\ \text{k}\Omega \text{ at } 25^\circ\text{C}$$

and

$$R_{ni} \simeq \frac{16}{0{\cdot}022^2} \simeq 33{,}000\ \text{k}\Omega \text{ at } 25^\circ\text{C}$$

Hence,

$$\frac{R_{ni}}{R_{nv}} = \frac{33{,}000}{126} = 262$$

Operating conditions do not seriously affect the noise performance of a FET stage, though it is advisable to choose a high g_{fs} unit and operate it near I_{DSS} where g_{fs} is maximal. This is because the channel Johnson noise is inversely proportional to g_{fs}, as has been shown by van der Ziel[8]

REFERENCES

1. Meindl, J. D. and Hudson. 'Low power linear circuits', *I.E.E.E. Journal.* Solid State Circuits, **SC-1**, No. 2 (December 1966).
2. Vogel, J. S. 'Nonlinear distortion and mixing processes in field-effect transistors', *Proc. I.E.E.E.*, **55**, 12 (December 1967).
3. Sherwin, J. S. 'Knowing the cause helps to cure distortion in FET amplifiers', *Electronics*, 99–105 (12 December 1966).
4. Watson, J. 'Biasing considerations in FET amplifier stages', *Electronic Engineering*, **40**, 489 (November 1968).
5. Evans, L. L. 'Biasing FETs for zero DC drift', *Electro-Technology*, No. 74, 93–96, (August 1964).
6. Watson, J. *Semiconductor Circuit Design*, Chapter 6. Hilger & Watts (1966).
7. Faulkner, E. A. 'The design of low-noise audio-frequency amplifiers', *Radio & Electronic Engineer*, **36**, No. 1, 17–30 (July 1968).
8. Van der Ziel, A. 'Thermal noise in field-effect transistors', *Proc. I.R.E.*, **50**, 1808–1812 (August 1962).

9

DIRECT-COUPLED AMPLIFIERS

9.1. Introduction

In many applications amplifiers are required having a frequency response extending down to zero frequency. Even in other cases, when the signal spectrum has a lower limit, DC amplifiers are preferred because they do not suffer from low frequency instability when feedback is applied, may have better transient response, and may be cheaper to build, in particular when monolithic integrated fabrication technology is adopted. These advantages are substantial: the only major disadvantage of the DC amplifier is that the extended low frequency response maximises the effect of $1/f$ noise.

The design of DC amplifiers presents two problems, namely interstage voltage level translation and drift. The first of these is relatively trivial. For junction FETs in the common source and common gate configurations and for enhancement mode insulated gate FETs in the common gate and common source configuration the DC levels at the input and output terminals of an amplifier stage are different, and a change of level will be required if stages are to be cascaded. However, the circuit arrangements must be such as to provide a shift of DC level with only a minimal loss of signal. Among the arrangements commonly used, Fig. 9.1(a) shows the simplest of all, a resistive potential divider, which is quite satisfactory provided that the bias supply (shown negative in this case) is large enough so that the voltage division ratio (and hence gain loss) in the coupling network is not too large. An alternative circuit, Fig. 9.1(b), replaces the lower resistor of the potential divider by a constant current source, in this case a bipolar transistor suitably biased. Because of its high incremental impedance, the constant current source greatly reduces signal loss. The diode D provides a measure of temperature compensation. A third possibility is the use of a zener diode, Fig. 9.1(c). Unfortunately, so-called zener diodes in fact operate in the

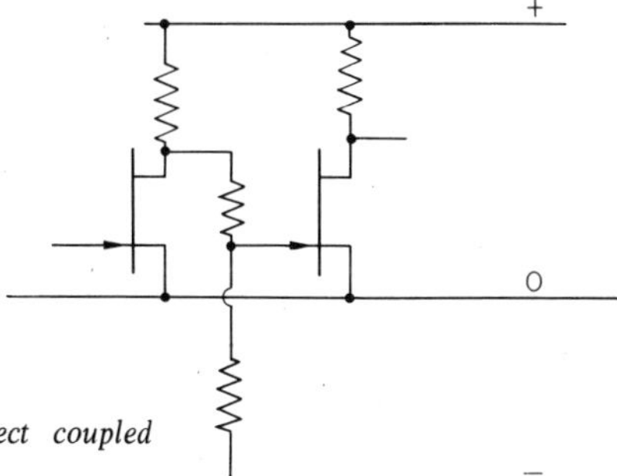

Fig. 9.1(a). A simple direct coupled amplifier

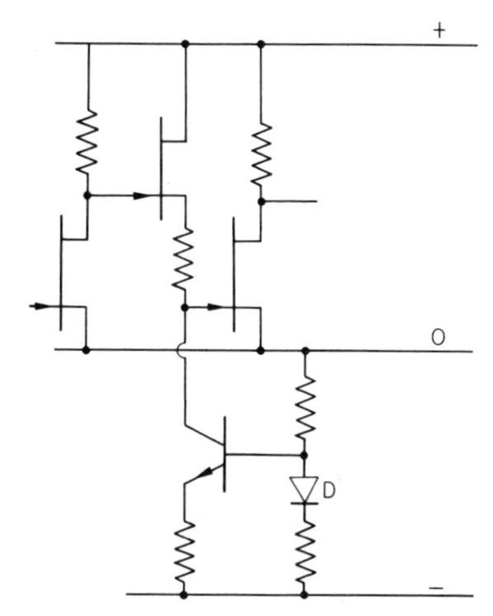

Fig. 9.1(b). Use of a constant current transistor to reduce loss of gain (the diode D *provides temperature compensation for the bipolar transistor)*

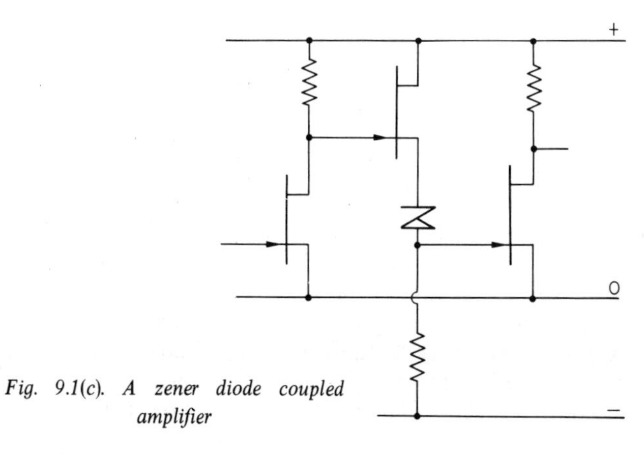

Fig. 9.1(c). A zener diode coupled amplifier

region of avalanche breakdown, for breakdown voltages much in excess of five volts. They are, therefore, very noisy and a zener diode coupling circuit would not be suitable for use in the very low level stages of an amplifier.

The problem of drift is a much more radical one, and presents the major design difficulty of DC amplifiers. Drift originates from three main causes:

(a) *Supply dependent drift.* Change in the supply voltages to the amplifier will, in general, cause a change in output. This is drift and is usually expressed as the voltage applied to the input terminals which would have produced the same change in output.
(b) *Component drift.* Components of the circuit, including FETs will change their properties slowly with time, due to ageing, even if maintained at a uniform temperature. This again may give rise to a change of output level, which can be expressed as an equivalent drift voltage at the input terminals.
(c) *Effect of temperature changes.* Since most electronic components, and in particular transistors, have properties which are temperature dependent, a temperature change can also result in amplifier drift.

Supply dependent drift can be minimised by careful attention to circuit design: this point will be referred to again subsequently. Its effects are usually reduced by careful attention to power supply design, so that power supply voltage changes are minimised. Component drift is minimised by careful attention to device fabrication technology so that drift rates are reduced, but it may also be minimised by compensation methods. Circuits are arranged so that components likely to drift are used in pairs in such a way that the drift due to one is in opposite sign to that due to the other, so ensuring partial or complete cancellation. Compensation techniques are even more widely used to overcome the effect of thermal drift, since in some cases the temperature dependence of the characteristics of, for example, pairs of FETs can be matched quite closely, thus it is in connection with thermal drift that compensation techniques will primarily be described.

The component which suffers from the most severe long term drift at the present time is the MOST insulated gate FET, as described in chapter 3. In particular the threshold gate voltage, at which drain current just begins, is subject to more or less pronounced drift, particularly at elevated temperatures, or if large signals are from time to time applied to the gate. This instability of characteristics is attributed to migration of impurity ions in the oxide insulating layer.

For this reason (and also because of its relatively high noise level) the MOST is very much less suitable for use in the early stages of DC amplifiers, at the present time, than junction FETs. Thus in the early part of this chapter discussion will be limited to the junction devices and the MOST will be considered later.

Temperature dependent effects are the largest cause of drift, in many cases, and can be countered in the following ways:

(a) by compensation
(b) by operating the amplifier at a fixed temperature
(c) by converting the signal from DC to AC and thus avoiding DC amplification altogether (this is the chopper amplifier approach) or
(d) by periodically resetting the zero of the DC amplifier in some automatic way, hence offsetting the effects of drift (the chopper-stabilised amplifier approach).

These techniques all have advantages and disadvantages and can be used either separately or in combination. In the following sections they will be considered in turn.

9.2. Compensation in transistor amplifiers

It will be instructive to begin by comparing the temperature dependent properties of bipolar and field-effect transistors. For a bipolar transistor in a simple common emitter amplifier configuration the collector current is given (Fig. 9.2(*a*)) to reasonable approximation by

$$I_C = h_{FE}\left\{\frac{E_B - V_{BE}}{R_B}\right\} + I_{CEO}$$

Here the temperature dependent transistor parameters are h_{FE}, V_{BE}, and I_{CEO}. For a typical transistor operating under normal conditions all three are in the same sense, so far as the effect on I_C is concerned. All three cause I_C to increase with increasing temperature.

For a junction field-effect transistor in the common source configuration, using Middlebrook's parabolic approximation to the large signal characteristics, Fig. 9.2(*b*).

$$I_D = I_{DSS}\left\{1 - \frac{V_{GS}}{V_P}\right\}^2 \tag{9.1}$$

where

$$V_{GS} = E_G - R_G I_G \tag{9.2}$$

In this case I_{DSS}, V_P and I_G are temperature dependent but whereas V_P and I_G vary in such a way as to cause I_D to increase with temperature, the variation of I_{DSS} is in the opposite sense. Under suitably chosen working conditions, indeed, the temperature dependence due to I_{DSS} can be made to balance exactly that due to V_P and I_G, hence resulting in zero temperature dependent drift.

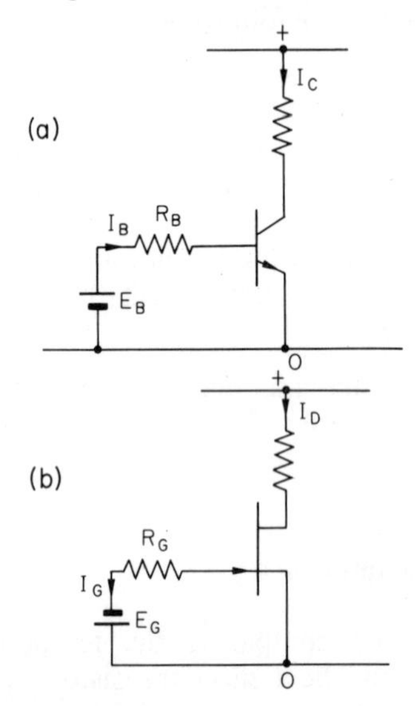

Fig. 9.2(a). A bipolar CE stage
(b). A FET CS stage

Thus in this respect the FET is radically different from (and superior to) the bipolar device.

Since all temperature dependent effects in bipolar devices are in the same sense, compensation can only be achieved by using two similar devices so arranged in the circuit that their temperature dependence produces opposite effects at the output. This mode of operation can also be used with FETs, but compensation in a different sense is also possible (autocompensation): a single FET may be operated at such a working point that its temperature dependent effects are of equal and opposite sign and hence balance out. Better still, both compensation techniques can be applied simultaneously, to achieve a very low level of drift.

Autocompensation, or, as it is often called, zero-drift operation, of a junction FET can have obvious advantages over the use of two transistors in a balanced circuit. The latter requires two transistors of identical characteristics, which may present difficulties unless they are fabricated side by side on a single silicon chip. Also techniques which use two transistors, even if they are on the same chip, are necessarily sensitive to temperature gradients which can produce differential temperature changes. The zero-drift operated FET is immune to this effect. In practice the thermal transient on switching on balanced amplifiers often has a large differential component, thus zero drift FET amplifiers can be expected to settle down more rapidly than balanced amplifiers during the 'warm-up' period. The next section, therefore will deal with the properties of a common source amplifier operated in this mode, using junction FETs.

9.2.1 THE ZERO-DRIFT CS AMPLIFIER

Consider a zero drift amplifier as in Fig. 9.2(*b*). To simplify matters it can be assumed that R_G is small enough that the product $R_G I_G$ can be neglected. The implications of this assumption will be considered later. By differentiation of equation (9.1)

$$\left(\frac{\partial I_D}{\partial T}\right)_{V_{GS}} = \frac{\partial I_{DSS}}{\partial T}\left(1 - \frac{V_{GS}}{V_P}\right)^2 - \frac{2I_{DSS}V_{GS}}{V_P^2}\left(1 - \frac{V_{GS}}{V_P}\right)\frac{\partial V_P}{\partial T} \qquad (9.3)$$

But

$$\left(\frac{\partial V_{GS}}{\partial T}\right)_{I_D} = \left(\frac{\partial I_D}{\partial T}\right)_{V_{GS}}\left(\frac{\partial V_{GS}}{\partial I_D}\right)_T = \frac{1}{g_{fs}}\left(\frac{\partial I_D}{\partial T}\right)_{V_{GS}}$$

and since, by differentiation of equation (9.1)

$$g_{fs} = -\frac{2I_{DSS}}{V_P}\left(1 - \frac{V_{GS}}{V_P}\right) \qquad (9.4)$$

it follows that the drift, defined as the rate of change of gate voltage with temperature required to keep the drain current constant, is given by

$$\left(\frac{\partial V_{GS}}{\partial T}\right)_{I_D} = -\frac{V_P}{2}\left(1 - \frac{V_{GS}}{V_P}\right)\frac{1}{I_{DSS}}\cdot\frac{\partial I_{DSS}}{\partial T} + \frac{V_{GS}}{V_P}\cdot\frac{\partial V_P}{\partial T}$$

$$= -\frac{\alpha}{2}(V_P - V_{GS}) + \frac{V_{GS}}{V_P}\lambda \qquad (9.5)$$

where

$$\alpha = \frac{1}{I_{DSS}}\cdot\left(\frac{\partial I_{DSS}}{\partial T}\right) \text{and } \lambda = \frac{\partial V_P}{\partial T}$$

Hence the drift will be zero if

$$(V_P - V_{GS})\frac{V_P}{V_{GS}} = \frac{2\lambda}{\alpha} = V^*, \text{ say} \tag{9.6}$$

where V^* is a constant having the dimensions of voltage.

$$V_P - V_{GS} = \frac{V^*}{1 + V^*/V_P} \tag{9.7}$$

Typically for silicon devices $\lambda = 2{\cdot}2$ mV/°C and $\alpha = 0{\cdot}5$ to $0{\cdot}7$ per cent per degree C. Taking the lower value for α, the constant V^* will be seen to be equal to about one volt. For FETs having pinch-off

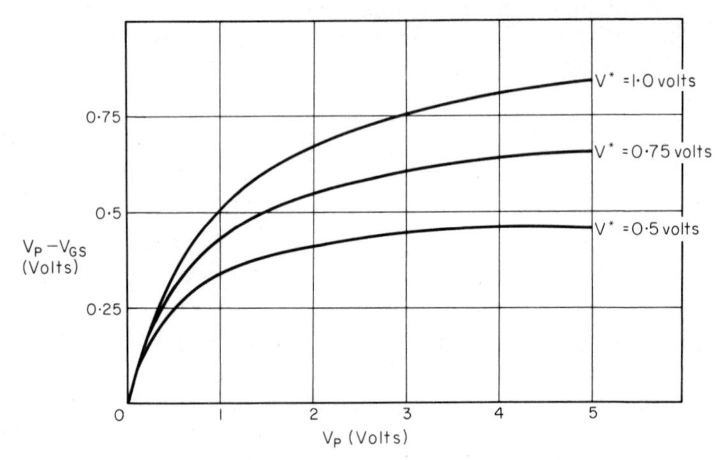

Fig. 9.3. $(V_p - V_{GS})$ *for zero drift at various* V_p *and* V^* *values*

voltages of the order of a few volts, therefore, the denominator of the right-hand side of equation (9.7) is near unity and the zero-drift bias point corresponds to V_{GS} about one volt (or less) smaller in magnitude than V_P. The general relationship between V_{GS} and V_P for various V^* is indicated in Fig. 9.3.

The coefficient λ is very nearly temperature independent, but this is not so for α. Assuming that all impurities are fully ionised at operating temperatures and that the channel dimensions remain substantially constant, the variation of I_{DSS} with temperature is due primarily to changes in charge carrier mobility. Mobility may be determined primarily by lattice or impurity scattering, but for the doping levels used in FETs and at room temperature the former predominates and results in α following a temperature law of the form

$$\alpha = -\frac{a}{T}$$

where a is a constant having a value between 1·5 and 2·5 for silicon.

A more convenient expression may be obtained in the form of a linear approximation by writing

$$\alpha = \alpha_0 \left\{1 - \frac{T - T_0}{T_0}\right\} \tag{9.8}$$

where α_0 is the value of α at a temperature T_0.

Equation (9.8) is a valid approximation provided that $(T - T_0) \ll T_0$ and is thus useful over a substantial range of temperatures about room temperature.

The expression for drift, therefore, can now be rewritten as

$$\left(\frac{\partial V_{GS}}{\partial T}\right)_{I_D} = -\frac{\alpha_0}{2}(V_P - V_{GS}) + \frac{T - T_0}{T_0}\frac{\alpha_0}{2}(V_P - V_{GS}) + \frac{V_{GS}}{V_P}\lambda$$

and if the value of $(V_P - V_{GS})$ is chosen to satisfy equation (9.7) at T_0 it follows that

$$\left(\frac{\partial V_{GS}}{\partial T}\right)_{I_D} = \frac{T - T_0}{T_0}\frac{\alpha_0}{2}(V_P - V_{GS}) - \frac{T - T_0}{T_0}\frac{V_{GS}}{V_P}.\lambda \tag{9.9}$$

Since λ is much the same for all silicon transistors (between -2 and -3 mV per degree C) the choice of transistor only affects the magnitude of drift away from T_0 in so far as it affects the ratio V_{GS}/V_P. Fig. 9.4 shows this ratio plotted against V_P for various V^*

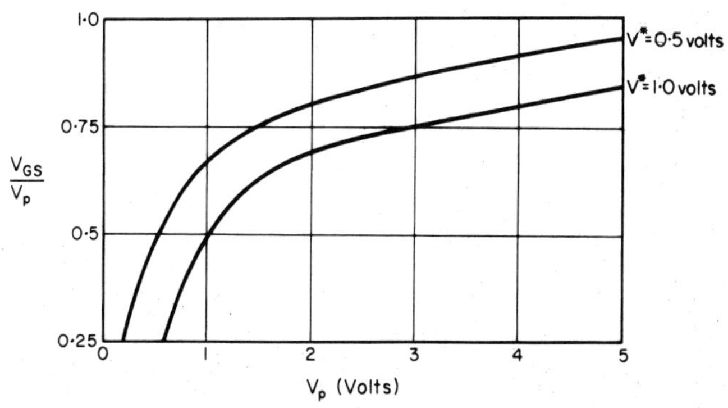

Fig. 9.4. Data of Fig. 9.3 re-plotted to give V_{GS}/V_P

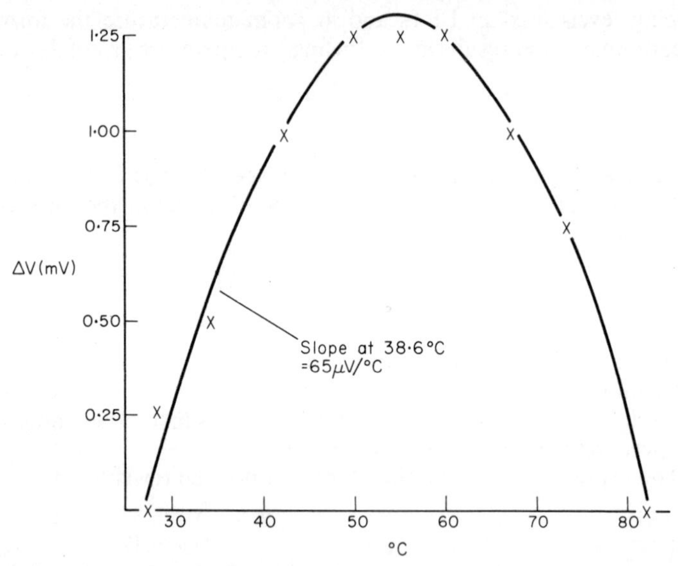

Fig. 9.5. The parabolic dependence of offset voltage on temperature for constant I_D

and it will be seen that the ratio, and hence the drift, will be lower for smaller values of V_P. Thus low pinch-off voltage transistors are at an advantage in this application.

The linear dependence of $\partial V_{GS}/\partial T$ with temperature offset from T_0 implies that V_{GS} plotted against T would have a parabolic form (Fig. 9.5).

$$V_{GS} = V_{GS}(0) - \lambda \frac{V_{GS}T_0}{2V_P}\left\{\frac{T - T_0}{T_0}\right\}^2 \tag{9.10}$$

A plot of drain current against temperature at fixed V_{GS} would have a similar generic form, since the fractional variation in I_D, and hence in g_{fs}, is small. This parabolic dependence was reported by Cobbold and Trofimenkoff[1].

9.2.2 EFFECT OF VARIATION OF BIAS POINT OR V_P

If at temperature T_0, V_{GS} varies from the correct value, as given by equation (9.7), by an amount ΔV_{GS} the resulting drift is no longer zero. Instead it will be (from equation (9.5) with $\alpha = \alpha_0$)

$$\left(\frac{\partial V_{GS}}{\partial T}\right)_{I_D} = \frac{\alpha_0}{2} + \frac{\lambda}{V_P}\Delta V_{GS} \tag{9.11}$$

substituting for λ from equation (9.6) (and writing $\alpha = \alpha_0$) this expression may be reduced to an alternative form as

$$\left(\frac{\partial V_{GS}}{\partial T}\right)_{I_D} = \frac{\alpha_0}{2}\frac{V_P}{V_{GS}}\Delta V_{GS} \tag{9.12}$$

Since V_P is typically a few volts whilst α_0 is just over one half per cent per degree the term $\alpha_0 V_P/2$ is of the order of ten millivolts per degree,

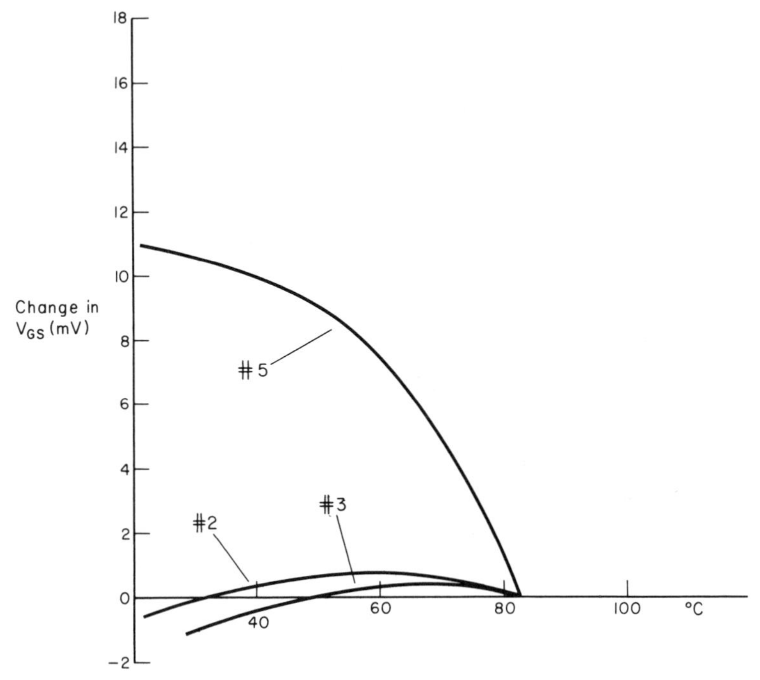

Fig. 9.6. Effect of change of transistor on the relationship between V_G *and* T, I_D *remaining constant*

so that to achieve a drift figure of, say, 10 microvolts per degree $\Delta V_{GS}/V_{GS}$ would need to be 0·1%. Thus the bias supply needs to be well stabilised but not impossibly so.

The form of equation (9.1) is sufficiently similar in respect of V_P and V_{GS} that for small changes in V_P, such that the ratio V_P/V_{GS} is not significantly altered, the drift caused if V_P alters by ΔV_P from the critical value for zero drift is well approximated by equations (9.11) and (9.12), substituting $-\Delta V_P\,.\,(V_{GS}/V_P)^2$ for ΔV_{GS}. Whereas in the case of V_{GS}, however, these equations imply not impossibly stringent conditions on the bias supply, in the case of V_P, for moderate values of drift a tolerance of a fraction of a per cent on V_P is indicated. It would be quite impossible to select field effect transistors to this tolerance on V_P: for a typical production spread the yield on selection would be a small fraction of one per cent and thus selection labour cost per unit would be prohibitive. Thus it will be necessary to adjust the values of V_{GS} if the FET is changed, to accommodate the change of V_P, and amplifiers using zero drift biasing will need individual adjustment and a thermal cycling test. Fig. 9.6 shows the effect of changing the FET on the dependence of V_{GS} on T for constant I_D.

9.2.3 THE COMMON DRAIN AMPLIFIER

For the CD configuration, Fig. 9.7

$$V_{\text{in}} = V_{GS} + RI_D$$

Hence

$$\left(\frac{\partial V_{\text{in}}}{\partial T}\right)_{I_D} = \left(\frac{\partial V_{GS}}{\partial T}\right)_{I_D} \tag{9.13}$$

but the drift in CD configuration is usually defined as

$$\left(\frac{\partial V_{\text{in}}}{\partial T}\right)_{I_D}$$

hence the expressions obtained for drift in the CS case remain valid in the CD case also, in particular equation (9.8). There is thus no direct drift advantage in this configuration. It can, however, be advantageous in one other respect. The CS amplifier is very sensitive to small changes in V_{GS}, and as indicated by equation (9.12), a small change in V_{GS}, which is supplied directly by an external bias supply, results in substantial drift. In the CD amplifier the bias voltage is not supplied directly, instead a negative (for n-channel FETs) supply V_{SS} is connected to the lower end of the source load resistor R_2. A change in V_{SS} then produces a consequent (but smaller) change in V_{GS}. Assuming that the output admittance of the CD stage is very nearly $-g_{fs}$ it follows that

$$\Delta V_{GS} = \frac{1}{1 + g_{fs}R_2} \Delta V_{SS}$$

and thus the drift consequent upon a change of V_{SS} is

$$\left(\frac{\partial V_{\text{in}}}{\partial T}\right)_{I_D} = \left(\frac{\partial V_{GS}}{\partial T}\right)_{I_D} = \frac{\alpha_0}{2} \frac{V_P}{V_{GS}} \frac{\Delta V_{SS}}{1 + g_{fs}R_2} \tag{9.14}$$

Since $g_{fs}R_2 \gg 1$ in most cases, the drift caused by a small change in V_{SS} is less than that caused by a similar bias change in the CS case.

The penalty that must be paid, however, for this easement is that the zero drift bias point is now dependent on I_{DSS} (as it was not in the CS case) and on R_2. The former is not too important since the value

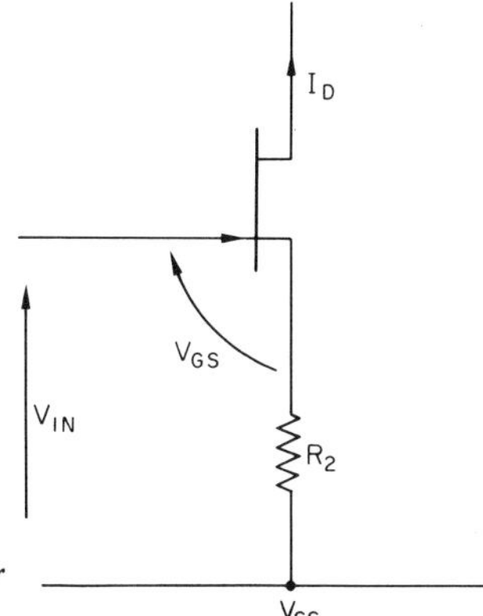

Fig. 9.7. A common-drain amplifier

of bias will anyway have to be adjusted individually for the transistor used, but the latter is significant since change in the value of R_2 due to ageing will adversely affect drift.

To estimate the magnitude of this effect it can be assumed that the change in I_D due to a change in R_2 is given to good approximation, by

$$\Delta I_D = \frac{g_{fs}\Delta R_2 I_D}{1 + g_{fs}R_2}$$

hence

$$\Delta V_{GS} = \frac{-I_D\Delta R_2}{1 + g_{fs}R_2}$$

and the drift is

$$\left(\frac{\partial V_{\text{in}}}{\partial T}\right)_{I_D} = -\frac{\alpha_0}{2}\frac{V_P}{V_{GS}}\frac{I_D\Delta R_2}{1 + g_{fs}R_2} \tag{9.15}$$

Since $I_D R_2$ is of the order of V_{SS}, this means that the fractional tolerance on R_2 will have to be at least as good as that on V_{SS}. Typically (Fig. 9.8) variation of R_2 by 1% will produce a serious degradation of drift performance, and stability of R_2 to $\pm 0{\cdot}1$% will be required. This is, however, not too difficult to achieve.

One interesting possibility is that R_2 may be replaced by a constant current source. In this case the dependence of drift on V_{SS} is eliminated, and if the current is given a slight positive temperature

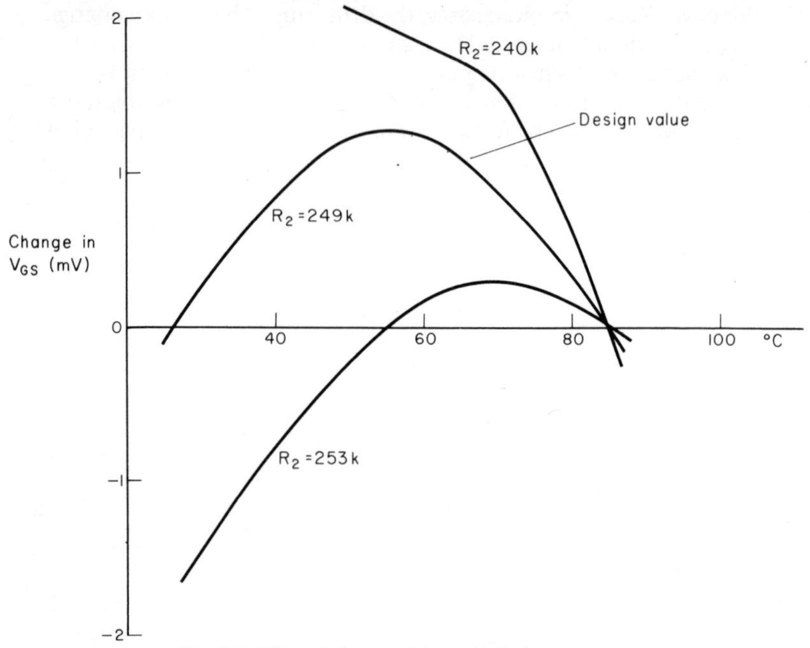

Fig. 9.8. Effect of change of R_2 *on drift characteristics*

coefficient the parabolic relationship between drain current and temperature no longer holds, and approximately zero drift conditions apply over a wide temperature range. Successful use of this technique has not, however, yet been reported.

9.2.4 THE LONG-TAIL PAIR CONFIGURATION

The most popular of all of the low drift amplifier configurations is, without question, the long-tail pair. Since drift is a common mode signal and to a first order the inclusion of a drain load resistor does not affect drift, each half of the long-tail pair has a performance identical with that of a CD amplifier. In particular equation (9.8) still holds. However, the drift of the two halves is in opposite sense at the output terminals and thus subtracts. If the two halves are identical the drift will be exactly zero at all temperatures for which the physical approximations implicit in this treatment hold good. This is not observed practically, since truly identical transistors cannot be made. Usually the zero drift temperature for the two units is not quite coincident. If the difference is ΔT, the drift is very nearly, from equation (9.9):

$$\left(\frac{\partial V_{G_1-G_2}}{\partial T}\right)_{V_{D_1-D}} = \lambda \frac{V_{GS}}{V_P}\left\{\left(1 - \frac{T - T_0 + (\Delta T/2)}{T_0}\right) - \left(1 - \frac{T - T_0 - (\Delta T/2)}{T_0}\right)\right\} = \lambda \frac{V_{GS}}{V_P}\frac{\Delta T}{T_0} \qquad (9.16)$$

Equation (9.16) demonstrates the principal advantage of the long-tail pair configuration. For a given value of $\Delta T/T_0$ (which depends only on the closeness of match of the units) the drift is constant, independent of temperature. The input offset voltage

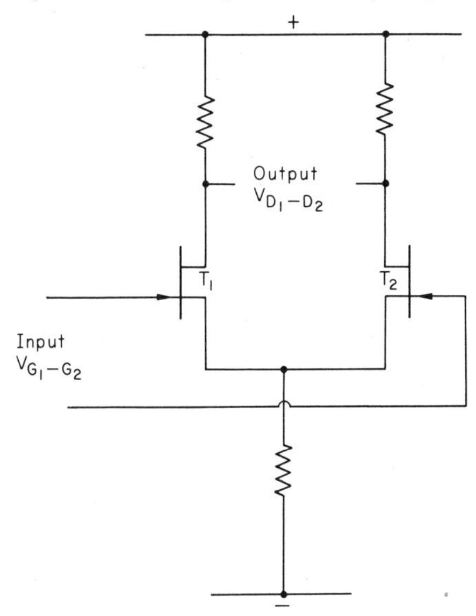

Fig. 9.9. A 'long-tail pair' amplifier

therefore is only linearly related to temperature and does not follow a parabolic law as for the single zero drift biased stage. Although the drift for the latter does fall to zero at T_0 it increases in magnitude rapidly both above and below this value, rapidly increasing above the constant, and usually low, value given by equation (9.16). Thus, whereas for the single stage the bias applied to the FET must be individually adjusted to give a value of T_0 in the middle of the working range, in the case of the long-tail pair provided that the units are well matched the drift will remain at a low constant value over a wide range of temperatures on either side of T_0, and thus individual bias adjustments are unnecessary. This is a powerful advantage of the long-tail pair and accounts for its wide popularity.

In order to determine the magnitude of the drift for a long-tail pair it is necessary to determine ΔT, the difference in zero-drift temperature for the two units. The usual reason why ΔT is not zero (i.e. the units are not identical) is a small difference of V_P. Since V_{GS} is the same for both units the zero drift condition of equation (9.7) will be satisfied for slightly different values of V^* and hence of α_0 and thus T_0. A small change of V_P can be shown (by an argument similar to that leading to equation (9.12)) to lead to drift given by

$$\left(\frac{\partial V_{GS}}{\partial T}\right)_{I_D} = -\frac{\alpha_0}{2}\frac{V_{GS}}{V_P}\Delta V_P \tag{9.17}$$

Hence, using equation (9.9) the difference in zero drift temperature is very nearly

$$\Delta T = -\frac{\alpha_0}{2\lambda} T_0 \Delta V_P$$

or

$$\frac{\Delta T}{T_0} = -\frac{\Delta V_P}{V^*} \tag{9.18}$$

Substituting this value into equation (9.16)

$$\left(\frac{\partial V_{G_1-G_2}}{\partial T}\right)_{V_{D_1-D_2}} = -\lambda\frac{V_{GS}}{V^*}\frac{\Delta V_P}{V_P} \tag{9.19}$$

Since V_{GS} is smaller for low pinch voltage transistors, evidently the smaller V_P the lower the drift for a given percentage tolerance on V_P. On the other hand, it becomes more difficult to hold the tolerance for very small V_P (one volt or less), thus small pinch-voltage units are at an advantage, but some compromise between conflicting requirements is involved. Drift rates down to 5 microvolts/degree C are obtained with well matched pairs indicating that $\Delta V_P/V_P$ is of the order of a small fraction of one per cent. Probably this result will not be greatly improved in the near future and thus represents the limit of what can be achieved at present by compensation techniques. Still further improvement must depend on the use of other methods, such as temperature regulation of the input pair of transistors or chopper stabilisation. Both techniques will be discussed subsequently.

9.2.5 INSULATED GATE DEVICES

So far only junction FETs have been considered. The insulated gate FETs, however, have closely similar properties in the present case, except that the values of α and λ are numerically different. Owen[2] has demonstrated, using an analytical treatment which differs only in detail from that in the preceding sections, that in this case also there is an approximately parabolic relationship between both drain current and offset voltage (referred to the gate) and temperature, and that consequently there exists a temperature at which the drift for individual units is zero. These results are a consequence of the fact that for the insulated gate devices (specifically MOSTs) the relationship between drain current and gate-source voltage is parabolic, to good approximation, following the relationship

$$I_D = \frac{\beta}{2}(V_{GS} - V_{th}) \tag{9.20}$$

which may be compared with the Middlebrook relationship for the junction gate device (equation (9.1)). Also the main temperature dependent effects are variations of charge carrier mobility in the channel and of gate channel contact potential, again as in the junction FET. The main difference is that the MOST is a surface device, and the mobility of charge carriers in the surface layer is much more dependent on device processing details than is the bulk mobility. Thus α can be modified somewhat, also its value is typically only half that for a junction device. By contrast λ is a little large, being at least 3 mV per degree, and hence V^* is likely to be at least 1·5 V, and may be more.

Because of the difficulties in processing a surface device it is at present not possible to match the V_{th} of a pair of MOSTs as closely as the V_P of a pair of junction FETs, and the voltage drift of a long-tail pair circuit is likely to be at least an order of magnitude worse. Against this must be set the advantage of even higher input impedance, which may be important in some cases. If the drift had been expressed as either current or power at the input terminals, rather than voltage, the MOST would appear superior to the junction FET. However, in many applications this is not of great significance.

A more serious problem with MOSTs arises from the instability of their parameters, particularly V_{th}. This problem has already been considered in chapter 4. At present, particularly when subjected to occasional large input signals, MOST DC amplifiers show long term drift levels, which are not temperature dependent, of the order of millivolts referred to the input gate. Long-tail pair circuits do not balance out this effect, since the polarity of signal is opposite at the two input gates and hence produces different 'hysteresis' effects. The effect is said to arise due to migration of ions in the oxide layer, a relatively very slow process. It is much affected by details of device fabrication, and is reduced by diffusing phosphorous into the oxide layer, or by overlaying the oxide with a layer of silicon nitride. The effect has not yet been entirely eliminated, however, in commercially available devices, and thus present MOSTs cannot be utilised in long-tail pair or zero-drift amplifiers unless long term drift of the order of millivolts can be tolerated. Chopper stabilised amplifiers, however, overcome this difficulty, and suitable techniques will be described subsequently.

Another problem in using MOSTs in high gain amplifiers arises from their relatively very high noise level. This again is a matter of details of fabrication and has been much improved in more recently available devices. Even so, the junction FET will be much superior except when the internal impedance of the signal source is very high, typically many megohms. The question of noise is discussed fully in chapter 4.

9.2.6 EFFECTS OF GATE CURRENT

So far the drift contributed by temperature dependent gate current has not been considered. As indicated in equation (9.2) the effect of this is to introduce a voltage drop $R_G I_G$ in series with the gate. Since gate current increases exponentially with temperature, the effect of gate current is to cause the drain current to increase with temperature. It is thus in the same sense as the effect of contact potential but in the opposite sense to the effect of charge carrier mobility variations. In the zero drift biased case the effects of contact potential and mobility variations are arranged to balance, thus the effects of gate current will be negligible if arranged to be small compared with the former. This leads to a condition on R_G, namely

$$R_G \ll \lambda \left\{ \frac{\partial I_G}{\partial T} \right\}^{-1}$$

But

$$I_G(T) = I_G(T_0) \exp \left\{ \frac{T - T_0}{a} \right\}$$

where a is a constant having the dimensions of temperature, typically 13 degrees for a silicon *pn* junction.

Hence

$$\frac{\partial I_G}{\partial T} = \frac{I_G}{a}$$

thus, for negligible drift due to gate current

$$R_G \ll \frac{a\lambda}{I_G} \tag{9.21}$$

For a small epitaxial diffused junction FET in silicon I_G might be 10 pA at 25°C and thus equation (9.21) would be equivalent to requiring R_G to be very small compared with 2·8 Gigohms. At higher temperatures the limit would fall, approximately by a factor of two for every 9 deg. C rise.

For an insulated gate device the gate current would be perhaps three or four orders of magnitude smaller and the upper limit on R_G would increase correspondingly. For this reason it is hardly ever necessary to consider MOST gate current, except perhaps in rare electrometry applications. Even for junction devices, the limit on gate circuit resistance can easily be met in all but a few cases.

REFERENCES

1. Cobbold, R. S. C. and Trofimenkoff, F. H. 'Theory and application of the field-effect transistor', *Proc. I.E.E.*, **111**, 1981–92 (December 1964).
2. Owens, A. R. and Perry, M. A. 'Temperature coefficient of drift of MOS transistors', *Electronics Letters*, **2**, No. 8, 309–10 (August 1966).

10

CHOPPERS AND ANALOGUE GATES

10.1. Chopper amplifiers

10.1.1 THE LOW-LEVEL MODULATOR, OR, CHOPPER FET

Many common transducers exist—such as strain gauges and thermocouples—which provide extremely low level output signals which vary very slowly with time. For all practical purposes, these signals can be treated as very small direct currents or voltages, and must therefore be processed by DC amplifiers. Because of the low level of such signals, the DC amplifiers involved must not be susceptible to inherent drift figures which are of comparable magnitude with these signals; that is, the noise figure near zero frequency must be very small. Apart from these data acquisition amplifiers, there are numerous applications for low drift amplifiers in other fields, including the operational amplifiers used in analogue computation.

Recent developments in microcircuit technology have made possible the close matching of two bipolar transistor structures which, when connected in the long-tailed pair configuration, exhibit differential drift voltages as low as 1 microvolt per degree centigrade referred to the input. These pairs normally form the input stages of monolithic operational amplifiers, and sometimes the actual substrate of a linear microcircuit of this type is self-heated and temperature controlled, which can reduce the drift even further. This order of voltage drift is quite close to that obtainable by chopper-type DC amplifiers, but at the time of writing, current drift figures for chopper amplifiers remain much superior. These have been quoted as typically 50 pA/°C for the better long-tailed pair input stages as compared with 0·5 pA/°C for good chopper amplifiers[1].

In addition to temperature drift, the long-term drift is of considerable importance in some applications, and here the chopper amplifier also remains superior, being capable of drifts of only a few microvolts over several months.

The block diagram for a simple chopper amplifier is given in Fig. 10.1. Here, the incoming signal is seen to be chopped into segments so that it may be injected into a normal, inherently drift-free AC amplifier. The output signal is then reconstituted by a demodulator chopper and an RC low-pass filter.

The most successful form of chopper from the drift point of view is the electromechanical types, but because it employs moving parts,

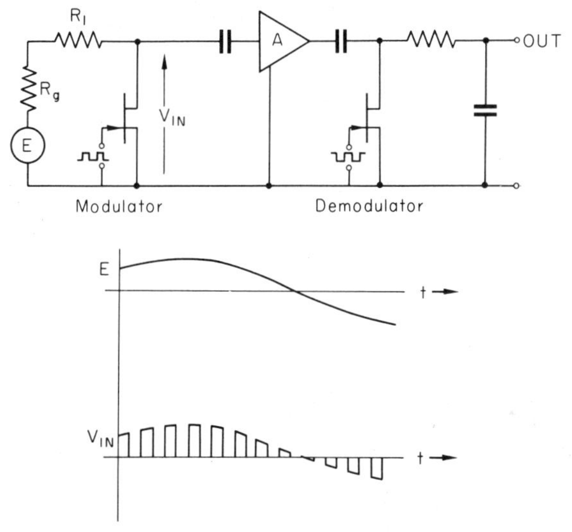

Fig. 10.1. Block diagram and typical input waveforms for chopper-type DC amplifier

not only is its maximum chopping frequency restricted to a few hundred hertz, but its life is strictly limited. In addition to these disadvantages, the electromechanical chopper is quite large, and requires a high driving power compared with solid state choppers.

The bipolar chopper transistor overcomes the aforementioned disadvantages, but unfortunately it has some severe drawbacks of its own. Notable among these is the existence of an offset voltage and current which appear at the collector due to the necessity of driving the base during the ON condition. Various circuit techniques have been developed to overcome this disadvantage[2,3] but the bipolar chopper amplifier remains unsuitable for signals in the microvolt region and lower.

The FET has a distinctly superior performance as a chopper owing to the absence of an offset voltage or current, but at the time of writing its performance is not quite comparable to that of the electromechanical unit.

10.1.2 THE FET SHUNT CHOPPER

Although the FET can be used as either a shunt or series chopper, the former is most commonly used in DC amplifier applications and the latter for analogue switching, which will be considered later in the chapter. For extremely high quality chopping, however, a combination of the two may be used. Fig. 10.2 shows the basic chopper input circuit, and the relevant portion of the FET output characteristic diagram is sketched in Fig. 10.3. Because the voltage and current levels involved are very small, the FET operates near the origin, well below pinch-off, and this means that

(*a*) the channel behaves as a pure resistance and

(*b*) if V_{DS} is reversed the chopper will still operate provided that the gate-channel junction does not become forward biased.

Notice that the characteristics all pass through the origin, which accounts for the lack of inherent offset voltage in the FET.

Consider the ON condition, the equivalent circuit for which is given in Fig. 10.4(*a*). If the parallel combination of $r_{DS(\text{on})}$ and R_L is R, then,

$$\frac{V_L}{E} = \frac{R}{R_g + R_1 + R} \tag{10.1}$$

Usually, $R_L \gg r_{DS(\text{on})}$, especially when R_L is the leakage resistance of an input capacitor to the following AC amplifier, so that,

$$\frac{V_L}{E} \simeq \frac{r_{DS(\text{on})}}{R_g + R_1 + r_{DS(\text{on})}} = \frac{1}{1 + \dfrac{R_g + R_1}{r_{DS(\text{on})}}} \tag{10.2}$$

Ideally, this equation should tend to zero, which implies that $(R_g + R_1)$ should be large and $r_{DS(\text{on})}$ small. Because R_1 is external to the source generator (and is often included to prevent the short-circuiting of this generator) it would appear simple to make R_1 large in comparison with $r_{DS(\text{on})}$. Unfortunately, this is inimical to optimum operation in the OFF condition, as will be shown.

As a numerical example, consider a chopper FET having an $r_{DS(\text{on})}$ of 50 Ω. (This will almost certainly be an *n*-channel type, for the mobility of holes is less than that of electrons, which means that

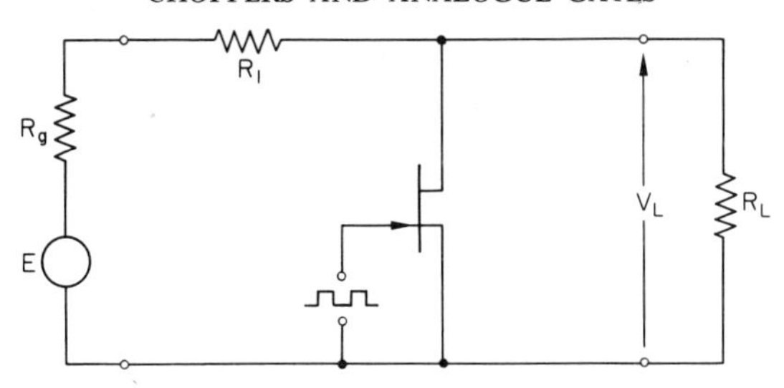

Fig. 10.2. FET shunt chopper

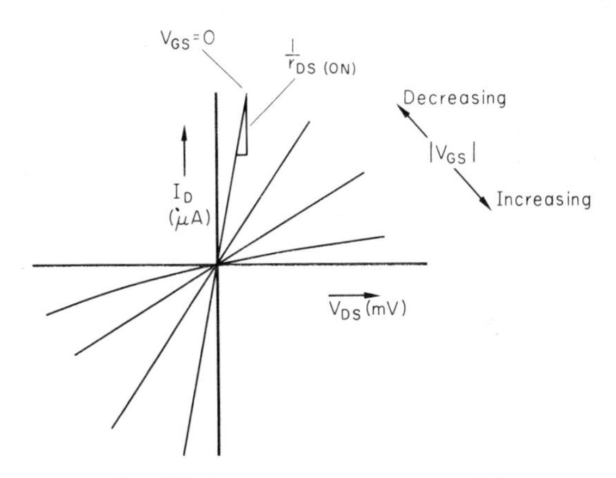

Fig. 10.3. Drain characteristics near origin

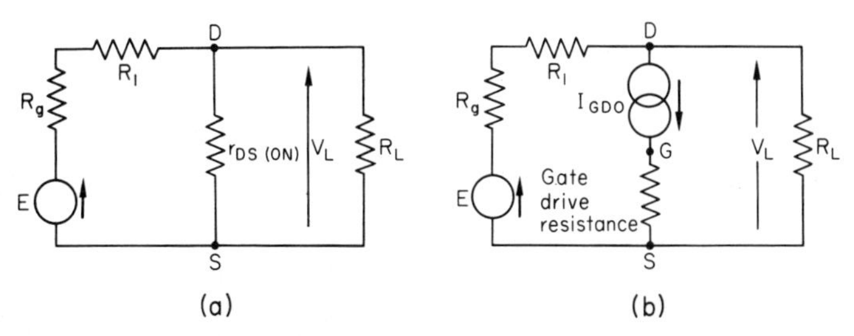

Fig. 10.4. Shunt chopper equivalent circuits.
(a) ON (b) OFF

$r_{DS(\text{on})}$ for p-channel types is in general the greater.) Letting $R_g = 1\ \text{k}\Omega$ and $R_1 = 10\ \text{k}\Omega$, equation (10.2) gives:

$$\frac{V_L}{E} = \frac{1}{1 + (11\,000/50)} = \frac{1}{221} \text{ or } 0{\cdot}45\,\%$$

This represents the ON error percentage.

When the FET is biased OFF, the equivalent circuit becomes that of Fig. 10.4(b). Here, it will be noticed that in place of the channel resistance (which has now become very large) a current generator has been included to take account of the drain–gate

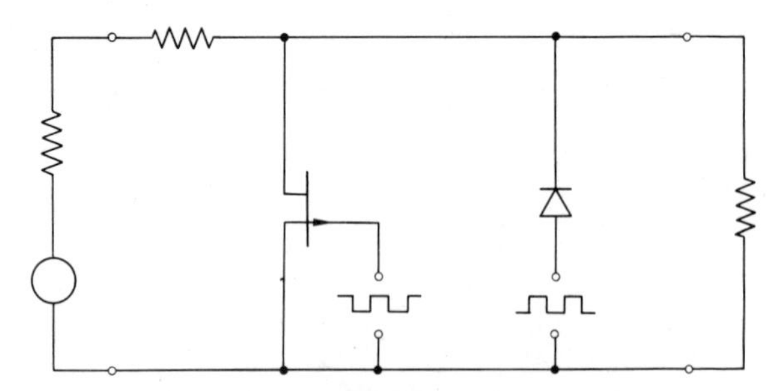

Fig. 10.5. Shunt chopper leakage compensation (After Barton[4])

leakage. This approximates to I_{GDO}, the inexactitude being due to the fact that the source is not open circuited, but is connected to the gate via the chopper drive impedance.

If r_{DS} and R_L are very large indeed, then

$$V_L \simeq E - I_{GDO}(R_g + R_1) \tag{10.3}$$

Using the values of R_g and R_1 previously quoted, and letting $I_{GDO} = 1$ nA,

$$V_1 = E - 11 \text{ microvolts}$$

Clearly, if E is to be of the order of a few tens of microvolts, this error is quite serious. The solution would be to either decrease R_1, which conflicts with the requirement of the ON condition; or choose a chopper FET with a lower gate leakage. It should also be remembered that I_{GDO}, being a reverse junction leakage current, is an exponential function of temperature.

It is possible to compensate for I_{GDO} by arranging to extract this amount of current from the drain during the OFF periods. Barton[4] suggests the form of circuit shown in Fig. 10.5, wherein a diode having a leakage current similar to I_{GDO} for the FET is supplied from the gate drive signal applied in antiphase. By this means a current I_{GDO} is extracted during each OFF period, this corresponding to the reverse-bias period of the diode. It would presumably make for better tracking with temperature if a matched FET were used in place of the diode, though this would be a somewhat expensive modification.

10.1.3 THE TRANSIENT PROBLEM

A square drive waveform is usually applied to the gate of a chopper FET so as to achieve fast turn-ON and turn-OFF. This means that the fast rise and fall of the leading and trailing edges of this square wave will cause transient spikes to be coupled from the gate to the drain of the FET. Fig. 10.6 shows the equivalent circuit of the FET

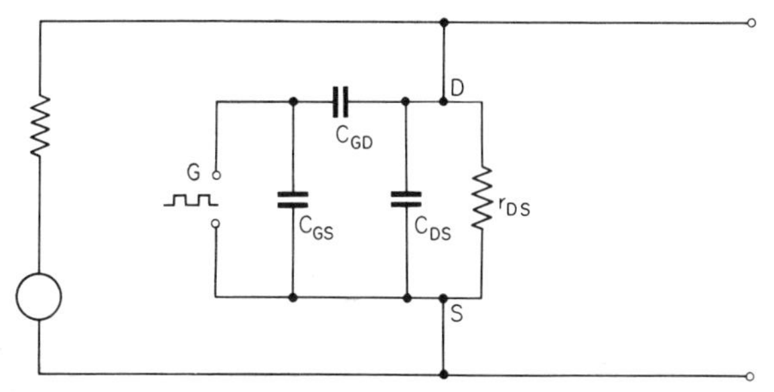

Fig. 10.6. Shunt chopper equivalent circuit including interelectrode capacitances

in the chopper configuration, and it will be seen that if the drive generator has a low internal impedance, C_{GS} has no effect, being simply a small capacitance in parallel with this drive generator. For the OFF condition, where r_{DS} is large, C_{GD} and C_{DS} form a voltage divider, and a proportion of the drive voltage will appear in the form of a spike across C_{DS} when the turn-ON edge appears. This will decay rapidly as r_{DS} falls. When the turn-OFF edge arrives, a spike in the opposite sense will appear, but now the low resistance r_{DS} is initially part of the voltage divider, so that until r_{DS} rises, the turn-OFF spike at the output will be quite small. The relevant waveforms are shown in Fig. 10.7.

It should be recalled that not only does r_{DS} vary as the drive voltage is applied, but so do the capacitances associated with the FET, and this situation means that it is invalid to use the small-signal equivalent circuit values for spike calculations. It is, however, reasonable to use these small-signal values to choose an FET having basically low interelectrode capacitances. Unfortunately, such an

FET would have a small-geometry construction, which would imply a rather high $r_{DS(on)}$. For these reasons, it is usual to design for a compromise between $r_{DS(on)}$ and interelectrode capacitance values, and it is in fact found that if the drive voltage is kept down to a little more than V_P for the FET, the spike generated can be considerably smaller than those exhibited by the bipolar transistor. This comes about in part because the transient response of the bipolar transistor is a function of charge storage effects in the base

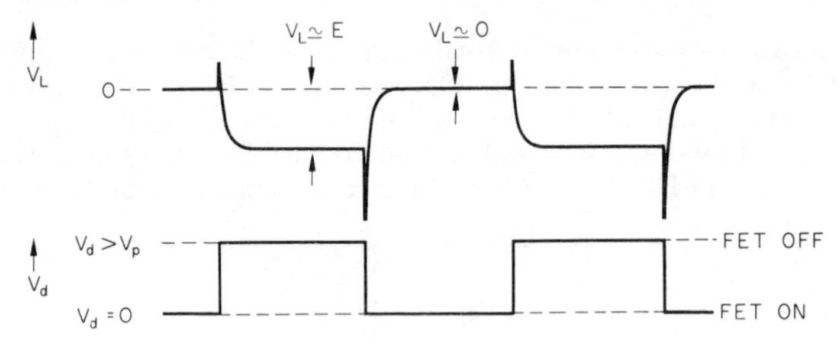

Fig. 10.7. Output and drive waveforms for p-channel *shunt FET chopper*

region in addition to interelectrode capacitances. An extensive analysis relevant to the IGFET shunt chopper is given by Dostal[5].

10.1.4 THE FET SERIES AND SERIES/SHUNT CHOPPER

In Figs. 10.2, 10.4 and 10.5, it is usual for R_L to be simply the very high leakage resistance of the input capacitor of an AC amplifier, as shown in Fig. 10.1. For circuits where this is true, it is mandatory to use a shunt chopper, for the series version, shown in Fig. 10.10 would have a floating drain. For this reason it is more usual to employ a series FET as a commutator, or analogue gate element rather than as a pure chopper. However, the series/shunt combination of Fig. 10.8 will operate into a capacitance, and it is in all ways superior to the shunt chopper. The two gates are driven in anti-phase so that the series FET is ON when the shunt FET is OFF and vice versa.

The series/shunt chopper succeeds in reconciling the conflicting requirements of high $(R_g + R_1)$ for the shunt ON condition and low $(R_g + R_1)$ for the shunt OFF condition. This is because R_1 is now replaced by the series FET, the channel of which presents a high resistance when the shunt FET is ON, and a low resistance when the shunt FET is OFF.

In addition to this, the transient spikes are reduced by the series/shunt combination, also because of the opposing resistance changes in the two channels.

The average value of the spikes appears as an apparent offset voltage, and Miles[6] has shown that, as expected, the series/shunt combination does much to minimise this offset. He also shows how further spike reduction can be attained by the judicious use of small trimming capacitors.

If the two FETs are complementary, a single-ended drive can be used, but in this case, care should be taken to establish that the p-channel device is very similar to the n-channel in terms of switching

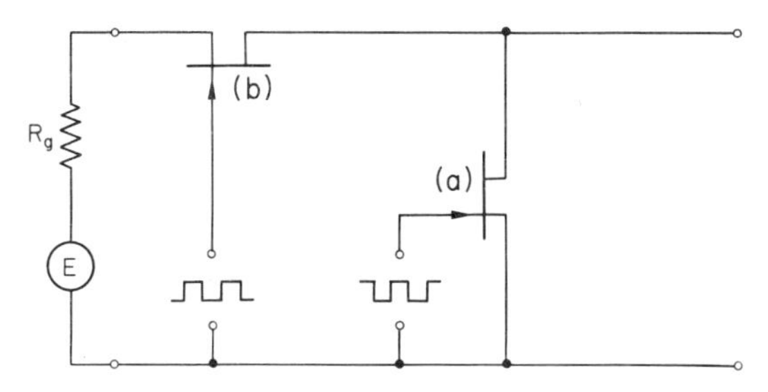

Fig. 10.8. Series/Shunt chopper using n-*channel FETs*

speed, otherwise a period may exist where the devices are both ON or both OFF. Normally, it is cheaper to use fairly well-matched n-channel FETs driven differentially from the antiphase outputs of a multivibrator, especially since such a drive is also useful for operating the demodulator FET.

10.1.5 DRIVE REQUIREMENTS

Providing that the input signal is small compared with V_P for the FETs used, the drive voltage(s) may be referred to the common line (i.e. may be singled-ended) for any of the chopper configurations discussed. In the interests of spike reduction, however, the drive voltage should be little greater than $|V_P| + |E_{max}|$. If E is to be significant compared with V_P, this implies a poor spike performance, and under these circumstances, a floating drive, or another means of solving the problem must be employed. Because a large value of E is more relevant to the commutator than to the pure chopper, techniques of drive isolation will be discussed in a later section.

The frequency of the drive waveform must normally be restricted to a few hundred hertz for the simple shunt or series chopper, because the relevant interelectrode capacitances must have time to discharge. However, for the series/shunt circuit, it will be found that this discharge time is much decreased because of the low channel resistances presented alternately to each FET. This means that a much higher chopping frequency is possible, and using IGFETs, this can be as high as several megahertz[7].

10.2. Analogue switching

10.2.1 THE ANALOGUE GATE

Unlike the digital gate, which transmits the state of the input signal—that is, ON or OFF—the analogue gate is required to transmit the actual input signal waveform, ideally without distortion. For example, a data logging assembly may monitor several

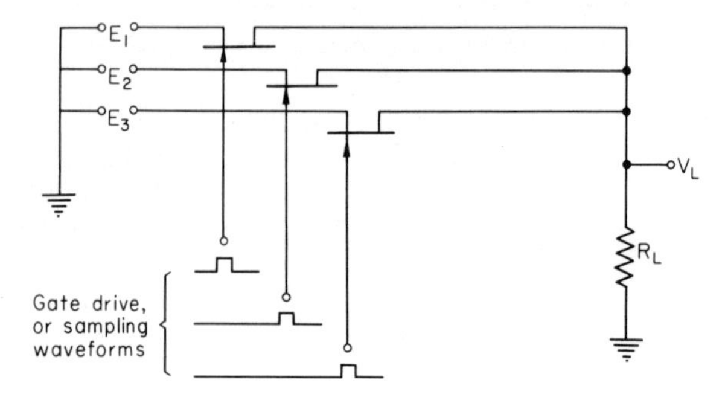

Fig. 10.9. Junction FET multiplexer or commutator

transducers, such as thermocouples, and print out the temperature readings consecutively. This commutation, or multiplexing, may be conveniently performed using an array of analogue-gate FETs driven by consecutively arriving pulses, as shown in Fig. 10.9.

The basic FET analogue gate is shown in Fig. 10.10, from which the errors in both the ON and OFF states may be obtained.

For the ON condition, if $R_L \gg r_{DS(\mathrm{on})}$ and R_g,

$$\frac{V_L}{E} = \frac{R_L}{R_g + r_{DS(\mathrm{on})} + R_L} \simeq 1 \tag{10.6}$$

For the OFF condition,

$$V_L \simeq I_{GDO} R_L \tag{10.7}$$

if $R_L \ll r_{DS(\text{off})}$, which is usual for an analogue gate circuit.

The gate drive waveform is shown as a square-wave, and this presents the usual spike breakthrough problems. If E is very small, the drive V_d can be held at a level little greater than V_P, making for spike minimisation; but for an analogue gate, as opposed to a

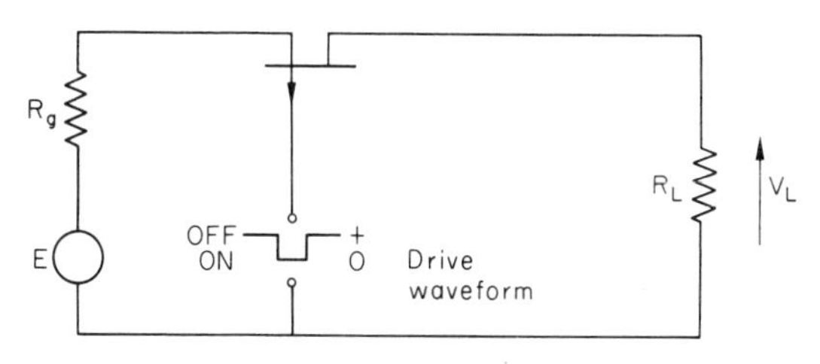

Fig. 10.10. Gate element

chopper, E is often large, which makes this technique unusable. Under these circumstances, working conditions can be tabulated as follows for the p-channel unit of Fig. 10.10:

(i) For the ON condition, where $V_d = 0$
- (a) If $E \to 0$, $r_{DS} = r_{DS(\text{on})}$, which is optimum.
- (b) If E goes significantly negative, V_{GS} increases and raises r_{DS}.
- (c) If E goes significantly positive, the junction may become reverse biased.

(ii) For the OFF condition, where $V_d > V_P$
- (a) If $E \to 0$, $r_{DS} \to \infty$, which is optimum.
- (b) If E goes significantly negative, the optimum condition remains.
- (c) If E goes significantly positive, V_{GS} may fall below V_P so that r_{DS} becomes finite.

A convenient and simple technique which obviates some of the disadvantages implied in the foregoing summary is shown in Fig. 10.11. Here, a diode has been connected in series with the gate drive, and a high value resistor connects the gate to the source. The new situation may be summarised as before:

(i) For the ON condition, where $V_d = 0$
- (a) If $E \to 0$, $r_{DS} = r_{DS(\text{on})}$, which is optimum.
- (b) If E goes significantly negative, V_{GS} increases and raises r_{DS}.
- (c) If E goes significantly positive, the diode cuts off, and

because its leakage current is negligibly small, V_{DS} remains zero, and r_{DS} remains at $r_{DS(on)}$.

(ii) For the OFF condition, the situation is as before.

If the two tabulations are compared, the conclusion will be reached that the *p*-channel analogue gate shown in Fig. 10.11 is suitable for a positive-going input signal such that $E \ngtr (V_d - V_P)$; but that only a small negative input can be accepted. However, the

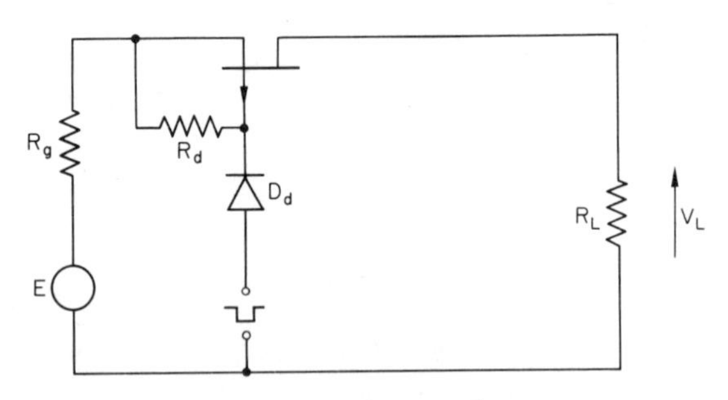

Fig. 10.11. Modified gate element

latter restriction may now be removed simply by making V_d *negative* rather than zero for the ON condition. Under these circumstances, the diode prevents forward current flow through the FET junction, and the resistor holds V_{GS} almost at zero, providing that the diode has a very low leakage current. The positive-going value of E must of course obey the inequality $E^+ \ngtr |V_d^-|$, where V_d^- is the negative-going excurşion of V_d.

When the FET is OFF, a small current will be fed back from the gate to the source via R_d and if R_g is low, this may be unimportant. However, if the drop across R_g due to this current is unacceptable (which is rare, for it cannot contribute to the output) then R_d must be made larger. This could lead to a slowing down in switching speed, and to obviate this, techniques for replacing R_d with another FET have been developed[8].

The foregoing discussion does, of course, apply equally well to *n*-channel devices with all polarities reversed. In fact, *n*-channel devices are in general to be preferred since they can exhibit lower values of $r_{DS(on)}$ and higher switching speeds.

Drive circuits using bipolar transistors are eminently suited to the analogue gate, and single packages containing both driver and one or more FETs are currently available[9].

10.2.2 THE IGFET CHOPPER

By virtue of their comparative ease of manufacture in large numbers on silicon slices, *p*-channel enhancement-type IGFETs are particularly suited to multiple arrays, including integrated choppers, Fig. 10.12, and multiplexers, Fig. 10.13(*a*).

Notice that in the case of the chopper, the gates of the two IGFETs are driven in antiphase from the gate and drain respectively

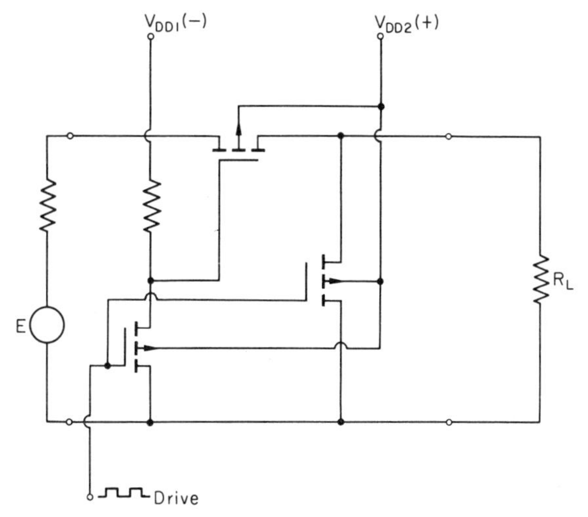

Fig. 10.12. Series/shunt integrated IGFET chopper including drive phase splitter

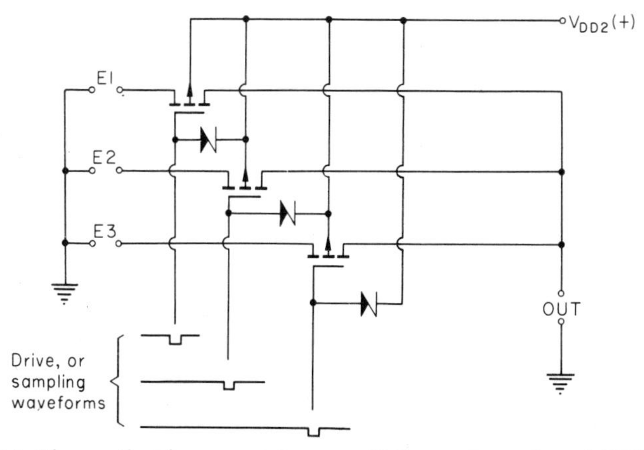

Fig. 10.13(a). Three paths of a commutator or multiplexer using p-channel *IGFETs integrated with Zener protectors*

of the driver IGFET. The drive waveform itself must be high enough to take the gates of the driver and shunt IGFETs from V_{th} to some value where $r_{DS(on)}$ becomes acceptably low. This implies quite a large excursion, but because V_{th} can be several volts, it also implies that signals of either polarity can be gated providing that the positive excursion does not exceed $|V_{th}|$. However, any positive input applied to the source would forward-bias the source-substrate junction, leading to clamping of that input. It is for this reason that the substrates have been taken to a fixed positive voltage V_{DD2} in Fig. 10.12. This voltage will normally be similar in magnitude (though opposite in sense) to V_{th}, and a side-effect of its application is that the value of $|V_{th}|$ is increased relative to that for a grounded substrate.

10.2.3 THE IGFET ANALOGUE GATE

Fig. 10.13(*a*) represents three units of a multi-path integrated *p*-channel enhancement IGFET analogue gate array. The insulated gate of each unit is protected by a Zener diode, and all the drains are commoned to one point.

By inspection of Fig. 10.13(*b*), which represents one path in this array, a series of statements regarding the working conditions can be made. Assuming that the load resistance is to be large, so that the voltage drop in R_g may be neglected, then:

(i) For the ON condition, where V_d is negative (V_d^-),

(a) If $E \to 0$, r_{DS} is defined only by V_d^-

(b) If E goes significantly negative (E^-), $V_{GS} = (V_d^- - E^-)$ and because V_{GS} must always be much more negative than V_{th} in order to hold the IGFET ON, this means that

$$V_d^- \ll (V_{th(max)} + E^-) \tag{10.8}$$

(c) If E goes significantly positive (E^+), it becomes additive with V_d^- and so helps to keep the IGFET hard ON.

(ii) For the OFF condition, where V_d is positive (V^+),

(a) If $E \to 0$, the IGFET is held hard OFF by V_d^+

(b) If E goes significantly negative (E^-), it becomes additive with V_d^+ and so helps to keep the IGFET hard OFF.

(c) If E goes significantly positive, (E^+),

$$V_{GS} = (V_d^+ - E^+)$$

and if $E^+ > V_d^+$ so that V_{GS} goes negative, it must not go more negative than $V_{\text{th(min)}}$, otherwise the IGFET may begin to turn ON. The necessary condition is therefore,

$$V_d^+ > (V_{\text{th(min)}} + E^+) \qquad (10.9)$$

Inequalities (10.8) and (10.9) define the excursion of V_d: it remains only to determine what voltage should be applied to the substrate, or body. There are three conditions to be fulfilled:

(i) $V_{DD_2} \geqslant V_d^+$, otherwise current will flow from the gate drive to the body through the gate-protection Zener diode in the *forward* direction.

(ii) $V_{DD_2} \geqslant E^+$, otherwise the source-body junction will be forward-biased and current will flow across it.

(iii) When E is negative, the voltage across the gate-protection Zener diode is $(V_{DD_2} - E^-)$, and this must not exceed the Zener breakdown voltage, which being the apparent gate breakdown voltage, is called BV_{GBS}. That is,

$$(V_{DD_2} - E^-) < BV_{GBS}$$

Summarising these conditions gives:

$$(E^+ \text{ or } V_d^+) \leqslant V_{DD_2} < (BV_{GBS} + E^-) \qquad (10.10)$$

To illustrate the use of the foregoing design expressions, consider a typical 5-channel p-type enhancement IGFET analogue gate array, the Silconix G114L, and let the input signal vary between ± 5 V.

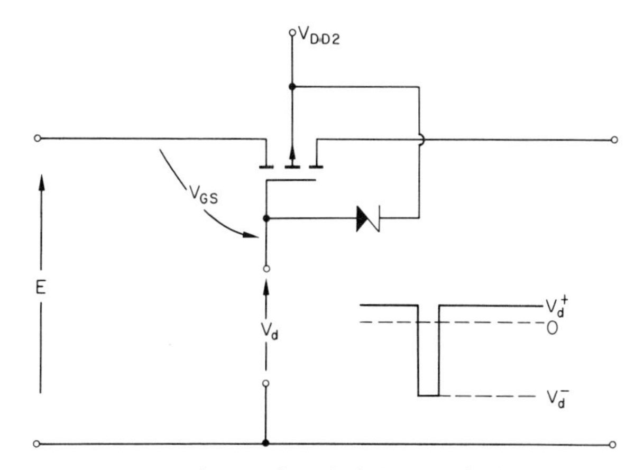

Fig. 10.13(b). One channel of IGFET multiplexer

For the G114L, V_{th} is between -2 and -6 V, so that inequality (10.8) gives:

$$V_d^- \ll (-6 - 5) = -11 \text{ V}$$

and inequality (10.9) gives:

$$V_d^+ > (-2 + 5) = +3 \text{ V}$$

Hence, an asymmetric drive of -18 and $+6$ V (for example) would be entirely satisfactory, and because BV_{GBS} for the G114L is -30 V (min), inequality (10.10) shows that V_{DD_2} could also be $+6$ V.

$$(+5 \text{ or } +6) \leqslant +6 < (30 - 5)$$

A drive circuit utilising supply rails of $+6$ and -18 V can be visualised, the $+6$ V rail being also taken to the G114L body.

Such devices are available commercially, and some, in monolithic microcircuit form[10], also perform an interfacing function between the analogue gate array and low-level bipolar logic. In fact, combination driver-gate modules are also manufactured, and offer the designer considerable freedom of choice in terms of both driver triggering levels and analogue gate switching combinations.

REFERENCES

1. Analog Devices Applications Manual for models 201 through 210.
2. Watson, J. *Semiconductor Circuit Design*, Chap. 6, Hilger & Watts (1966).
3. Hutcheon, I. C. and Summers, D. 'Low-drift transistor chopper-type DC amplifier', *Proc. I.E.E.*, **107-B**, 451 (1960).
4. Barton, K. 'The FET used as a low-level chopper', *Electronic Engineering*, **37**, 80–83 (February 1965).
5. Dostal, J. S. 'Analysis of the capacitance error of a MISFET chopper modulator', *Electronic Engineering*, **39**, 476 (October 1967).
6. Miles, J. F. 'MOS transistors and their applications', *Mullard Technical Communications* **9**, No. 83 54–63 (September 1966).
7. Bergersen, T. 'FETs in chopper and analog switching circuits', *Motorola Applications.* Note AN-220 ( ).
8. Shipley, M. 'Analog switching circuits use field-effect devices', *Electronics* (28 December 1964).
9. Gulbenk, J. and Prosser, T. F. 'How modules make complex design simpler', *Electronics* (28 December 1964).
10. Evans, A. D. 'ICs end the driver gap', *Electronic Design* (25 *October* 1966).

11

THE FET AS A VARIABLE RESISTOR

11.1. Introduction

In the previous chapters the field-effect transistor was considered in applications in which it is used as an amplifier or a switch. Because of its very favourable properties it is replacing the bipolar transistor for many purposes. These include very high power gain and input impedance, low noise, speed and, as a switch, zero voltage offset.

However, the field-effect transistor has a further property which is almost unparalleled in other circuit devices. Operated under suitable conditions it can exhibit the properties of an ohmic resistor, so far as the circuit between source and drain terminals is concerned, but the value of the resistance is electrically controlled, being determined by the potential difference between the source and gate terminals. The exploitation of this variable resistance property in circuit applications has already been the subject of publications, and there can be little doubt that it will play a significant role in the future development of electronic circuit technology, not least because it is the only variable linear passive element available for use in micro-circuits.

In order that a FET shall behave like an ohmic resistor two conditions must be fulfilled:

(a) the potential difference between the gate region and the conducting channel must be such that no significant gate current flows and

(b) the depletion region of the reverse-biased gate to channel *p–n* junction must at no point along the axis of the channel extend so far as to close or 'pinch-off' the channel.

If these two conditions are met then the source-to-drain channel will show approximately ohmic properties, although there will be some departure from a linear voltage-current relationship as circuit conditions approach those at which pinch-off occurs.

Typical drain characteristics of a junction field-effect transistor (Type C82) are shown in Fig. 11.1 and an enlarged plot of the region near the origin where the channel is not pinched-off is shown in Fig. 11.2. This is sometimes referred to as the triode region of the device characteristics, although as will be seen it bears very little resemblance to the characteristics of a thermionic triode. The relationship between drain current and voltage is almost linear, provided that the gate-to-channel junction remains reverse-biased, and also that the channel pinch-off conditions are not too closely

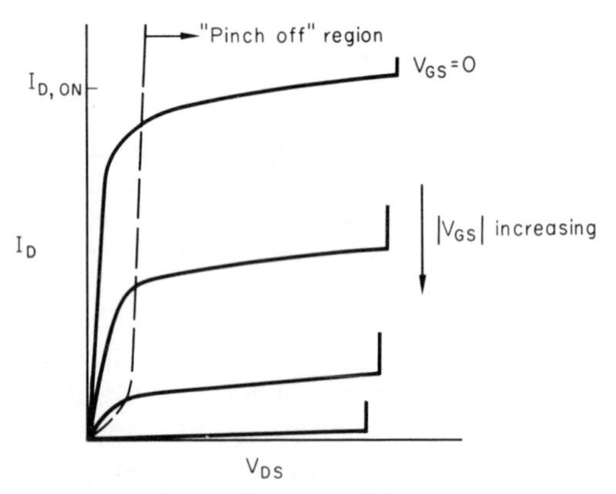

Fig. 11.1. FET drain characteristics

approached. There is no voltage offset; that is to say, the drain-to-source voltage is zero when the drain current is zero.

The value of resistance measured between drain and source will depend upon the voltage between gate and source terminals of the device, increasing with the reverse bias applied to the gate junction due to the increasing penetration of the depletion region into the conducting channel, and hence the reduction of its effective cross-sectional area. The exact law relating channel resistance to gate control voltage depends critically on the structural details of the device considered. Usually there is an approximately exponential increase in resistance as reverse gate bias increases, followed by a much sharper increase in resistance as pinch-off conditions are approached. The latter effect is rather undesirable, since it occurs at a gate voltage which varies widely from unit to unit. It also results in a marked increase in control sensitivity which may complicate design and sets a limit to the resistance variation range which can be used. Development of transistor structures for variable resistance use will probably aim at 'remote pinch-off' types having an extended range of near exponential dependence of resistance on control voltage.

Most FET structures developed for amplifier and switching applications have a prismoidal untapered channel having plane parallel gate junctions (or metal gates) on opposite sides of the channel. In this case the mathematical model developed by

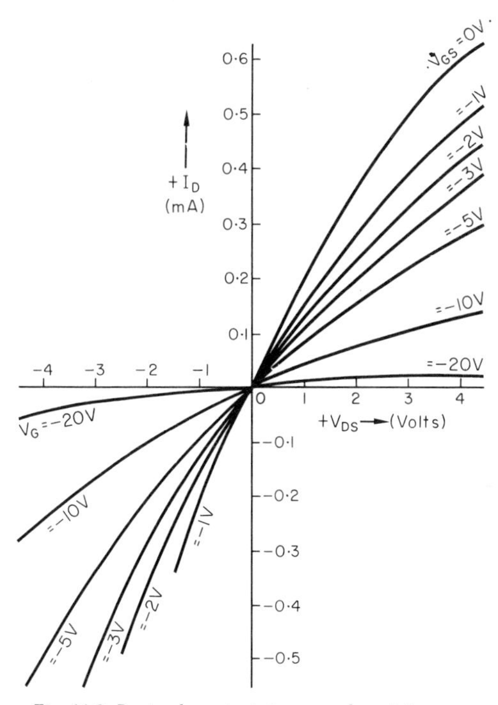

Fig. 11.2. Drain characteristics near the origin

Shockley[1], or the alternative and rather simpler charge controlled treatment due to Middlebrook[2], applies, and leads to a resistance law of the form

$$r_{DS} = \frac{r_0}{1 - \{V_{GS}/V_P\}^{\frac{1}{2}}} \tag{11.1}$$

in the Shockley case, or

$$r_{DS} = \frac{r_0}{1 - (V_{GS}/V_P)} \tag{11.2}$$

in the Middlebrook case.

The two expressions give results which do not differ very greatly and actual double diffused or epitaxial diffused junction FETs and MOSTs have resistance laws which fall between the limits predicted by the two expressions. They are not, however, very useful in variable resistance applications, since as V_{GS} approaches V_P the rate of variation of r_{DS} with control voltage becomes very rapid. Thus, the useful range of r_{DS} is limited to no more than about a 10:1 ratio. The effect can be seen clearly if the drain–source resistance is plotted on a logarithmic scale against gate–source voltage (Fig. 11.3).

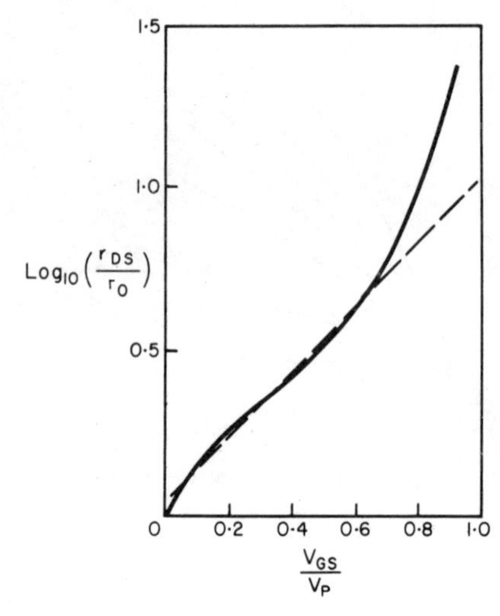

Fig. 11.3. The relationship of r_{DS} to V_{GS} as predicted by Shockley's model (The broken line represents an exponential relationship

The broken line on the same graph represents an exponential relationship. It will be seen that for $V_{GS}/V_P > 0{\cdot}7$ (at which value the drain–source resistance has increased only by a factor of slightly over six, relative to its value at zero gate–source voltage) the relationship diverges sharply from the exponential approximation.

For the present application it would be desirable for the sharp rise in resistance as V_{GS}/V_P tends to unity (which is due to the pinching off of the conducting channel) to be less marked, thus extending the range of usable resistance variation. Transistors may be produced by adopting a modified device geometry; one such structure, based on an alloyed junction, is shown in Fig. 11.4. Transistors of this type are commercially available, and other structures, utilising different fabrication techniques, which would yield comparable properties, are also possible. It is to be expected that the alloyed structure will rapidly become obsolete, even for the present purpose, since it is incompatible with modern planar technology. Also alloyed devices cannot be incorporated in integrated circuits and are slightly noiser and less reliable than their planar counterparts.

Possible structures using modern fabrication technology which would yield the desirable 'remote pinch-off' FETs have been

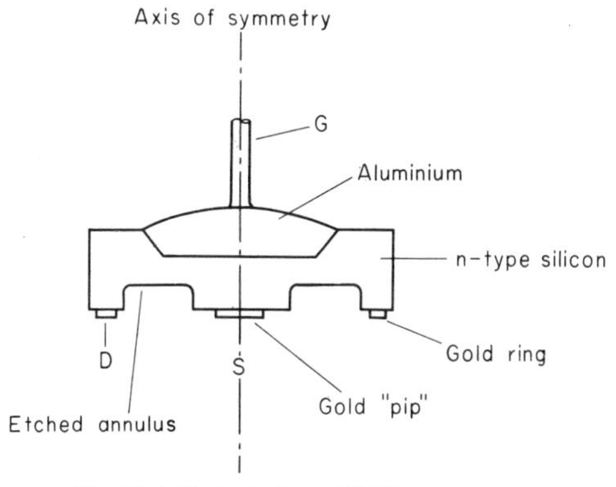

Fig. 11.4. Etched-alloyed FET structure

described[3], and commercial availability of devices of this type can be expected in the near future.

The variation of r_{DS} with gate–source voltage observed experimentally for a commercial alloyed unit is shown in Fig. 11.5. It is apparent that the law of dependence of r_{DS} upon gate–source voltage may be reasonably well-represented by an expression of the form

$$r_{DS} = r_0 \exp(\eta V_{GS}) \tag{11.3}$$

where r_0, η are constants.

Transistors of the same nominal pinch-off voltage have approximately the same value of η but will vary markedly in r_0. Even with the modified geometry of this type of transistor a pinch-off effect is observed, and at large gate–source voltages, a more rapid than exponential variation of drain–source resistance occurs. However, equation (11.3) remains a good approximation for more than a decade resistance change, and even beyond the resistance deviates much more gently from the exponential approximation than in the case of a transistor obeying equation (11.1) and (11.2). The results shown in Fig. 11.5 were obtained using an AC bridge to measure the resistance (at a frequency $\omega = 10^4$ rad/s) taking care to keep the alternating voltage between source and drain less than 0·3 V r.m.s. Since the source–drain channel is not perfectly ohmic, the applied voltage is kept small so that the true slope resistance at the origin may be measured. The consequences of the nonohmic nature of the conducting channel are investigated further later in this chapter.

The FET cannot be regarded as a pure resistance, since the channel is shunted by a small stray capacitance, leading to the

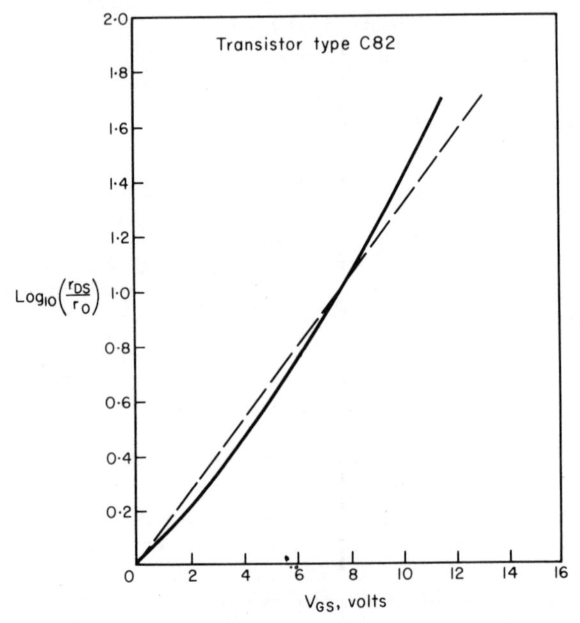

Fig. 11.5 Resistance versus control voltage characteristic for a 'remote pinch-off' FET

equivalent circuit of Fig. 11.6. The effect of this shunt capacitance is more serious at high frequencies, and will be described later in connection with particular applications.

All resistors generate a certain amount of noise, principally Johnson (thermal) noise. This effect is also present in the FET, used as a variable resistor. However, there is a further source of noise in the transistor, namely a component which has a power spectral density function which obeys a $1/f$ law, (see chapter 4). This noise source, which is probably associated with surface effects, typically becomes of significant magnitude, relative to Johnson noise, below about 200 Hz and will therefore be important in many audio applications. As will be shown, circuits using FETs can operate at quite high signal voltage levels and, consequently, can yield satisfactory signal-to-noise ratios despite the excess noise generated.

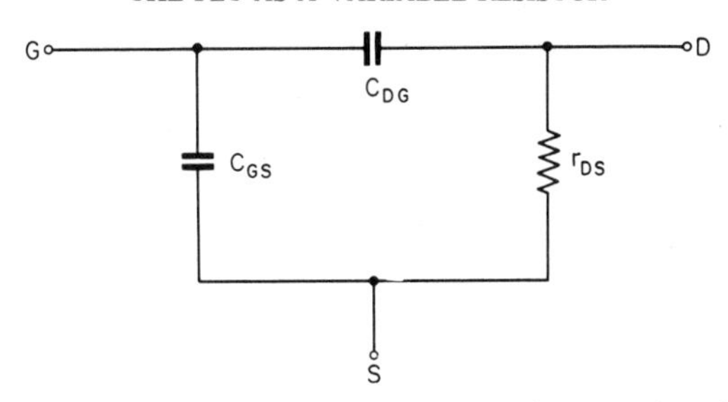

Fig. 11.6. Equivalent circuit for a variable resistance made FET. The value of r_{DS} *is a function of* V_{GS}

The importance of $1/f$ noise depends very much on the value of the lower cut-off frequency of the system and on the particular transistor used. Insulated gate (MOST) transistors at present available have inferior characteristics in this respect.

11.2. Temperature effects

The effects of temperature in specific circuit applications will be discussed later However, in all cases, provided that the resistance in the gate circuit of the FET is not so large that the effects of gate current need be taken into account, the temperature dependence of the channel resistance r_{DS} depends on two factors:

(i) Variation of contact potential between gate and channel This is equivalent to a change in the effective voltage applied to the gate terminal, and is of such a sense as to cause a decrease in channel resistance with increasing temperature. It is equal in magnitude to about 2·2 mV/deg. C, for a silicon junction-gate unit.

(ii) Variations of carrier mobility. The mobility of charge carriers falls with increasing temperature, causing an increase in the magnitude of channel resistance. In silicon junction FETs the effect varies between about 0·5 and 0·7% per deg. C

To derive an expression for the variation of applied gate–source voltage needed to offset the effects of temperature, equation (11.3) must be rewritten in a form which explicitly includes contact potential. Writing Φ for the latter, equation (11.3) may be amended to

$$r_{DS} = r_0 \exp(\eta |V_{GS} + \Phi|) \quad (11.4)$$

The theory of variable resistance networks derived in succeeding sections will be based on equation (11.3) rather than the more precise form of equation (11.4). However, ϕ is only a fraction of a volt and the magnitude of η, for the type of FET best suited to these circuits, is 0·1 or less, so that the consequent error is very small.

By differentiation of equation (11.4) r_{DS} will not vary with temperature if

$$0 = \frac{\mathrm{d}r_0}{\mathrm{d}T}\exp(\eta|V_{GS} + \phi|) + \eta r_0\frac{\mathrm{d}V_{GS}}{\delta T} + \frac{\mathrm{d}\phi}{\mathrm{d}T}\exp(\eta|V_{GS} + \phi|)$$

or, expressed as a condition on the rate of change of the magnitude of the gate bias with temperature to keep r_{DS} constant,

$$\left|\frac{\mathrm{d}V_{GS}}{\mathrm{d}T}\right| = \frac{\alpha}{\eta} + \frac{\mathrm{d}\phi}{\mathrm{d}T} \quad (11.5)$$

where

$$\alpha = \frac{1}{r_0}\cdot\frac{\delta r_0}{\delta T}$$

Practically, the reverse gate bias will need to decrease in magnitude by an amount varying between zero and a few tens of millivolts per degree. Note that whilst α is always positive, for an n-channel device η is negative and $\mathrm{d}\Phi/\mathrm{d}T$ (since the voltage is of such sense as to make the channel more negative than the gate) is positive. The two terms in equation (11.5) are thus of opposite sense, the first predominating for normal values of η.

The effect of temperature changes may be reduced by putting the FET in a series-parallel network of non-temperature sensitive resistors (Fig. 11.7) in which case the total resistance, R_T, of the network is given by

$$R_T = \frac{R_2(R_1 + r_{DS})}{R_2 + R_1 + r_{DS}} \quad (11.6)$$

and the variability of the combination is reduced, relative to r_{DS}, in the ratio

$$\frac{\partial R_T}{\partial r_{DS}} = \frac{R_2}{R_2 + R_1 + r_{DS}} \quad (11.7)$$

The dependence of R_T on gate control voltage is, of course, reduced in the same ratio as its dependence on temperature. Thus the gate voltage swing must be increased to achieve a similar variation of R_T to that which would have been obtained had R_1 and R_2 not been used. However, in many applications where the value of the resistance must be held accurately, only a small range of variation is required. Further reduction of drift is possible if R_2 has a negative temperature coefficient. The main critical factor in all compensation methods of this kind is that the compensation component must be at the same temperature as the FET. Thus, little power should be dissipated in either and the mechanical design should be such that thermal conductance is much higher

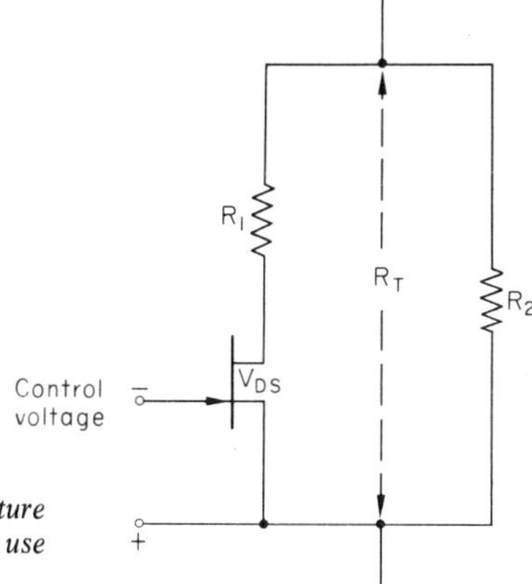

Fig. 11.7. Reduction of the temperature coefficient of the variable resistor by use of a series-parallel network

between these two than to other components. Given this requirement, and writing

$$\frac{1}{R_2}\frac{\partial R_2}{\partial T} = -a$$

by differentiation of equation (11.6) and equating

$$\frac{\partial R_T}{\partial T} = 0$$

the condition on the value of a for zero overall temperature coefficient of R_T is

$$a = \frac{R_2}{(R_1 + r_{DS})^2} \cdot \frac{\partial r_{DS}}{\partial T}$$

A temperature coefficient of this magnitude can easily be attained if R_2 consists of a thermistor and an ohmic resistor in series.

As a further alternative, R_1 could be arranged to have a negative coefficient, and the value for zero overall coefficient can readily be shown to be b, where

$$b = -\frac{r_{DS}}{R_1} \cdot \left(\frac{1}{r_{DS}}\frac{\partial r_{DS}}{\partial T}\right)$$

In this case satisfactory temperature coefficients may be obtained from parallel combination of thermistors and ohmic resistors. In both methods of compensation, however, difficulties arise due to the fact that the temperature coefficient of a thermistor is itself dependent on temperature, and so compensation of the transistor characteristics can only be perfect at one temperature. Thus if very close compensation is desired (say, reduction of the apparent temperature coefficient by an order of magnitude) the working range of temperature is certain to be restricted.

11.3. Voltage controlled attenuators

One of the most obvious applications of the FET variable resistor is in the design of voltage controlled variable attenuators. Such an attenuator can be realised in many forms, including a bridge network with the field-effect transistor as one branch, or embodying

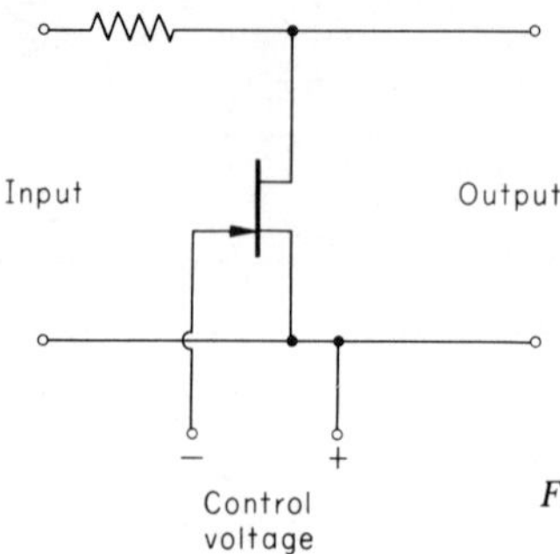

Fig. 11.8. 'Voltage divider' attenuator, using a FET

the device in the feedback network of an amplifier, but the simplest is that in which the transistor forms one arm of a resistive potential divider (Fig. 11.8), and it is this form which will be considered here. The field effect transistor will be assumed to obey the law of equation (11.4), and where a sharp pinch-off types is used there will be some error as the pinch voltage is approached. The effect is small, however, as will be shown.

The attenuation is given by

$$\frac{E_{out}}{E_{in}} = A = \frac{r_0 \exp(\eta V_{GS})}{R' + r_0 \exp(\eta V_{GS})}$$

where R' is the value of the fixed branch of the potential divider, or

$$A = \frac{\exp(\psi)}{1 + \exp(\psi)} \tag{11.8}$$

where

$$\psi = \eta V_{GS} + \log\left(\frac{r_0}{R'}\right) \tag{11.9}$$

If the attenuator is preceded and followed by amplifiers of zero output impedance and infinite input impedance, respectively, the attenuation of the output signal due to the insertion of the network may then be expressed in decibels as

$$L = 20 \log_{10}\left\{\frac{\exp(\psi)}{1 + \exp(\psi)}\right\} \tag{11.10}$$

This relationship is plotted in Fig. 11.9 for the case where r_0/R' is chosen so that $L = -20$ dB when $V_{GS} = 0$, and also taking $\eta = 0{\cdot}135$. Experimentally observed points for an actual attenuator with R' chosen for an initial attenuation of 20 dB are superimposed.

It will be noted that the experimentally observed variation of attenuation expressed in decibels is more linearly related to control voltage than the approximate theory predicts. The convex upwards curvature of the characteristic at high control voltages is partially offset by the more rapid variation of drain resistance with control voltage at higher values where pinch-off is approached. The

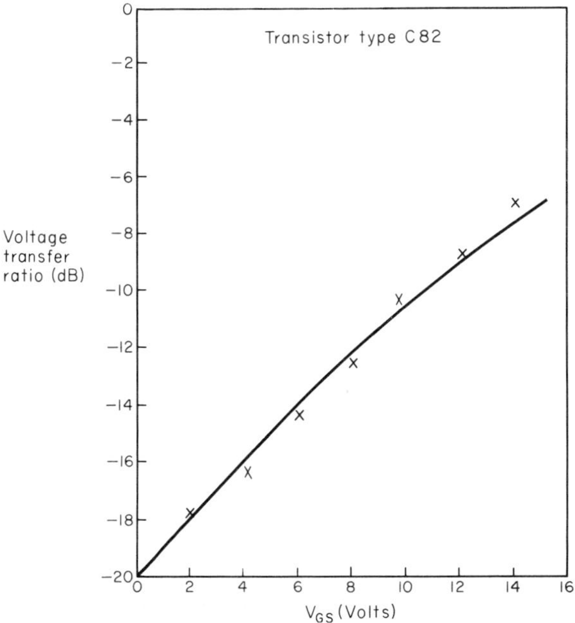

Fig. 11.9. Comparison of measured (×) and calculated attenuation for the network of Fig. 11.3, the latter assuming an exponential dependence of r_{DS} on V_{GS}

characteristics of three different attenuators using the same transistor but different values of fixed series resistor, and hence yielding different values of attenuation at zero gate bias, are compared in Fig. 11.10. For lower values of series resistor the convex upwards curvature is more noticeable, while at intermediate values the curvature is to some extent offset by the faster than exponential increase in the transistor resistance, resulting in a relatively linear

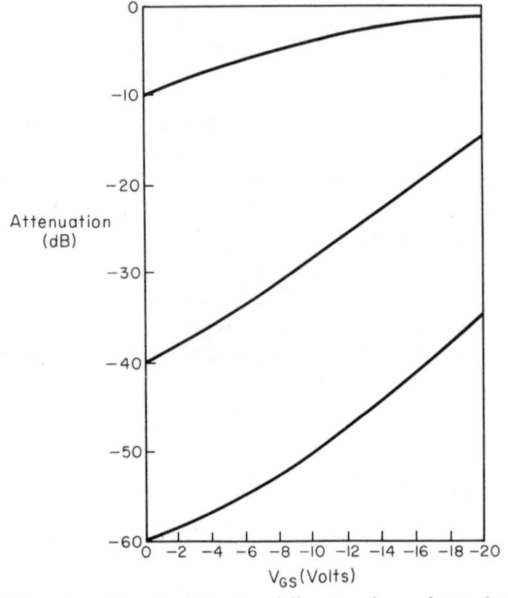

Fig. 11.10. Attenuator characteristics for different values of zero-bias attenuation

scale. Where the series resistor has a still larger value, resulting in even higher initial attenuation, the curvature is concave upwards; since in this case, the effect of more rapid variation of drain resistance dominates.

The utility of a voltage controlled attenuator such as this, having a linear decibel attenuation to control voltage characteristic, would be particularly marked in audio-frequency equipment where it will yield a linear subjective loudness relationship over a substantial range. It should also prove of value in carrier amplifiers designed to have a logarithmic response.

The range of attenuation which can be obtained depends on the tolerable attenuation when r_{DS} is at its maximum value, since this sets the value of R'. Using a remote pinch-off transistor, r_{DS} may be variable over a range of 50 to 1, or even more. If R' is large compared with r_{DS} then A varies over the same range. However, if R' is smaller the range of A is less. Writing

$$\frac{(r_{DS})_{max}}{(r_{DS})_{min}} = m$$

Then

$$\frac{A_{max}}{A_{min}} = \frac{(r_{DS})_{max}}{(r_{DS})_{min}} \frac{R' + (r_{DS})_{min}}{R' + (r_{DS})_{max}}$$

but

$$(r_{DS})_{max} = \frac{A_{max}}{1 - A_{max}}$$

hence

$$\frac{A_{max}}{A_{min}} = m \left[1 - A_{max} + \frac{1}{m} A_{max} \right]$$

or, very nearly,

$$= m[1 - A_{max}] \text{ since } m \gg 1 \qquad (11.11)$$

Alternatively, in terms of A_{min}

$$\frac{A_{max}}{A_{min}} = \frac{m}{1 + (m - 1) A_{min}} \qquad (11.12)$$

Clearly, if

$$(m - 1) A_{min} \ll 1$$

the ratio approaches closely to its maximum value, namely m.

The actual value of m which may be utilised, as indicated above, is typically of the order of 50 for an attenuator intended to operate over the audio-frequency range. The value of minimum source–drain resistance is determined by the dimensions and construction of the transistor, whereas maximum usable resistance depends primarily upon the maximum frequency at which the attenuator is required to retain its accuracy, since the resistive path between source and drain is also shunted by a capacitor. This point will be discussed further later.

The rate of change of attenuation with control voltage applied to the gate electrode may be obtained by differentiation of equation (11.8). Differentiating first with respect to ψ.

$$\left(\frac{\partial A}{\partial \psi}\right)_{r_0} = \frac{\exp(\psi)}{\{1 + \exp(\psi)\}^2}$$

hence

$$\frac{1}{A}\left(\frac{\partial A}{\partial \psi}\right)_{r_0} = \frac{1}{1 + \exp(\psi)} = 1 - A$$

and thus since

$$\left(\frac{\partial \psi}{\partial V_{GS}}\right)_{r_0} = \eta$$

$$\frac{1}{A}\left(\frac{\partial A}{\partial V_{GS}}\right)_{r_0} = \eta(1 - A). \tag{11.13}$$

Thus the sensitivity to control voltage diminishes as the gate is driven toward pinch-off (negatively for an *n*-channel device), but at low values of A tends to η, which it will approach closely, provided that the minimum value of A is small.

For the transistors used to obtain the attenuator characteristics shown in Fig. 11.9, a rate of variation of attenuation with control voltage, applied to the gate, of about one decibel per volt is obtained. This is affected hardly at all by the value of fixed series resistor chosen, and hence the value of A_{min}. The value of the characteristic gradient would be affected, however, if transistors having a different η were substituted. The value of η should vary directly with the resistivity of the material used to fabricate the transistor and inversely as the cross-sectional dimension of the channel. However, values of η much larger than that of the transistor relevant to the characteristics of Fig. 11.6 are of limited use, since pinch-off voltage is inversely proportional to η, and as will be shown subsequently, units of reasonably large pinch-off voltage are desirable in order that the attenuator may handle signals of reasonable magnitude without excessive distortion. As an alternative, a more rapid attenuation characteristic may be obtained by using two or more attenuator sections in cascade. If the sections are separated by suitable buffer amplifiers, such as FETs in the common drain configuration, the overall attenuation is just the product of the attenuation of the individual sections. When sections are cascaded directly without the use of intervening amplifiers, there is some loading of each section by that following, which results in additional attenuation; however, the effect is often small.

11.3.1 DISTORTION AND NOISE IN ATTENUATOR NETWORKS

The FET, even when the drain voltage is restricted to avoid pinch-off, is not a truly ohmic device and, in consequence, there is some harmonic distortion of the attenuated signal. However, two circumstances can cause the distortion to increase considerably. The first, which occurs principally when the reverse bias applied to the gate is small, arises when the drain potential moves sufficiently negatively (in the case of an *n*-channel device), so that the gate potential difference is in the sense of forward rather than reverse bias and of such a magnitude that appreciable conduction occurs through the gate-to-channel junction.

The apparent resistance measured at the drain then drops sharply with a consequent large increase in attenuation by the network.

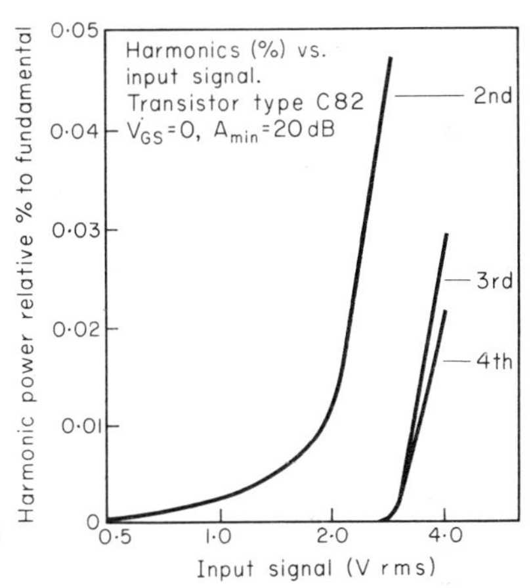

Fig. 11.11. Distortion arising from gate conduction

Since the conduction occurs at one peak of the waveform, it results in substantial harmonic distortion, particularly, second-harmonic. The effect is noticeable for quite small signals when the gate-to-source voltage is zero (Fig. 11.11). For small applied signal voltage, there is little distortion, and that is principally second-harmonic. However, as the signal level is increased, the point is reached where significant gate conduction occurs, and there is then a very sharp increase in second-harmonic distortion and also in higher harmonics. As might be expected for a silicon device, the effect becomes noticeable when the forward bias on the gate-to-channel junction is of the order of half a volt.

The other cause of excess distortion arises at the other extreme of bias conditions, when the drain-to-gate potential difference becomes sufficiently large for pinch-off effects to occur. Distortion due to this cause is not likely to be very marked in the case of remote pinch-off FET types. It is further reduced by the fact that pinch-off effects occur primarily when the gate is strongly negatively biased (n-channel). Under these conditions the resistance of the transistor may be of the same order as the series resistor (R'); and there will be little attenuation of the signal, so that the nonohmic nature of the drain–source channel will introduce only very slight distortion. Experimentally, it is difficult to demonstrate serious pinch-off distortion with such attenuators. Thus, provided that the sum of the direct control voltage applied to the gate and the peak value of any waveform applied to either of the other terminals is not

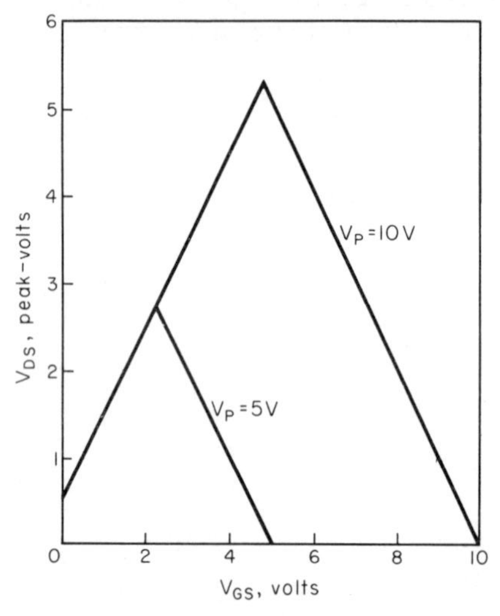

Fig. 11.12. Permitted operating regions for ohmic behaviour. The device behaves as an ohmic variable resistor provided that the peak drain voltage is such that the working point is within the triangle appropriate to the pinch-off voltage of the device in use

sufficient (either in magnitude or sign) either to cause gate conduction or channel pinch-off, it may be assumed that for practical purposes the device is ohmic.

Fig. 11.12 shows the limit on peak values of V_{DS}. The devices will behave in an ohmic manner only below the contours shown. The superior signal handling capacity of high pinch-off voltage (V_P) devices is evident.

Within this region of quasi-ohmic operation there is still some residual non-linearity of the current-voltage relationship for the FET channel. Typically, in the case of remote pinch-off transistors total harmonic power is at least 40 dB down on the fundamental, and is thus negligible in all but the most critical applications. The distortion is almost wholly second harmonic. Sharp pinch-off devices may be worse by 10 or even 20 dB.

Insulated-gate FETs, such as the MOST, have negligible gate currents, whatever the polarity of the gate bias up to the point at which the gate insulation fails. In the case of transistors of this type, therefore, only the pinch-off effect boundary is significant. However, only MOSTs which have sharp pinch-off at voltages not high enough to be very useful in the present application are generally available at present.

As well as distortion within the ohmic working domain, excess noise contributed by the FET acting as a variable resistor also exists, but in the same way may usually be neglected. The noise introduced is slightly greater than that which would result from an ideal resistor, but in many cases the difference is no more than a fraction of a decibel. Since the FET variable resistor is capable of operation with several volts of signal applied to it, the very small noise voltages can usually be left out of consideration.

11.3.2 EFFECTS OF TEMPERATURE ON ATTENUATORS

The general effects of temperature on field-effect transistors operating as variable resistors have already been considered (section 11.2). Provided that circuit arrangements are made to cause the gate voltage to vary with temperature in accordance with equation (11.5) the attenuation of the network will be constant. Alternatively, if the gate voltage is held constant, the rate of change of attenuation with temperature is given by

$$\frac{1}{A}\cdot\left(\frac{\partial A}{\partial T}\right)_{V_{GS}} = \frac{1}{A}\cdot\left(\frac{\partial A}{\partial V_{GS}}\right)_{T}\cdot\left(\frac{\partial V_{GS}}{\partial T}\right)_{A} = (1 - A)\left\{\alpha + \eta\frac{\partial \Phi}{\partial T}\right\} \quad (11.14)$$

The second term is small (except with FETs of only one or two volts pinch-off), and thus to good approximation.

$$\frac{1}{A}\cdot\left(\frac{\partial A}{\partial T}\right)_{V_{GS}} = (1 - A)\,\alpha \quad (11.15)$$

or, in decibels per degree

$$\frac{\partial L}{\partial T} = 8{\cdot}7\,(1 - A)\,\alpha \quad (11.16)$$

Since α is typically 6×10^{-3} per degree, it will be seen that typical temperature coefficients are of the order of 0·05 dB per degree. This is likely to be negligible in applications where the attenuation by the network is large but more significant in applications where only very small attenuation is required. In these cases the use of a series-parallel network of non-temperature sensitive resistors to decrease the apparent temperature dependence of the variable resistor is indicated, as explained in section 11.2. Alternatively the series resistor R' can be made to match the temperature coefficient of the FET, for example by replacing it by a second FET operated with fixed bias.

11.3.3 HIGH FREQUENCY EFFECTS

Signal-frequency voltages cannot be allowed to appear at the gate of the transistor, otherwise the attenuation of the network will be modified by feedback effects. Thus, the gate terminal can be taken to be at a low impedance to the source terminal so far as signal-frequency currents are concerned. Both the gate–drain and the drain–source capacitances effectively shunt the output terminals of the attenuator. The sum of these two is typically from a few picofarads to a few tens of picofarads with devices of the types described, although, of course, they are both strongly dependent on the dimensions of the device.

The time constant at the output of the attenuator is seen to be (assuming that the network is driven by a signal source of negligible internal impedance):

$$\tau = (C_{GD} + C_{DS})\, R' A_0$$

and, hence, the angular frequency at which the output from the attenuator has dropped by three decibels may be shown to be

$$\omega_0 = \frac{1}{R'(C_{GD} + C_{DS})\, A_0} \qquad (11.17)$$

where A_0 is the attenuation at very low frequency.

This expression might be taken to imply that the cut-off frequency is lowest when the gate of the transistor is at its most extreme negative bias, giving the maximum value of A. However, the gate–drain capacitance is voltage dependent, decreasing with increasing gate reverse bias, and this offsets the fall in ω_{co} due to increasing A. Since the law relating C_{GD} and V_{GS} depends on the nature of the junction (whether abrupt or gradual) and since the effect on ω_{co} is modified by C_{DS}, the effect presents certain complexities in computation, but in general, does not cancel out variation in attenuation due to change of A to any marked extent.

As an example of the typical values of cut-off frequency which may be obtained with currently available transistors, the attenuator whose characteristics have been plotted in Fig. 11.6 has a value of R' of 90 kΩ, and at maximum gain, the output has fallen by 3 dB at 347 kHz signal frequency. Higher cut-off frequencies could be obtained with the same transistor by the use of a lower value of R' with, however, an increase in the value of A_{min} and, hence, a reduction in the available range of attenuation. For example, if R' were reduced by a factor of ten, the cut-off frequency would increase by a slightly smaller factor (since maximum gain would increase somewhat).

Substantially higher working frequencies may be attained with transistors having appreciably lower value of r_0 or of interelectrode capacitances. Such transistors (fabricated by double diffusion) are available, but usually have a sharp pinch-off characteristic at a rather low pinch-off voltage, limiting the signal handling capacity of the attenuator.

11.3.4 OTHER ATTENUATOR CONFIGURATIONS

The simple resistive potential divider is not the only possible voltage-controlled attenuator configuration. Among many other possibilities two are particularly worth mentioning. A bridge type of attenuator is of interest because it can, at least in principle, give an unlimited decibel attenuation at balance and even in practice can be made to better 60 dB insertion loss under these conditions. The bridge can be either in transformer or four-arm form (Fig. 11.13(*a*) and (*b*)); however, the latter has the disadvantage that the input and output ports do not share a common terminal, and thus the circuit is best followed by a differential amplifier. Monolithic differential amplifiers are now so inexpensive, however, that this arrangement will generally be preferred to the use of a transformer.

For a four-arm bridge the output voltage e_0 is given by

$$e_0 = \left\{\frac{r_{DS}}{R_1 + r_{DS}} - \frac{R_2}{R_2 + R_3}\right\} e_i$$

Hence, the attenuation, A, is

$$A = \frac{R_3 r_{DS} - R_1 R_2}{(R_1 + r_{DS})(R_2 + R_3)} \qquad (11.8)$$

In a case of special interest $R_1 = R_2 = R_3 = R$ in which event

$$A = \frac{r_{DS} - R}{2(R + r_{DS})}$$

For $r_{DS} > R$ the output voltage is in phase with the input, for $r_{DS} < R$ the input and output are in antiphase and when $r_{DS} = R$ the output is zero. Near this latter condition the attenuation is large and given approximately by

$$A = \frac{1}{4}\left\{\frac{r_{DS}}{R} - 1\right\}$$

Under these conditions

$$\frac{1}{A}\left(\frac{\partial A}{\partial V_{GS}}\right) = \frac{1}{4R}\left(\frac{\partial r_{DS}}{\partial V_{GS}}\right)\frac{4}{(r_{DS}/R) - 1} = \frac{1}{r_{DS} - R}\left(\frac{\partial r_{DS}}{\partial V_{GS}}\right)$$

Evidently, the fractional rate of increase of A with V_{GS} becomes very large as r_{DS} tends to R. Thus near balance sensitivity of the attenuator

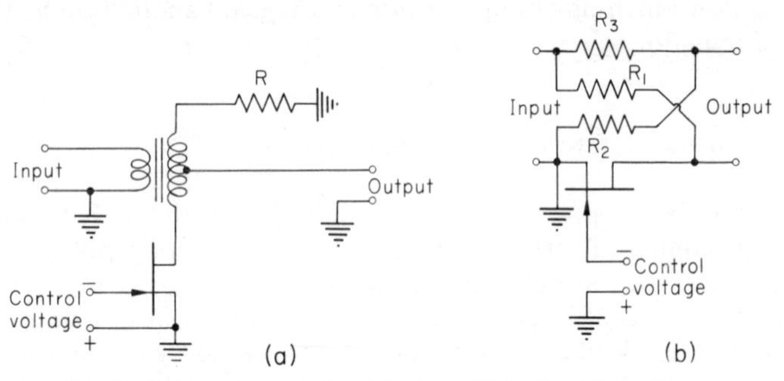

Fig. 11.13. Transformer (a) and four-arm (b) bridge attenuators

to control voltage becomes very large indeed, but unfortunately the rate of change of attenuation with temperature also becomes large since

$$\frac{1}{A}\left(\frac{\partial A}{\partial T}\right)_{V_{GS}} = \frac{1}{r_{DS} - R}\cdot\left(\frac{\partial r_{DS}}{\partial V_{GS}}\right)_T\cdot\left(\frac{\partial V_{GS}}{\partial T}\right)_{r_{DS}}$$

and $(\partial V_{GS}/\partial T)$ is, as in the previous case, a constant characteristic of the transistor and typically a few millivolts per degree. Because of the high temperature dependence of the attenuation figure, bridge attenuators can rarely be used near balance, and hence show little advantage over voltage-divider types.

Variable attenuation can be achieved in yet another way, by using the FET in a negative feedback network; for example, in the emitter lead of a CE bipolar transistor amplifier (Fig. 11.14) or the source of a CS FET. To take the former case, the gain of such an amplifier is very nearly

$$A_V = -\frac{R_L}{r_{DS}} \tag{11.19}$$

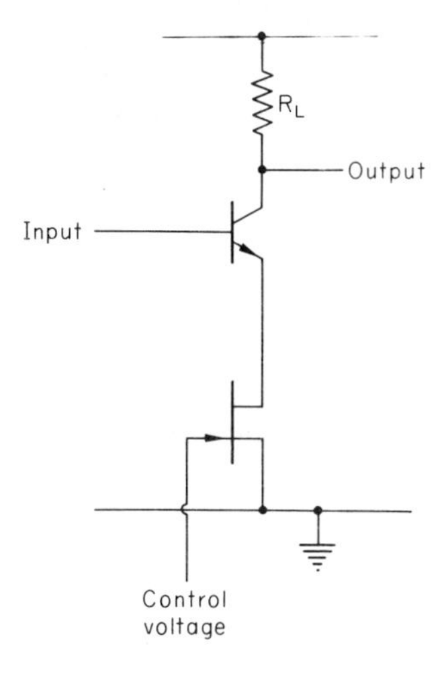

Fig. 11.14. A variable gain amplifier, using a single bipolar transistor

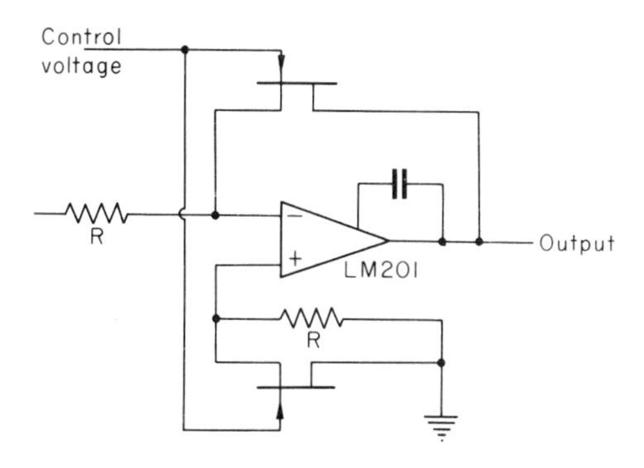

Fig. 11.15. Voltage-controlled high-gain amplifier based on a monolithic circuit

provided that this value is not large and that R_L is much smaller than $(h_{ob})^{-1}$ It will be seen that $(A_V)_{max}$ can be equal to or greater than unity without reducing the ratio of maximum to minimum voltage transfer ratio, which in this case is always just m. The advantage is, however, achieved at the cost of including a transistor, although this can contribute gain.

In an alternative version, the variable resistance FET controls feedback not simply over a single transistor, but rather over an integrated amplifier. A suitable circuit is given in Fig. 11.15. Neither end of the FET channel is at earth potential, but since the amplifier input terminal is a virtual earth (summing) point, the signal voltage impressed in series with the gate–source control voltage is negligibly small. The voltage gain of the amplifier with feedback is

$$A_V = -\frac{r_{DS}}{R}$$

and the second FET connected to the non-inverting input terminal (which should be identical to the gain control FET) avoids the introduction of a control-voltage-dependent DC offset, as the gain control FET provides part of the return path for the amplifier input terminal bias current. This precaution can be omitted if the feedback circuit is AC coupled, or if an FET input stage amplifier having negligible input bias current is used.

11.4. Phase shifting networks

Field-effect variable resistors can be used in many other circuits besides attenuators. One important class is circuits for shifting the phase of a sinusoid.

A particularly simple class of phase-shifting network consists of a resistor and either a capacitor or an inductor Due to technological problems associated with the use of inductive components, consideration will here be limited to circuits composed of a FET acting as a variable resistor, and a capacitor Simple lead and lag circuits appear in Figs. 11.16(*a*) and (*b*). Alternative forms are possible, as in Fig. 11.16(*c*), where the signal source is a current generator.

For the lead-network of Fig. 11.16(*a*) the ratio of output voltage to input voltage is

$$\frac{e_{out}}{e_{in}} = \frac{r_{DS}}{r_{DS} + (1/j\omega C)} \qquad (11.18)$$

where ω is the angular frequency of the signal, and C is the value of the capacitor.

Hence phase shift ϕ given by the network is

$$\phi = \arctan \frac{1}{\omega C r_{DS}} \tag{11.19}$$

$$= \arctan \frac{1}{\omega C r_0} \exp(-\eta V_{GS}) \tag{11.20}$$

assuming an exponential resistance-voltage relationship, as before By suitable choice of C the phase advance at zero gate bias may be set at any required value greater than zero and less than 90°. When

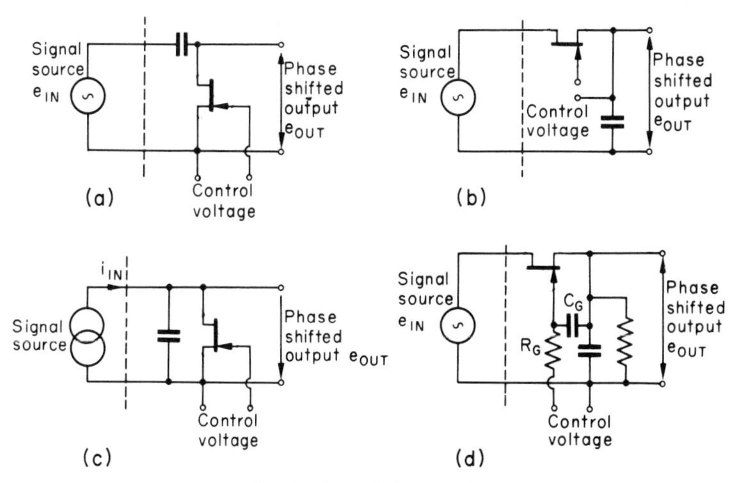

Fig. 11.16. Simple phase-shift networks using FETs
(a) *a phase advance network.*
(b) *a phase retard network*
(c) *an alternative phase retard network*
(d) *a variant of (b) in which the control voltage is not 'floating' relative to earth*

a reverse bias is applied to the FET gate the phase lead is reduced, falling to a low non-zero value as the bias approaches the pinch-off value. Fig. 11.17 shows a series of graphs of ϕ against V_{GS} for a number of different values of initial phase shift. For small angles of phase shift the dependence is initially linear, but becomes less rapid as the phase shift is reduced below approximately one-third of its initial value, and in this region the divergence between experimental results and the predictions of equation (11.20) become significant. As the phase shift is varied there is also a change in magnitude of the output voltage from the circuit, in fact, taking the modulus of both sides of equation (11.18) and using equation (11.19)

$$\frac{|e_{out}|}{|e_{in}|} = \cos\phi \tag{11.21}$$

The corresponding lag network, Fig. 11.16(*b*), has closely similar properties, the angle of lag being given by

$$\phi' = -\arctan(\omega C r_{DS}) \tag{11.22}$$

$$= -\arctan\{\omega C r_0 \exp(\eta V_{GS})\} \tag{11.23}$$

This circuit is, however, occasionally inconvenient in use because the common terminal of the input and output ports of the network is usually required to be earthed, whilst the control voltage is also

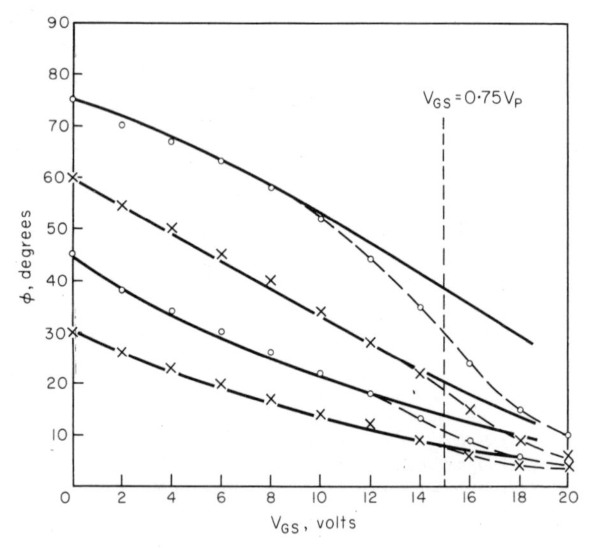

Fig. 11.17. Experimentally observed dependence of phase shift on V_{GS} *(point and broken line). The full line is the relationship of equation (11.20)*

usually most conveniently available relative to earth (or some other fixed) potential. This is incompatible with the circuit of Fig. 11.16(*b*), which requires the control voltage to be floating with regard to the earth line. Although as shown in Fig. 11.16(*d*), this difficulty is quite easily overcome, it may be more convenient to use a different lag circuit if the signal can be derived from a source of high internal impedance, essentially a current generator. In this case the circuit of

Fig. 11.16(*c*) can be used, the values of lag angle given in equations (11.22) and (11.23) still applying. The main difference between the two circuits, apart from the greater convenience of applying control voltage in the second, is that whereas for the first the magnitude of the voltage transfer ratio

$$\frac{|e_{\text{out}}|}{|e_{\text{in}}|} = \frac{1}{(1 + \omega^2 C^2 r_{DS}{}^2)^{\frac{1}{2}}} = \cos \phi \qquad (11.24)$$

as in the case of the lead network, for the second circuit the ratio of output voltage to input current (the transfer impedance) is equal in magnitude to

$$\frac{|e_{\text{out}}|}{|i_{\text{in}}|} = \frac{R_{DS}}{(1 + \omega^2 C^2 r_{DS}{}^2)^{\frac{1}{2}}} = r_{DS} \cos \phi \qquad (11.25)$$

This dependence of output voltage on r_{DS} is a disadvantage for certain applications.

Experimentally, the results shown in equations (11.20) and (11.23) are well reproduced, subject to two limitations. The FET drain–source circuit has been represented by a pure resistance, r_{DS}. In fact it is shunted by a stray capacitance which may be in the range from a picofarad to a few tens of picofarads, depending on the details of device design. Thus the results quoted are inaccurate if the value of the capacitor C is not large compared with this stray capacitance, particularly when r_{DS} takes on large values. This effect is most serious at high frequencies where low values of C are used to obtain a given phase shift. Maximum frequency of operation is obtained with transistors having low values of stray capacitance or low values of r_{DS} (permitting higher values of C). At present diffused devices are commercially available having values of r_{DS} as low as 50 Ω, combined with a shunt stray capacitance of only 10 pF, permitting operation up to about 3 MHz without significant change in performance.

The second factor limiting the accuracy of the agreement between predicted and experimentally observed variation of phase shift with V_{GS} is due to failure of the channel resistance to obey the simple exponential law, as assumed. This is particularly obvious at higher values of V_{GS} where the channel begins to pinch-off and its resistance increases much more rapidly than predicted.

The simple phase-shift networks described may be used by themselves, or may be combined in various ways to produce more complicated networks having a variety of applications. Some of these applications will now be described.

11.4.1 CASCADED NETWORKS

The range of phase shift at a single frequency which can be conveniently obtained with the simple networks so far described is not more than 70° (i.e. 10° to 80°). Where the phase angle is required to vary over a larger range (assuming constant signal frequency) several similar circuits may be cascaded. The phase shift of a series of cascaded networks is not simply the sum of that which they would have produced separately, due to the effect of loading on earlier networks by those following. For example, in the case of a phase-advance network having two identical cascaded sections

$$\frac{e_{out}}{e_{in}} = \frac{j\omega Cr_{DS}\left\{\dfrac{j\omega Cr_{DS}(1 + j\omega Cr_{DS})}{1 + 2j\omega Cr_{DS}}\right\}}{\left\{\dfrac{j\omega Cr_{DS}(1 + j\omega Cr_{DS})}{1 + 2j\omega Cr_{DS}}\right\}\{1 + j\omega Cr_{DS}\}} \tag{11.26}$$

From this expression the phase shift at a given frequency can be calculated if C and r_{DS} are known. For this particular network the effects of loading are zero when the nominal phase shift per section is 0°, 45° or 90° and rises to a maxima of about $11\frac{1}{2}$° midway between these points. The loading effect has been shown (to good approximation), for the case of two identical sections, to give a total phase shift equal to θ, where

$$\theta = 2\arctan\frac{1}{\omega CR} + 11{\cdot}5\sin 4\arctan\frac{1}{\omega CR} \text{ in degrees} \tag{11.27}$$

The effect of loading is given by the second term and may be seen to be small. It may be still further reduced by using a sequence of networks in which the value of the capacitor is successively reduced. This, however, is at the cost of reducing the total range of phase shift available. In some applications (for example, where signal levels are rather small) buffer amplifiers may be interspersed between successive phase shift networks, and in this case the effect of loading is eliminated.

11.4.2 A PHASE-SHIFT OSCILLATOR

A simple phase shift oscillator may be constructed (Fig. 11.18) using three cascaded lead-networks to produce a phase shift of 180°, so permitting oscillation when connected as a feedback loop over an amplifier which produces phase inversion. In this case the total phase shift to be produced is constant, and variation of bias applied to the field-effect transistor varies the frequency at which oscillation occurs.

For a three-mesh oscillator of this type the oscillation frequency is given by

$$f = \frac{1}{2\pi\sqrt{(6Cr_{DS})}} = \frac{\exp(-\eta V_{GS})}{2\pi\sqrt{(6Cr_0)}} \quad (11.28)$$

where the symbols are as above, and subject to the condition that the forward path voltage gain of the amplifier must be not less than 29. This very simple circuit achieves a frequency range of 30:1 with

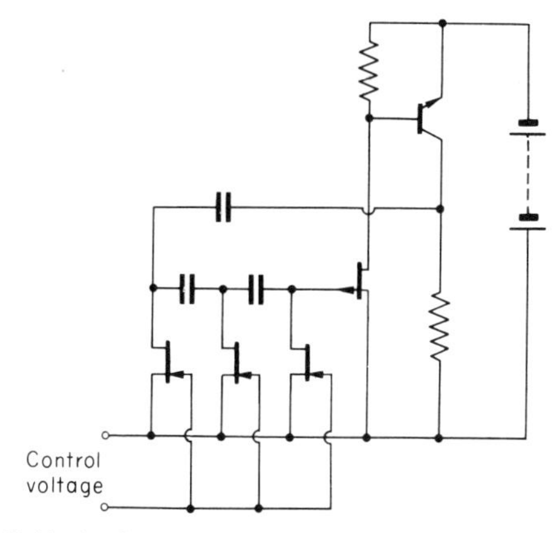

Fig. 11.18. A voltage-controlled variable frequency oscillator

good agreement with equation (11.28) over the upper decade of the frequency range. With the FETs used in the variable resistance mode in this case (type C85) the lower decade of frequency variation was more rapid than predicted by equation (11.28) due to the onset of pinch-off effects. Provided that the loop gain is set sufficiently low, good waveform may be obtained. However, a practical oscillator would embody some form of a.g.c. to adjust the loop gain, and maintain stable, relatively distortion-free oscillations. With modern transistor structures, having low r_0, there seems to be no reason in principle why oscillators of this type should not operate well above the audio range.

11.4.3 FREQUENCY INDEPENDENT PHASE SHIFT

In the above networks the phase shift has been calculated as a function of frequency. Many systems operate on fixed frequencies and in this case the simple networks described will yield fixed values of phase shift. There is, however, an important field of application for networks which yield a constant phase-shift irrespective of frequency. This result can be achieved, using FETs, over a limited frequency range. Consider the circuit shown in Fig. 11.19. The system will automatically adjust the value of resistance of the FET

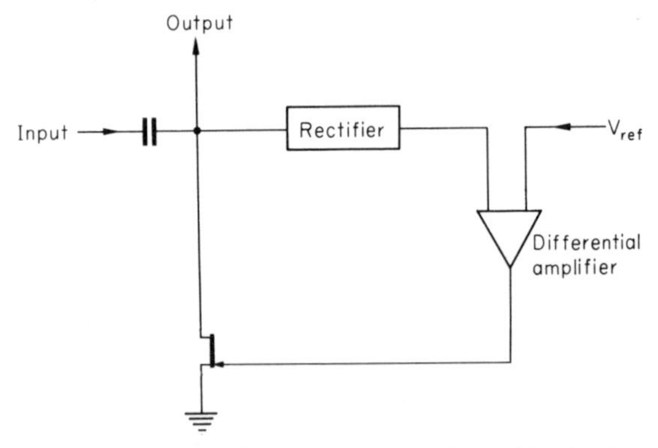

Fig. 11.19. A system which produces constant phase shift of a sinusoid

so that the rectified value of the signal voltage obtained from the phase-advance network is almost equal to V_{ref}, which is assumed smaller than the magnitude of the input signal. Thus, assuming that the peak value of the input signal (assumed sinusoidal) is V, then the phase-shift can be shown to be

$$\phi''' = \arccos\left\{\frac{V_{ref}}{V}\right\}, \text{ assuming } V_{ref} < V \qquad (11.29)$$

This is unsatisfactory as it stands, because the phase shift obtained is dependent on V.

However, V_{ref} may itself be derived by means of a resistive potential divider and rectifier from the input signal, in which case

$$\phi''' = \arccos(a) \qquad (11.30)$$

where a (less than unity) is the voltage division ratio (assuming perfect rectification). Thus the phase shift of the network is constant, independent of frequency, and is set by a simple resistive network.

In practical terms, the properties of the circuits are fairly complex, partly because the assumptions made are too simple. In fact, the phase shift obtained is not constant, but depends upon a number of factors. However, by careful design, a close approach to constancy can be made which would be sufficient to render the circuit useful in a variety of applications.

An example of a practical circuit for a phase shifter of this kind is shown in Fig. 11.20. To avoid loading the phase-shift network, FET buffer amplifiers of unity gain and very high input impedance are used in front of the rectifiers, which are diode circuits of the peak-responding type. A two-loop lead phase-shift circuit is used, embodying remote pinch-off FETs. With the value of a adjusted to give 50° of phase shift, for example, the phase-shift remained within

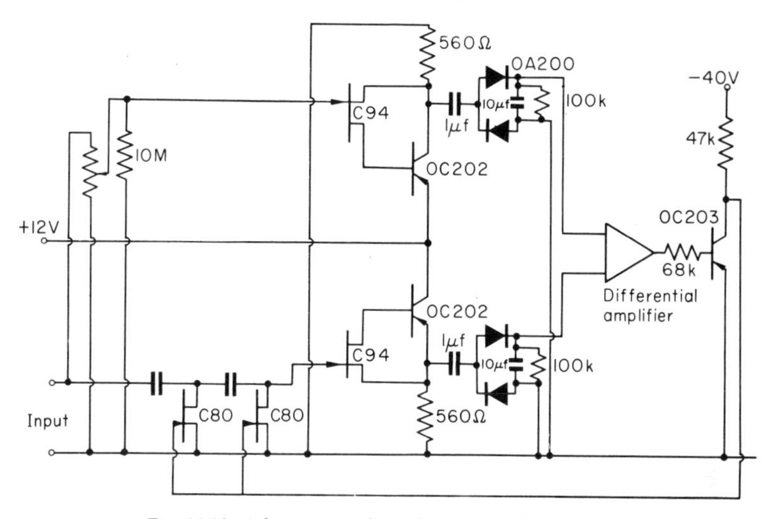

Fig. 11.20. A frequency independent phase advance circuit

±1° from 2 kHz to beyond 10 kHz. The system is sensitive to signal waveform, since the harmonic content of the shifted waveform is enhanced by the phase advance circuit, but the effect is not large and can usually be neglected for up to 10% distortion, provided that a full-wave type of rectifier is used.

11.4.4 TEMPERATURE EFFECTS

For the phase shifting circuits, as for the attenuators, equation (11.5) gives the change of gate voltage required to balance the effects of temperature change and hence keep the phase shift unchanged.

Rate of change of gate voltage needed to keep the phase shift constant despite temperature changes is the most convenient mode for expressing drift magnitude in cases where the gate voltage can be adjusted by a feedback system. However, in other cases the gate voltage remains constant and the phase-shift changes. Under these circumstances the drift rate is more conveniently expressed as $(\delta\phi/\delta T)\, V_{GS}$ where ϕ is the phase shift. Since, using a well known result of calculus,

$$\left(\frac{\partial V_{GS}}{\partial T}\right)_\phi \left(\frac{\partial \phi}{\partial V_{GS}}\right)_T \left(\frac{\partial T}{\partial \phi}\right)_{V_{GS}} = -1$$

and from equation (11.20) for a phase-lead network of one section

$$\left(\frac{\partial \phi}{\partial V_{GS}}\right)_T = \frac{\eta}{2} \sin(2\phi),$$

it follows that for a network of this type

$$\left(\frac{\partial \phi}{\partial T}\right)_{V_{GS}} = \frac{\alpha}{\lambda} + \left(\frac{\partial \phi}{\partial T}\right) \frac{\eta}{2} \sin(2\phi) \qquad (11.31)$$

Similar expressions can be obtained for other phase shifting networks. Typical drift values range from zero up to a small fraction of a degree of angle per deg. C.

11.5. A variable *Q*-factor filter

A circuit widely used in active *RC* filter networks is an amplifier with a feedback path containing a twin-T network, as shown in Fig. 11.21. The normal design assumes

$$N = M = 2 \qquad (11.32)$$

where N, M are as defined in the figure, in which case, provided that the network operates into a virtual open circuit, the locus of the

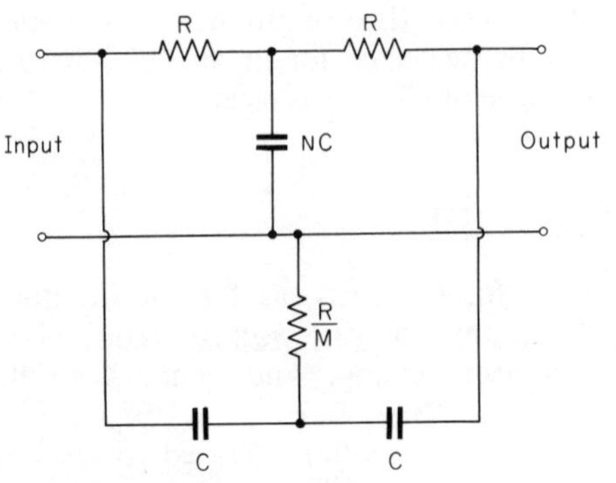

Fig. 11.21. The twin-T feedback network

vector representing the ratio of voltage V_2 to V_1 (where these are as shown in the diagram) is a circle, as in Fig. 11.22(*a*), the feedback voltage passing through a minimum at

$$\omega_0 = \frac{1}{RC} \tag{11.33}$$

and the gain of an amplifier with such a network in a negative feedback path is shown in Fig. 11.22(*b*) which will be seen to follow

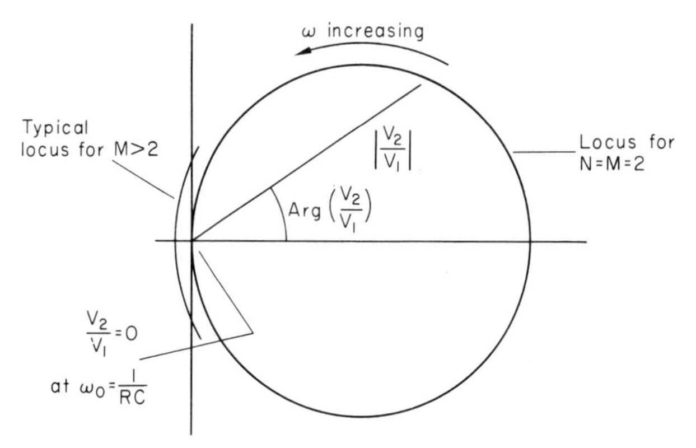

Fig. 11.22(a). Voltage transference of the feedback network

the general form of the response curve of a resonant circuit, the gain peaking at ω_0 to a value A_V equal to the gain of the amplifier without feedback. A 'Q-factor' for such an amplifier can be defined as

$$Q = \frac{\omega_0}{\Delta\omega}$$

where $\Delta\omega$ is the difference between the two frequencies at which the gain is $1/\sqrt{2}$ of that at the peak. It can be shown that

$$Q = \frac{A_V + 1}{4} \tag{11.34}$$

Provided that $N = M = 2$

If, however, $M > 2$, the locus of the feedback ratio V_2/V_1 cuts the imaginary axis on Fig. 11.22(*a*) so that at ω_0 there is a small output from the twin-T network in antiphase to V_1, that is, a small positive feedback. This will cause the value of the amplifier gain to be larger than A_V at ω_0, with a corresponding increase of Q-factor above the value given by equation (11.34). A sufficient increase in M will cause the circuit to oscillate (corresponding, from some points of view, to infinite Q-factor). Conversely, if M is slightly less than 2

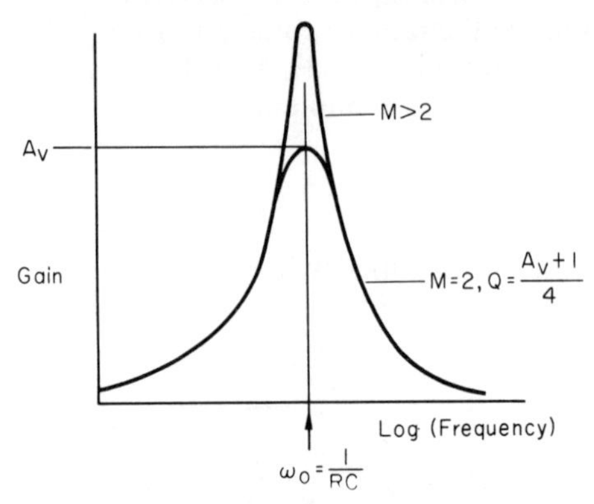

Fig. 11.22(b). Forward path transfer properties of the amplifier with feedback

there will remain a little negative feedback at frequency ω_0, so that Q will be reduced. Thus by variation of M, Q can be varied over a wide range, up to the onset of oscillation. It can be shown that if M is not very different from two, then

$$\frac{\partial Q}{\partial M} = \frac{Q^2}{4}$$

A suitable circuit would use the FET to shunt a fixed resistor of value $R' > \frac{1}{2}R$ (Fig. 11.23), so that Q could be varied on either side

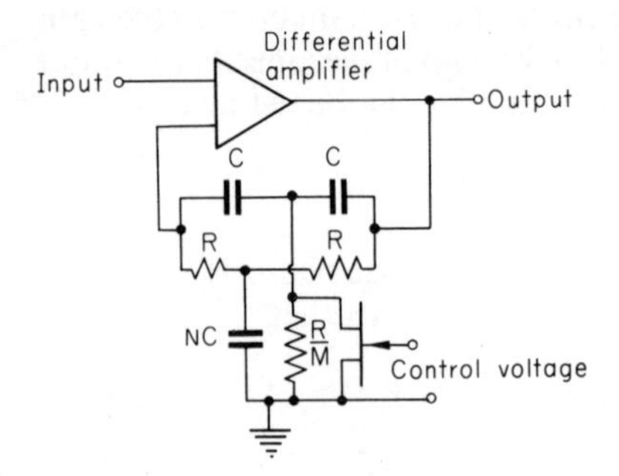

Fig. 11.23. A 'controlled-Q' active filter

of $\frac{1}{4}(A_V + 1)$. Similar Q control circuits can easily be devised for other active RC filters; for example, the Wien bridge circuit. For present purposes both the twin-T and the Wien bridge circuits have the advantage over other active filters that Q-factor can be altered by adjustment of a single resistive element without substantial effect on the resonance frequency.

Other active filters can be made voltage controllable both in respect of bandwidth and cut-off frequencies by exploitation of the voltage controlled resistance characteristics of the FET but usually several identical units which 'track' in resistance (as control voltage varies) will be required.

11.6. Analogue computer applications

There are important potential applications of the FET in analogue computers, particularly in multipliers and non-linear function generators. One multiplier is based on Ohm's law: if the magnitude of a resistance is proportional to one variable and the current through it to the other, then the voltage across it is proportional to the product. A possible circuit is that of Fig. 11.24. A known standard current I flows through an FET and the potential difference

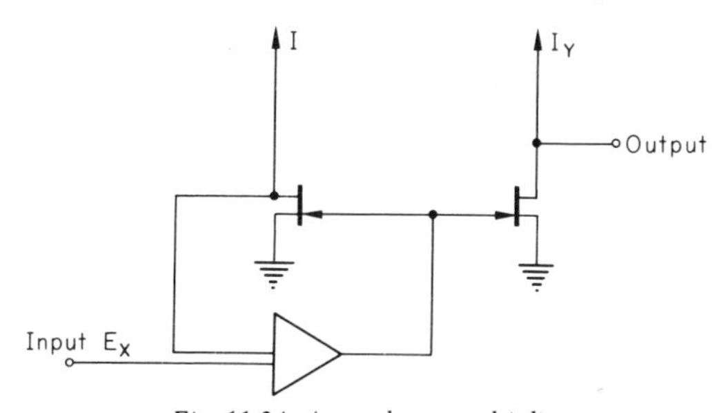

Fig. 11.24. An analogue multiplier

developed across it is compared with the voltage analogue E_x of one of the variables to be multiplied. The output of the differential amplifier which compares the two voltages is connected to the gate of the FET in such polarity that the resistance of the FET is adjusted to bring the two inputs to the amplifier very close to equality. When this is the case, writing R'_D for the resistance of T_1

$$R'_D = \frac{E_x}{I} \tag{11.35}$$

The gate of T_1 is connected to that of a similar FET T_2 through which a current I_y passes, which is proportional to the second variable. Thus if R''_D is the resistance of T_2 the output voltage will be

$$E_{out} = I_y \,.\, R''_D \tag{11.36}$$

But if both FETs follow the same law relating resistance to gate voltage

$$R''_D = kR'_D \tag{11.37}$$

where k is constant. Hence,

$$E_{out} = \frac{k}{I} I_y \, . \, E_x \qquad (11.38)$$

to good approximation, and the system will thus act as an analogue multiplier. It should be noted that the transistors need not be identical provided that they have the same law of variation of resistance with V_{GS}, that is, the same gate cut-off voltage. The actual values of resistance at the same applied gate voltage may be quite different.

Circuits of this kind may be made insensitive to temperature changes if the two transistors, T_1 and T_2, are kept at the same temperature provided that, as specified, the variation of the resistance of the two units follows the same law: they need not be matched in terms of value at a particular temperature. Substitution of transistors, similarly, will only affect the constant k. A divider may be constructed by a simple modification of this circuit: E_x is replaced by a fixed voltage E, and I by a current I_x proportional to the first variable. Equation (11.35) then becomes

$$R'_D = \frac{E}{I_x}$$

and equation (11.36) is unaltered. Thus equation (11.38) becomes

$$E_{out} = kE \frac{I_y}{I_x} \qquad (11.39)$$

Some epitaxial-diffused sharp pinch-off FETs obey the resistance law of equation (11.2) quite closely and this may be exploited to design another multiplier. Writing this relationship as

$$g_{DS} = g_0 \left(1 - \frac{V_{GS}}{V_P}\right)$$

If a voltage V is applied to the transistor a current I will flow given by

$$I = g_0 V - \frac{g_0}{V_P} \, . \, V_{GS} V$$

$$= I_0 - a \, . \, V_{GS} V \qquad (11.40)$$

where I_0, a are constants.

The application of this equation to an analogue multiplier is obvious, and a fast, but not very accurate, multiplier or divider can easily be designed.

A different family of non-linear function generators can be based on the relationship of r_{DS} and V_{GS} given in equation (11.3). The generation of an exponential function, by passing a fixed current through an FET, applying a voltage to the gate proportional to the exponent and taking the required function as V_{DS}, is obvious. By a simple feedback system, logarithmic functions could also be obtained, and by combining exponential and logarithmic function generators, systems could be built up which multiply or raise variables to higher powers, including fractional powers. However, the utility of circuits of this kind is limited by the poor agreement between the characteristics of available transistors and equation (11.3). Probably more types of remote pinch-off FET will become available in due course, however, in which case circuits of this kind will merit consideration.

11.7. Comments

In the preceding chapters the use of the FET to replace bipolar transistors or thermionic devices in conventional applications was described, but the circuits quoted in the immediately foregoing sections, in which the FET acts as a variable ohmic resistor, are quite new, since no other component has these properties (except the indirectly heated thermistor and the calistor, both of which have a very long response time). It is consequently quite possible that new applications will appear in future, not covered in this chapter, as designers become familiar with the properties of below pinch-off FETs. In the field of microelectronics, the 'below pinch-off' FET is, in fact, the only kind of variable resistor available which can be integrated with the rest of the electronics, and it is to be expected that it will consequently find wide application there. Substantial progress must, however, depend on the evolution of better FET structures having the desirable 'remote pinch-off' characteristic, with a near exponential dependence of channel resistance on control voltage over a considerable range.

REFERENCES

1. Shockley, W. 'A unipolar field-effect transistor', *Proc. IRE*, **40**, 1365–1376 (1952).
2. Middlebrook, R. D. 'A simple derivation of FET characteristics', *Proc. I.E.E.E.*, **51**, 1146 (August 1963).
3. Morgan, A. N. 'The FET as an electronically variable resistor', *Proc. I.E.E.E.*, **54**, 892–893 (1966).

12

HF AMPLIFIERS AND FREQUENCY CONVERSION

12.1. High-frequency models for FETs

In chapter 5 equivalent circuits were discussed which were applicable for both junction and insulated gate FETs over the widest frequency range. At radio frequencies, and indeed even at the higher audio frequencies, the shunt conductance across the gate–drain and gate–source capacitance can be neglected: however, the very small drain–source capacitance, usually neglected at low frequencies, becomes significant at the very highest operating frequencies. Similarly the loss-angle of the gate–drain and gate–source capacitors must be allowed for at the upper extreme of the working range by including a small value resistor in series with them in the equivalent circuit model. Thus a typical high-frequency equivalent circuit is as shown in Fig. 12.1(*a*). Although such an equivalent circuit is a perfectly valid model for the FET, it is complex; so much so, in fact, that it is of little practical value in circuit analysis. A topologically simpler representation becomes essential.

The same problem occurs in bipolar transistor circuit analysis, and the solution adopted is to replace the relatively complex equivalent circuit based on 'real' (i.e. frequency invariant) components by a much simpler form in which the properties of the elements depend upon frequency. The commonest approach of this type is to specify the FET in terms of the y-parameters, principally because these are particularly easily measured using the well-known GR admittance bridge. Although generally related to the description of an active two-port in terms of a two-by-two admittance matrix, the y-parameters can equally legitimately be interpreted as defining a topologically simple equivalent circuit consisting of only four branches, as in Fig. 12.1(*b*) (for the common-source case). The simplicity of such a model makes circuit analysis far more straightforward than if, say, Fig. 12.1(*a*) were used. However, a price must be paid for this, which is simply that the circuit equations derived in this way will be in terms of the *y*-parameters, which are themselves frequency-dependent. This is particularly inconvenient if analytical

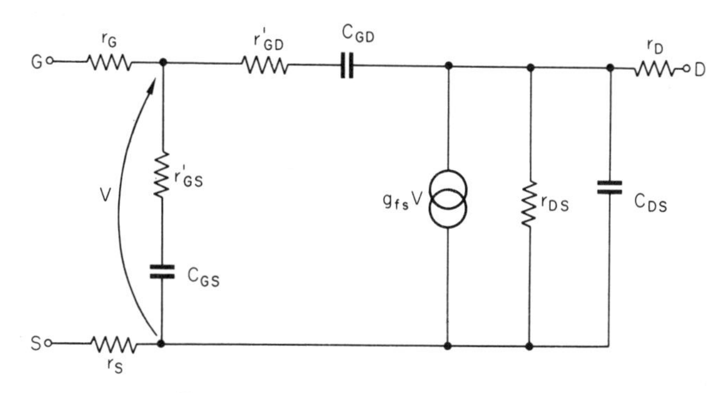

Fig. 12.1(a). Equivalent circuit of a FET

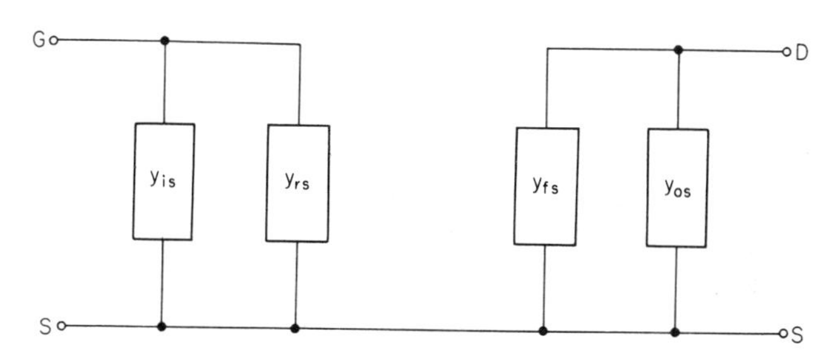

Fig. 12.1(b). y-*parameter equivalent circuit*

expressions are required giving the parameters of the circuit being analysed over a wide range of frequencies.

However, there are two cases where the difficulties associated with the *y*-parameter approach are not encountered, namely in numerical treatments of response (using a computer) where the values of the *y*-parameters at a number of different frequencies can be stored, and in the case of narrow-band amplifiers, for which the values of the *y*-parameters can be treated as constants (or at worst as simple linear functions of frequency) within the small frequency range of interest. This chapter will primarily be concerned with the latter case.

To appreciate the magnitude and significance of the high-frequency effects encountered in field-effect devices, it will be convenient to compare the actual *y*-parameters of a good high frequency FET with those which would result from the simplified equivalent circuit generally used at audio frequencies, as shown in Fig. 12.2. The y-parameters for this model can be derived by simply applying the relationships

$$\left.\begin{aligned} y_{is}e_{GS} + y_{rs}e_{DS} &= i_G \\ y_{fs}e_{GS} + y_{os}e_{DS} &= i_D \end{aligned}\right\} \qquad (12.1)$$

where e_{GS}, e_{DS} are the small alternating gate–source and drain–source voltages respectively,

and i_G, i_D are the alternating gate and drain currents.

By inspection of Fig. 12.2

$$\left.\begin{aligned} y_{is} &= j\omega(C_{GS} + C_{GD}) = j\omega C_{is} \\ y_{rs} &= j\omega C_{GD} = j\omega C_{rs} \\ y_{fs} &= -g_{fs} - j\omega C_{GD} = -g_{fs} - j\omega C_{rs} \\ y_{os} &= g_{DS} + j\omega C_{GD} = g_{os} + j\omega C_{rs} \end{aligned}\right\} \qquad (12.2)$$

In Fig. 12.3 the real and imaginary parts of the y-parameters are plotted against frequency for an actual RF junction FET (Siliconix 2N5397) and the approximations corresponding to equations (12.2) are indicated. The deviations will be seen to occur at the upper frequency extreme; in this case they become significant above 500 Mhz. This frequency can be determined quite simply by inspection of the y-parameter plots for any FET. Below it the simple equivalent circuit of Fig. 12.2 may be used, or alternatively a y-parameter

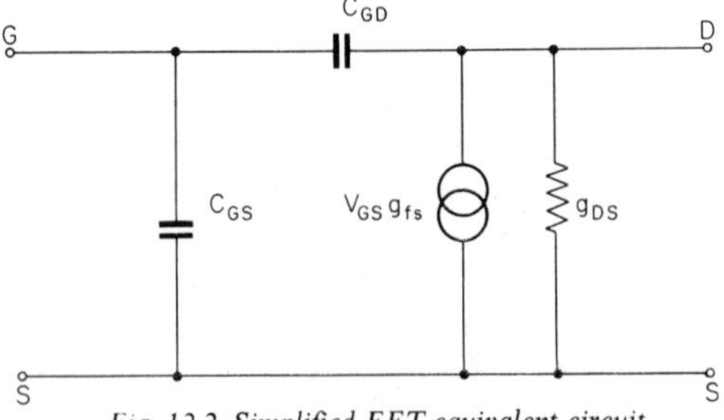

Fig. 12.2. Simplified FET equivalent circuit

model using the parameter values given by equations (12.2). At higher frequencies, since the equivalent circuit of Fig. 12.1(a) is too complex to be of much use, a y-parameter approach is adopted, using empirically determined values for the y-parameters at each frequency for which a circuit calculation is to be performed.

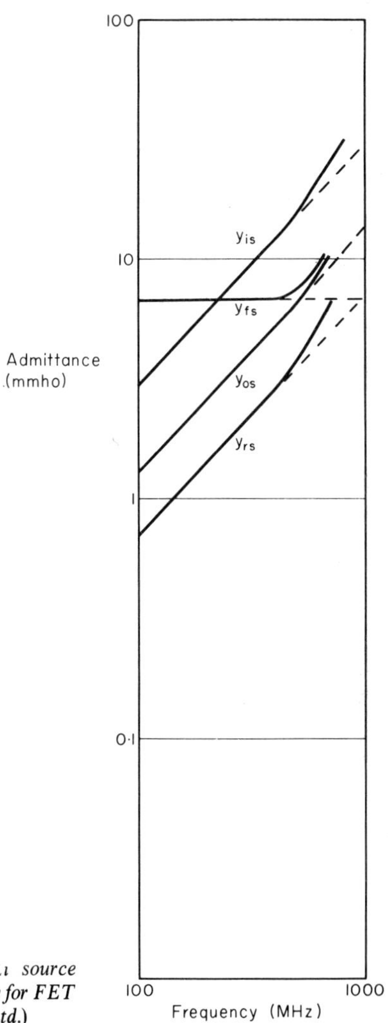

Fig. 12.3. A plot of commc.ı source y-parameters against frequency for FET Type 2N5397 (Siliconix Ltd.)

The values of the FET y-parameters are, of course, working-point dependent, but much more predictably so than in the case of bipolar transistors. The only exception to this is g_{os}, which falls with increasing V_{DS} and V_{GS}, but not according to a simple law, However, in many cases g_{os} is anyway small compared with external conductance connected between drain and source terminals, and may be ignored. By contrast, g_{fs} falls approximately linearly with V_{GS},

$$g_{fs} = g_{fso} \cdot \left| \frac{V_{GS} - V_P}{V_P} \right| \tag{12.3}$$

and the law of dependence of C_{GS} and C_{GD} on applied voltage depends on the nature of the device. For a junction FET, both capacitance values fall as an inverse square-root function of the applied voltage, in keeping with the usual law for a sharp junction *pn* diode capacitor. Provided that the device is not allowed to operate into the enhancement region, the range of capacitance is not too large in normal operation, usually, in fact, less than 2:1. By contrast the capacitance values for an MOS device are only very slightly voltage-dependent. This is because the capacitors are essentially of the MOS type, and thus relatively stable. For some purposes (particularly when neutralisation is employed) this can be an advantage.

12.2. Narrow-band amplifiers

A very simple narrow band amplifier is as shown in Fig. 12.4, which assumes that the input signal comes from a source of high internal impedance. The voltage gain of the amplifier is a maximum at the frequency at which the impedance of the resonant circuit in the drain connection is a maximum, and falls on either side.

The gate circuit is also tuned to resonance, which further improves the ratio of resonant frequency gain to that remote from resonance, and also, by resonating the input capacitance of the amplifier, avoids signal loss which would otherwise result from the relatively high input admittance. In this circuit, biasing and power-supply arrangements are omitted for simplicity, but would, of course, follow an entirely conventional pattern. Choice of DC operating conditions is as for any other amplifier (whether small signal or large) and is covered in chapter 6. Typical biasing arrangements will be shown later in this chapter when examples of complete amplifiers are described.

It will also not be necessary to dwell at great length on the design of resonant coupling networks for amplifiers of this type. An important difference between field-effect and bipolar transistors makes design of interstage circuits much simpler in the former than the latter case. Whereas the bipolar transistor has only a finite power gain, typically perhaps 40 dB or so, that of the field-effect device is virtually infinite, due to the very high input resistance of the device, provided that it is operated in the common-source or common-drain configuration, and also that the operating frequency is not so very high that the loss angle of the interelectrode capacitances need be taken into account.

Subject to these conditions, the power dissipated at the input circuit of the FET is negligible and the interstage coupling circuit can be allowed to dissipate all the signal power delivered to it by the preceding stage. This is in sharp contrast to the situation with bipolar transistors, in which considerable drive power is required, and consequently design is dominated by the need to secure effective power transfer between stages.

Thus tuned circuits, used as coupling elements can, in FET applications, have a working Q-factor virtually identical with their unloaded Q, provided that the output resistance of the preceding stage is sufficiently high relative to the equivalent resistance of the LC circuit at resonance. The Q-factor of circuits can consequently be chosen virtually solely on bandwidth considerations, whilst the required stable gain is determined by resonant impedance and hence the choice of L/C ratios, tapping points (if used) or transformation ratios for double-wound circuits. This effective separation of gain and bandwidth (within limits) as independent parameters greatly simplifies design, compared with the bipolar case.

Similarly, a straightforward approach is possible to stability problems. There is very little low-frequency internal feedback in field-effect devices operating in the CS mode, so that stability problems arise (as with thermionic valves) only as a result of feedback capacitance, specifically C_{GD}. Where problems of stability arise, the solution is generally found in terms of device technology, by reducing C_{GD}. Contemporary dual gate MOS pentodes have values of C_{GD} of 0·01 pfd or lower for g_{fs} values of the order of 10 millimhos, so stability problems are not very serious until the VHF working frequencies are reached. As an alternative, neutralisation procedures are possible, and will be described subsequently: however, they are not of great practical importance. Usually, it is not acceptable to have to adjust individually the neutralising capacitors used in narrow band amplifiers, consequently neutralisation can only be used to give a very approximate cancellation of feedback effects, and thus a marginal improvement in stability. Exceptions to this are amplifiers operating at the upper VHF and UHF frequencies (say above 150 MHz) where at present devices with C_{GD} low enough to give satisfactory unneutralised operation in the CS configuration are not available, and consequently either the amplifier must be in CG configuration (with consequent loss of power gain) or individually adjusted neutralisation circuits must be used (or even both) in extreme cases.

Over the last few years, very elegant theories of the problems of stability in tuned amplifiers have been developed, in particular by Linville and his collaborators[1]. All this work can, of course, be applied to field-effect transistors, but it must be appreciated that the devices have such very different terminal properties, and specifically such very different typical y-parameter values (as inspection of equation (12.2) shows) that the results obtained by applying bipolar transistor theory must be interpreted with great care. Central to this theory is the criterion that establishes whether a transistor is stable when arbitrary values of source and load impedance are connected to it. For a CS FET amplifier this would be:

$$g_{is} - \frac{\mathscr{R}(y_{fs} \cdot y_{rs})}{2g_{os}} - \frac{|y_{rs}y_{fs}|}{2g_{os}} > 0 \qquad (12.3)$$

where $\mathscr{R}$ implies 'the real part of . . .'

Since for a field-effect device g_{is} is substantially zero, except at the upper limit of operating frequency, this condition cannot generally be met. However, this does not imply that the FET is much more difficult to stabilise than the bipolar device. In fact the condition of arbitrary source and load impedances at the terminals of the device is a very much more stringent one for the FET (where it includes, for example, the case of a resonant circuit of virtually infinite Q-factor connected to the gate) than for the bipolar device, so that whilst it is good practice to ensure that inequality (12.3) is met by any transistor used in a bipolar amplifier, it is impossible to make this stipulation for field-effect devices. Instead a modified condition is used in which g_{is} is supplemented by G_G, the equivalent external conductance between the input terminals giving

$$G_G + g_{is} - \frac{\mathscr{R}(y_{fs} \cdot y_{rs})}{2g_{os}} - \frac{|y_{rs}y_{fs}|}{2g_{os}} > 0 \qquad (12.4)$$

as the stability condition when the load impedance is arbitrary but the generator admittance has a real part G_G and an arbitrary imaginary part.

Further since

$$y_{fs} = g_{fs} + j\omega C_{GD}$$

provided that

$$\omega \ll \frac{g_{fs}}{C_{GD}} \tag{12.5}$$

then y_{fs} is almost perfectly real. Similarly, y_{rs} is almost perfectly imaginary. Thus the product $(y_{fs} \cdot y_{rs})$ is also imaginary, subject to this condition on frequency. The stability criterion can thus be rewritten in simplified form, and in doing so it is almost always possible in practice to make the further simplifying assumption that g_{is} is much less than G_s, in which case the criterion becomes

$$G_s - \frac{\omega C_{GD}\, g_{fs}}{2g_{os}} > 0 \tag{12.6}$$

subject to the condition of inequality (12.5).

This criterion is appropriate, and must be satisfied when the resonant circuit connected to the drain of the FET (either directly or through a transformer) is of very high loaded Q-factor. More commonly g_{os} does load the drain circuit but there are also losses in the resonant circuit itself which cannot be ignored. If the loaded Q-factor is Q' and the resonant circuit directly coupled to the drain as in Fig. 12.4, has an inductance L' and capacitance C', then the equivalent shunt conductance is $(Q'\omega L')^{-1}$ to good approximation, provided that Q' is greater than about 5, when the stability criterion can be written as

$$G_s - \frac{\omega^2 Q' L' C_{GD}\, g_{fs}}{2} > 0 \tag{12.7}$$

Usually a condition on the maximum value of gain is more useful for design purposes, and remembering that the voltage gain A'_V at resonance is very nearly $(g_{fs}\omega L' Q')$, the condition becomes

$$G_s - \frac{\omega C_{GD} |A'_V|}{2} > 0$$

or

$$|A'_V| < \frac{2G_s}{\omega C_{GD}} \tag{12.8}$$

This is a particularly useful form for the stability criterion and can serve as a convenient starting point for an amplifier design. It remains valid, as can very easily be shown, if the drain circuit is transformer coupled to its load, or a double-tuned transformer is used. When, as is commonly the case, a resonant circuit is connected between gate and source of the amplifier (Fig. 12.4) having the parameters L, C, Q, then at resonance

$$G_s = \frac{1}{\omega L Q} = \frac{\omega C}{Q}$$

to good approximation if $Q \geqslant 5$ and hence the stability condition is

$$|A'_V| < \frac{2C}{QC_{GD}} \tag{12.9}$$

which is also most convenient, practically. Strictly, this condition only guarantees stability at the resonant frequency of the gate circuit; it can easily be shown that if stable at this frequency with arbitrary load, the stage will be stable at all other frequencies.

Note that the criteria of inequalities (12.6), (12.8), and (12.9) only guarantee that the circuit is stable. They do not imply, however,

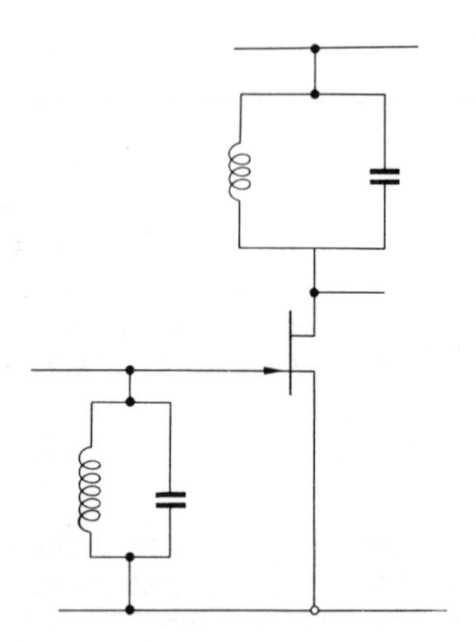

Fig. 12.4. A primitive narrow-band amplifier

that no regeneration effects at all occur, and hence modification of the shape of the amplifier pass-band will take place compared with that predicted by a simple theory which ignores feedback effects. In particular, this will be very marked if the stability criterion is only just met. For a simple amplifier (Fig. 12.4) the effect is to raise the apparent power gain below the centre frequency and depress it above, although the effect on voltage gain will be less marked because of the shunt form of the feedback.

Analytical treatments of regeneration effects are possible but extremely unwieldy. This is a case where the numerical approach is preferable, using an AC circuit analysis technique such as the IBM 'ECAP' (Electronic Circuit Analysis Programme), for which complete software is available applicable to any computer with FORTRAN capability. Calculation of the transfer properties of narrow band amplifiers using this method presents no outstanding difficulties, and a good approach is to arrange the programme to calculate the properties of the stage for a number of values of voltage gain less than the critical one at which instability occurs.

12.3. Typical CS amplifier circuits

A typical unneutralised CS amplifier is shown in Fig. 12.5 (omitting the components indicated in broken lines). However, unless the voltage gain is very restricted, this configuration can only be used at relatively low frequencies. For example, by use of equation (12.9)

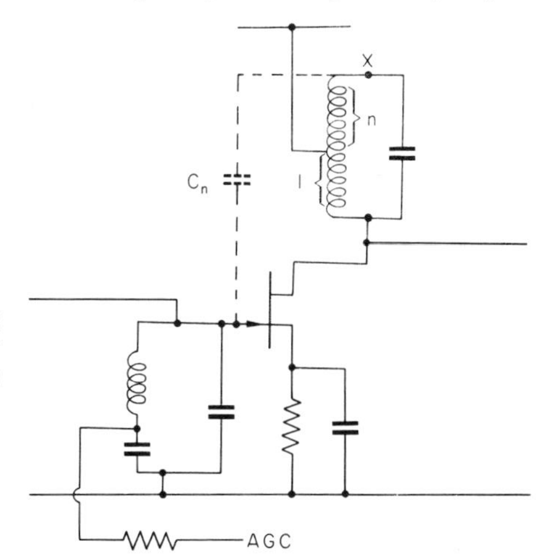

Fig. 12.5. A more practical narrow band amplifier, incorporating reverse AGC and neutralisation (shown in broken line)

assuming that the FET is a 2N3819 for which C_{GD} is 3 pF, if the Q-factor of the resonant circuit is, say, 100, then even for C as large as 1000 pF, the maximum stable value of A_V is as low as 6·7. This type of amplifier is therefore restricted to low gain, wide bandwidth applications, particularly at the lower frequencies where large values of C can be used.

The use of a tapped inductor in the drain resonant circuit, as shown, allows the effective impedance at resonance presented to the FET to be reduced without the need for an excessively low L/C ratio and hence inconvenient component values. If the output is taken from the point X, instead of from the drain terminal as shown, the resonant circuit may be used as an autotransformer, with a step-up ratio 1 : n. The apparent voltage gain to the output terminal (A'_V, say) is then n times that to the drain of the FET, and hence the stability condition of equation (12.9) becomes

$$|A'_V| < \frac{2nC}{QC_{GD}} \tag{12.10}$$

Thus an arrangement of this kind permits a higher gain to be realised. Instead of an autotransformer a double wound component

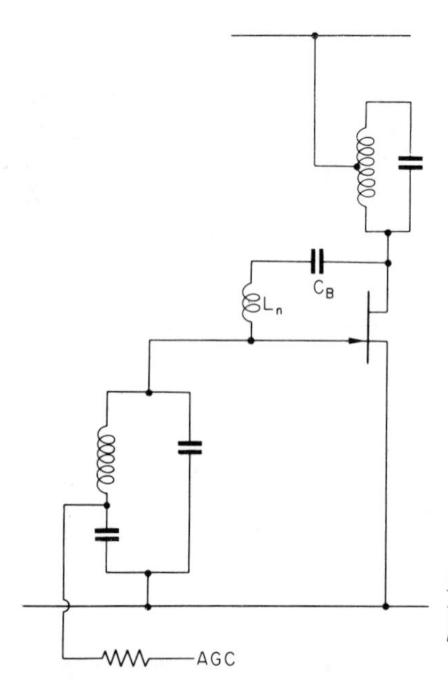

Fig. 12.6. Simple neutralising circuit with an inductor connected, via a DC blocking capacitor, between the drain and gate of the FET

may be used, with the primary either tuned or untuned. For air cored or iron-dust slug cored coils it is usually difficult to get even an approximation to perfect transformer action for $n > 5$, and above this ratio little further improvement is obtained.

If the circuit is neutralised, since it is the effective value of feedback capacitance which must be used to replace C_{GD} in the stability condition, the situation is eased, and typically even when the neutralisation is not individually adjusted for the FET in use, the gain can be increased by a factor of at least two without risk of instability. Some neutralising circuits which have been used at various times are indicated in the accompanying Figs. 12.5 to 12.7.

The use of an inductor (normally made variable by means of a variable tuning slug) connected, via a DC blocking capacitor, between the drain and gate of the FET (Fig. 12.6) has the merit of simplicity, but the frequency dependence of the reactance of the inductor is the inverse of that of the feedback capacitor C_{GD}, thus neutralisation is only correct at a single frequency given by

$$\omega = \frac{1}{\left\{L_n \cdot \frac{C_{GD}C_B}{C_{GD} + C_B}\right\}^{\frac{1}{2}}} \tag{12.11}$$

where L_n is the neutralising inductance, and C_B is the DC blocking capacitance. This neutralisation scheme can obviously be used only

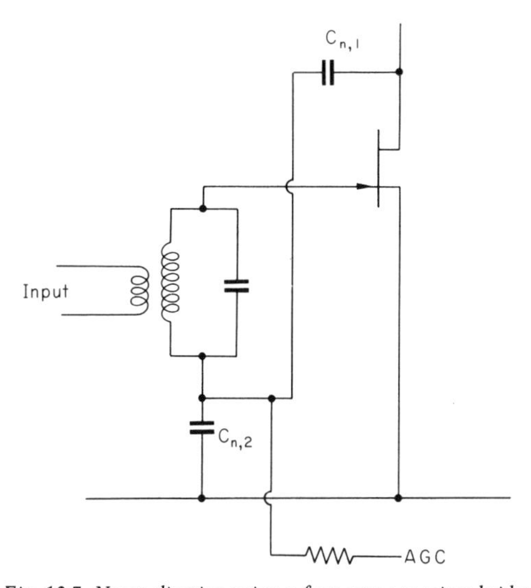

Fig. 12.7. Neutralisation using a four-arm capacitor bridge

in narrow band synchronously tuned amplifiers, and its application is therefore limited.

Of wider application is the Hazeltine circuit (shown in broken line in Fig. 12.5) in which neutralisation is effected by means of an additional feedback capacitor, C_n, driven from a point in antiphase to the drain and connected back to the gate. The circuit can be seen as, in effect, a transformer bridge, with the tapped drain resonant circuit acting as the transformer and C_{GD} and C_n as the two capacitor arms which are balanced. The condition for neutralisation is

$$C_n = \frac{C_{GD}}{n} \tag{12.12}$$

Because both the feedback and neutralisation admittances derive from capacitors they have the same frequency dependence, and thus the adjustment is not, in principle, frequency dependent. In fact, some frequency dependence does occur, for two reasons. The first is that the tapped inductor does not have its two parts 100% coupled, so that it cannot be regarded as a perfect transformer, and the relationship between the EMFs driving current through C_n and C_{GD} is slightly frequency dependent. Secondly, the assumption that C_{GD} is the only source of internal feedback is over-simplified, since it ignores the (small) feedback effect due to bulk silicon resistance in series with the FET source. However, analytical treatments of the frequency dependence are too complicated to be of much use, and here again numerical treatments, using computer techniques, which are based on measured y-parameters, in the frequency range of interest, for both FET and the drain circuit transformer, yield more tractable results.

Another circuit variant, which avoids the need for a transformer or tapped inductor, also uses a bridge type of circuit, but in this case a four-arm capacitor bridge (Fig. 12.7). The balance condition in this case is

$$\frac{C_{n,1}}{C_{n,2}} = \frac{C_{GD}}{C_{GS} + C_{st}} \tag{12.13}$$

where C_{st} is the total circuit stray capacitance from gate to earth.

Provided that C_{st} is kept small and circuit construction is such that it is reasonably constant this is a particularly attractive neutralising arrangement, since the ratio C_{GD}/C_{GS} will not vary very widely between different FETs of the same nominal type, even although the actual value of both capacitors may be widely variable. Unfortunately, the ratio is, however, dependent on working point.

The dependence of interelectrode capacitance on working point is one of the most important factors limiting the usefulness of neutralisation as a means of decreasing feedback effects. In this respect MOS devices are much superior to junction FETs, since the voltage dependence of interelectrode capacitance is more than an order of magnitude less. Even here, however, variability of capacitances from device to device limits the accuracy with which feedback effects can be balanced out. All neutralisation can do is to yield a reduction in the apparent magnitude of C_{GS}, with an increase in the stable gain. Typically this will amount to no more than $2\times$ or $3\times$.

12.4. Cascode amplifiers

A more radical approach to the design of stable FET amplifiers is to replace the single FET in CS configuration by a cascode pair

(Fig. 12.8). It is, of course, possible to analyse an amplifier of this type from first principles, but the pair of transistors connected in cascode configuration (CS followed by CG, Fig. 12.9) can simply be regarded as a new active two-port, and its y-parameters can be

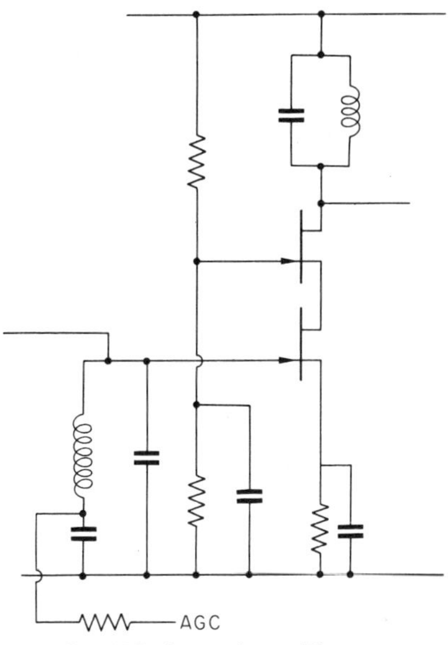

Fig. 12.8. A cascode amplifier

calculated in an entirely straightforward manner. The parameters of the combination are then,

$$\left.\begin{aligned} y_{ic} &= y_{is} - \frac{y_{rs}y_{fs}}{y_{ig} + y_{os}} \\ y_{rc} &= \quad - \frac{y_{rs}y_{rg}}{y_{ig} + y_{os}} \\ y_{fc} &= \quad - \frac{y_{fs}y_{fg}}{y_{ig} + y_{os}} \\ y_{oc} &= y_{og} - \frac{y_{rg}y_{fg}}{y_{ig} + y_{os}} \end{aligned}\right\} \qquad (12.14)$$

where the s-subscript y-parameters denote the common source parameters, and those having a g-subscript relate to the common gate.

Assuming that the simple pi-form equivalent circuit of Fig. 12.3 can be used as an approximation to the FET, the values of the common gate parameters can be approximated as

$$y_{og} = g_{DS} + j\omega C_{GD}, \qquad y_{rg} = g_{DS}$$

$$y_{fg} = -g_{fs} + g_{DS}, \qquad y_{ig} = -g_{fs} + g_{DS} + j\omega C_{GS}$$

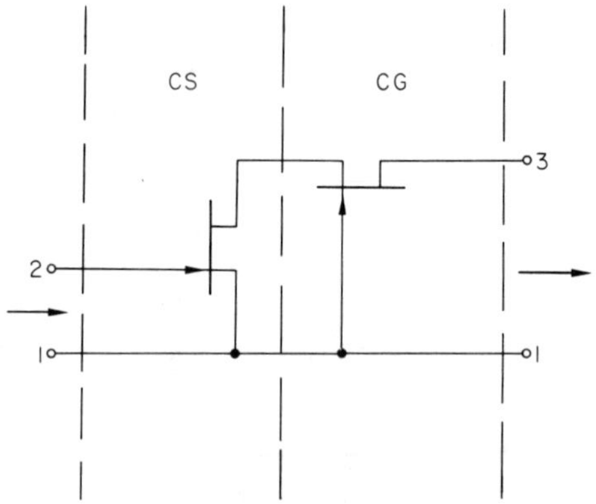

Fig. 12.9. A cascode pair as a two-port

However in most applications further approximation is permissible. Usually

$$|g_{DS}| \ll |g_{fs}|$$

hence to good approximation

$$\left.\begin{aligned} y_{og} &= g_{DS} + j\omega C_{GD}, \qquad & y_{rg} &= g_{DS} \\ y_{fg} &= -g_{fs}, & y_{ig} &= -g_{fs} + j\omega C_{GS} \end{aligned}\right\} \qquad (12.15)$$

Substituting into equations (12.15) the y_g values and also y_s values from equations (12.2), values of the cascode y-parameters can be obtained. Again approximations derived from the fact that

$$y_{ig} \gg y_{os}$$

are always permissible.

Thus

$$y_{ic} = j\omega(C_{GS} + 2C_{GD})$$

$$y_{rc} = j\omega C_{GD} \cdot \left\{\frac{g_{DS}}{g_{fs}}\right\}$$

$$y_{fc} = g_{fs}$$

$$y_{oc} = j\omega C_{GD}$$

to fair approximation.

Evidently the CS–CG cascode combination may be represented by an equivalent circuit as in Fig. 12.10, identical in form with that for a CS amplifier, but with element values modified as shown, and with an output capacitance replacing g_{DS}. The increase in input and output capacitance is of little significance, since it can be offset by a change in the values of the external tuning capacitors which form part of the two resonant circuits. Of more importance

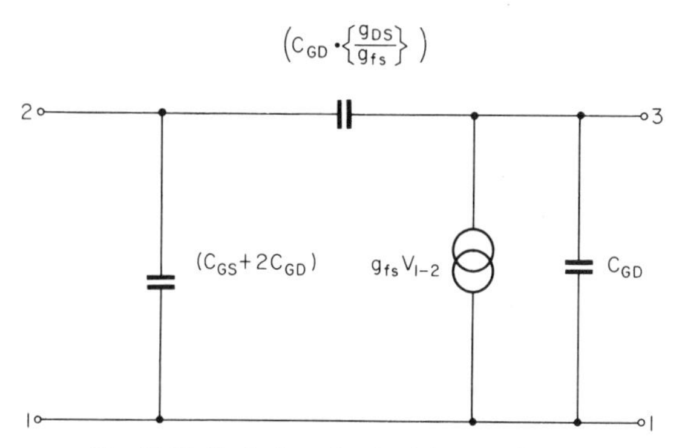

Fig. 12.10. Equivalent circuit of the cascode pair

is the dramatic reduction in feedback capacitance, in the ratio $|g_{DS}/g_{fs}|$. This ratio (which is invariably much less than unity and may be less than 10^{-3}) would in theory be the inverse of the ratio by which stable gain could be increased. However, the actual improvement obtained is not so large, since small terms in the equivalent circuit, ignored in this analysis, are in fact significant. Practically, feedback capacitance, or, more conveniently, y_{rs}, must be measured for the particular cascode pair in use, computed values, even with a more sophisticated basis of calculation than that indicated above, being insufficiently accurate.

Early cascode amplifiers using FETs invariably used two devices externally connected in cascode configuration, and in the case of junction devices this is still the case. For MOS devices, however, integrated cascode pairs are now available, and in some devices the source connection to the CG MOST and the drain of the CS unit are not brought out. Instead the device is simply fabricated as a channel with source and drain at either end and two gates, the gate nearer the source acting as the input terminal (control gate) and that nearer the drain (shield gate) being held at a fixed potential. The combined device can be regarded as a new type of five-electrode (including the substrate) FET, or alternatively as an integrated cascode package. Typical construction is shown in Fig. 12.11. Devices of this type have very favourable properties for use in narrow band amplifiers: for example type 248 BFY (Mullard) has $y_{fs} = 12$ millimho at $I_D = 10$ milliamps, with $C_{rs} = 25$ femto-farads (10^{-15} farad), and a noise figure of 2·7 dB at 200 Mhz. The

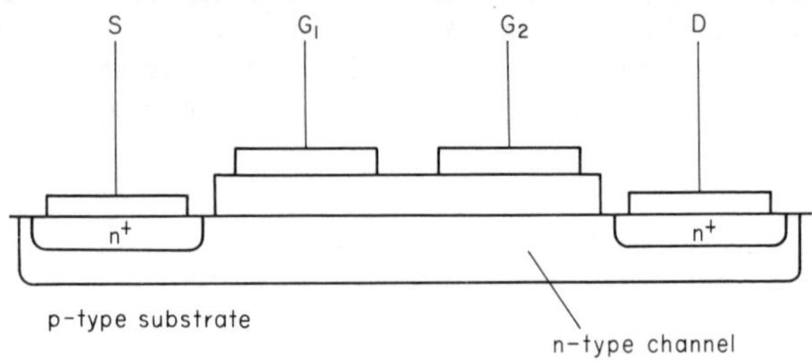

Fig. 12.11. An integrated MOS cascode pair

comparison with a good thermionic pentode is striking. Probably devices of this type will replace all others in narrow band RF and IF amplifiers.

The DC characteristics of cascode pairs (or equivalent integrated devices) are slightly different from those of a single device. As indicated already, the gate of the upper (CG) device is held at a fixed potential (shield gate, in the case of an integrated unit). To take the case of a cascode pair of junction gate devices, assuming that both devices are identical and that the chosen drain current is I_D (obviously, the same for both devices) then the gate–source bias voltage for the lower (CS) device is just

$$V_{GS} = V_P\left[1 - \left\{\frac{I_D}{I_{DSS}}\right\}^{\frac{1}{2}}\right]$$

where V_P is the pinch voltage, here assumed positive for *p*-channel devices and negative for *n*-channel.

However, in order that the CS stage shall not bottom it is necessary that

$$\left|\frac{V_{DS,1}}{V_P}\right| \geqslant 1 - \left[1 - \left\{\frac{I_D}{I_{DSS}}\right\}^{\frac{1}{2}}\right]$$

$$\geqslant \left\{\frac{I_D}{I_{DSS}}\right\}^{\frac{1}{2}} \qquad (12.16)$$

where $V_{DS,1}$ is the drain–source voltage of the CS (lower) stage.

The gate–source voltage for the CG amplifier will be the same as for the CS, hence the gate of the upper stage will be at a potential

$$V_{G2,S1} > V_P\left[2\left\{\frac{I_D}{I_{DSS}}\right\}^{\frac{1}{2}} - 1\right] \qquad (12.17)$$

relative to the source of the CS stage.

The total supply voltage must be sufficient to ensure that the CG stage also is not bottomed hence if the supply voltage is E (negative for p-channel, positive for n-channel).

$$\left|\frac{E}{V_P}\right| > 2\left\{\frac{I_D}{I_{DSS}}\right\}^{\frac{1}{2}} \qquad (12.18)$$

This assumes that a separate bias supply (for example from the AGC) is available to bias the CS FET. If this is not so the total supply must be increased to allow the CS stage bias to be developed across a resistor in the source connection, in which case

$$\left|\frac{E}{V_P}\right| > 1 + \left\{\frac{I_D}{I_{DSS}}\right\}^{\frac{1}{2}} \qquad (12.19)$$

The above expressions relate to depletion-mode (e.g. junction gate) devices. For MOSTs, which operate in the enhancement mode

$$V_{GS} = V_{\text{th}}\left[1 + \frac{1}{V_{\text{th}}}\left\{\frac{2I_D}{\beta}\right\}^{\frac{1}{2}}\right]$$

Hence

$$V_{DS,1} \geqslant \left\{\frac{2I_D}{\beta}\right\}^{\frac{1}{2}} \qquad (12.20)$$

and

$$E \geqslant 2\left\{\frac{2I_D}{\beta}\right\}^{\frac{1}{2}} \qquad (12.21)$$

At a given drain current level, in both cases the effective 'bottoming' voltage for the cascode configuration is at least twice that for a single triode device, thus at moderate values of supply voltage the power efficiency of cascode (or equivalent integrated) amplifiers is poorer. However, in small signal amplifiers, which are the major application of the cascode configuration, this is unimportant.

12.5. Intermodulation and mixing

It is convenient to treat the topics of intermodulation and mixing together, since they are both aspects of the non-linearity of the amplifier. A mixer is a special type of non-linear amplifier in which incoming signal and AC power from a local oscillator are intentionally intermodulated in order to generate an intermediate frequency signal. By their nature mixers must necessarily be non-linear, but design concentrates on ensuring that the type of non-linearity is only that required to ensure correct mixing action, and ideally outputs resulting from non-linearity other than the required product of signal and local oscillator should not occur. A linear amplifier, by contrast, is required simply to amplify an incoming signal, and should ideally have no non-linearities whatsoever, so that if more than one signal is simultaneously applied to the input no intermodulation products should be present in the output.

Consider a voltage-controlled amplifier having an input $v_1(t)$ applied to it and producing an output current $i_2(t)$. In general, and neglecting the possibility of internal feedback, the transfer properties of the amplifier may be represented by a polynomial

$$i_2(t) = a_0 + a_1 v_1 + a_2 v_1^2 + a_3 v_1^3 + \text{etc.} \tag{12.22}$$

In general the input voltage $v_1(t)$ may be regarded as the sum of n different voltages, thus

$$v_1(t) = \sum_{r=1} v_{1,r}(t) \tag{12.23}$$

An amplifier would have all the $v_{1,r}$ representing different input signals, and a mixer would have all but one of the $v_{1,r}$ input signals with the additional one, say $v_{1,1}$, derived from the local oscillator. To take the case of sinusoidal inputs, it is permissible to write

$$v_{1,r}(t) = v'_{1,r} \cos(\omega_r t) = \frac{v'_{1,r}}{2}\{e^{j\omega_r t} + e^{-j\omega_r t}\}$$

and this can be substituted into equation (12.23) and hence into (12.22). The resulting equation (12.22), however, when fully expanded is rather complex, thus it will be convenient to consider it term by term. The term in a_0 is simply a constant and can be neglected, in considering intermodulation effects, as can the term in a_1, which simply represents the linear response to the input signals. The lowest order term of interest from the present point of view is that in a_2, which may be expanded as

$$a_2 v_1^2 = a_2 \sum_{r=1}^{n} \sum_{s=1}^{n} v_{1,r}(t) \,.\, v_{1,s}(t)$$

$$= a_2 \sum_{r=1}^{n} \sum_{s=1}^{n} \frac{v'_{1,r} v'_{1,s}}{4} \{e^{j\omega_r t} + e^{-j\omega_r t}\}\{e^{j\omega_s t} + e^{-j\omega_s t}\}$$

$$= a_2 \sum_{r=1}^{n} \sum_{s=1}^{n} \frac{v'_{1,r} v'_{1,s}}{4} \mathrm{e}^{j(\omega_r + \omega_s)t} + \mathrm{e}^{-j(\omega_r + \omega_s)t}$$

$$+ a_2 \sum_{r=1}^{n} \sum_{s=1}^{n} \frac{v'_{1,r} v'_{1,s}}{4} \mathrm{e}^{j(\omega_r - \omega_s)t} + \mathrm{e}^{-j(\omega_r - \omega_s)t}$$

Thus the square law term introduces AC components into the output current of frequencies $|\omega_r + \omega_s|$ and $|\omega_r - \omega_s|$ (since the possibility of negative frequency need not be considered here) for all r and s from 1 to n. These sum and difference frequencies are, of course, the intermodulation products.

In the case of the term in a_3, the cube law results in three pairs of exponential terms being multiplied together, with the consequent production of alternating terms having frequencies $|\omega_q + \omega_r + \omega_s|$, $|\omega_q - \omega_r + \omega_s|$, $|\omega_q + \omega_r - \omega_s|$, and $|\omega_q - \omega_r - \omega_s|$, where q, r and s take values from 1 to n. Similarly the term in a_4 produces components having frequencies given by $|\omega_p \pm \omega_q \pm \omega_r \pm \omega_s|$ where p, q, r and s take values from 1 to n, and so on for higher terms.

Intermodulation products of this kind are objectionable if they are at a frequency such that they are within the pass-band of the amplifier subsequent to the stage in which they arise, and hence can interfere with the wanted signal. In the case of an amplifier, all the second and higher order terms, can, at least in principle, produce interfering signals of this kind, thus all the coefficients a_2, a_3, a_4 and so on should be as small as possible. However, in this case a large value of a_2 may not matter very much, since if two frequencies have a sum or difference equal to the mid-band frequency of the amplifier then one or both must be remote in value from the mid-band frequency. Thus great attenuation by the filter preceding the amplifier will occur, provided that this filter is fairly narrow band. Such an argument does not apply to the a_3 term. Thus although for a linear amplifier the a_1 term should be maximised and all other terms made as small as possible, in the narrow band case a large value of a_2 will not be a serious disadvantage.

In the case of a mixer, the required IF output is at a frequency equal to the difference between the local oscillator and incoming signal frequencies, and is thus generated by the a_2 term in the transfer polynomial. The a_0 and a_1 terms generate no useful output but also do not contribute to spurious intermodulation products, which are due to terms in a_3, a_4 and so on. Thus a mixer should have large a_2 and all other terms negligible, although non-zero a_0 and a_1 will only cause some loss in conversion efficiency and will not generate unwanted IF signals.

Field-effect transistors have nth power law transfer characteristics where n is close to two. In the MOS case, assuming a standing DC gate–source bias of V_B and an input AC signal $V_1(t)$, the resulting drain current is given by

$$2I_D = \beta(V_B + V_1 - V_{th})^n \tag{12.24}$$

for all $(V_B + V_1 - V_{th}) > 0$

If the value of V_B is chosen so that

$$V_B - V_{th} = 0$$

and also a MOST is used which has n very near to 2, the response of the device is very nearly a perfect square law, for one half cycle of the input voltage $V_1(t)$. This might be thought to be ideal for the case of a mixer, where $V_1(t)$ could be the sum of the local oscillator and incoming signal voltages. However, operation at this point has disadvantages, in particular the transfer conductance is low, and low intermodulation is not obtained, partly because of the fact that square law behaviour is only obtained for $(V_B + V_1 + V_{th}) > 0$, (no drain current flows if the gate voltage is less than V_{th}) with the consequent need to introduce higher order terms into the transfer characteristic to take account of this, and partly because the deviations of the drain current from the value predicted by the simple square-law equation become quite severe near the threshold gate voltage. Thus it is usual practice to bias the device so that there is appreciable standing drain current. Under these conditions, assuming that $n = 2$ in equation (12.24) and writing $(V_B - V_{th})$ as V_F,

$$2I_D = \beta V_F^2 + 2\beta V_F V_1 + \beta V_1^2$$

provided $|V_1| < |V_F|$

Hence, in terms of equation (12.23)

$$\begin{aligned} 2a_0 &= \beta V_F^2 \\ 2a_1 &= 2\beta V_F \\ 2a_2 &= \beta \\ a_3, a_4, \text{etc.} &= 0 \end{aligned} \tag{12.25}$$

It might be supposed that for a linear amplifier V_F would be chosen such that

$$2V_F \gg V_1$$

whilst for a mixer the inequality would be reversed, however this is not the case, since the drain current does not follow the parabolic law for $V_F < V_{th}$. More correctly the $I_D - V_{GS}$ relationship is represented by the non-analytic expression.

$$2I_D = \beta(V_{GS} - V_{th})^2 \,.\, u_1(V_{GS} - V_{th})$$

where

$$u_{-1}(x) = 0 \qquad \text{for } x < 0$$

$$u_{-1}(x) = 1 \qquad \text{for } x > 0$$

It can easily be shown that excursions into the cut-off region result in sharp deterioration in intermodulation performance. Thus both amplifiers and mixers are usually biased about half way between zero bias and pinch-off (for depletion mode and junction devices) or at a voltage in the middle of the range over which g_{fs} is a linear function of V_{GS}, for enhancement mode devices, and the total peak V_{GS} swing is restricted to avoid excursions into the cut-off region.

12.6. Automatic gain control

Often electrical means of varying the gain of amplifiers is required, for example, in AGC systems. The most common approach is to vary g_{fs}, to which gain is usually proportionate, by varying the bias point. This is sometimes called reverse AGC, since for junction devices the gate remains reverse biased. The gain can be varied in this way over a substantial range (usually more than 20 dB, and in some cases as much as 40 dB) however, very low gain involves biasing to very low drain currents and, unfortunately, level of intermodulation effects rises as the bias voltage approaches the pinch-off (threshold) value. This may not matter too much, since this low-gain condition only tends to arise when the wanted incoming signal is relatively large, and thus intermodulation effects are less critical. However some attention has been given to the design of FET structures which do not exhibit a sharp cut-off of drain current at a particular value of gate voltage, and thus are less prone to this trouble. Such structures are quite closely related to those, described in chapter 11, which are being developed for variable resistance applications. The majority of amplifiers, however, still use 'sharp cut-off' FETs.

So far as junction gate devices operated in the CS configuration are concerned, an alternative mode of gain control is also possible, in which the gate-channel junction is forward biased, causing gate current to flow. Although increasing forward bias increases g_{fs}

slightly, the principle effect is greatly to reduce the input resistance of the FET, hence loading the signal source preceding it and reducing the overall signal level at the gate. Although this mode of operation gives a wider range of gain variation and has been advocated, for example, by Farell[2], it does suffer from certain serious disadvantages. The first is that when the gate is driven into conduction the loading on any filter circuit connected to the gate is severe, with corresponding loss of selectivity. Also due to the severely non-linear relationship between gate voltage and current, there is a sharp increase in intermodulation with this mode of operation. For these reasons the forward bias AGC mode of operation will not be considered further here.

In any AGC system alteration of the voltage applied to the FET electrodes will cause some change in interelectrode capacitance. This can cause de-tuning of resonant circuits connected to the device. The effect is made worse because of Miller effect through the feedback capacitance C_{DG}. This results (assuming no neutralisation) in the FET presenting a complex input impedance, as explained above in connection with problems of stability. The imaginary part of this input impedance modifies the resonant frequency of the gate circuit. Unfortunately the effect varies directly both with C_{DG} and with the voltage gain of the stage. The only solution to this problem is to use a cascode amplifier (or equivalent dual gate MOST) which reduces the effective feedback capacitance to negligible proportions. So far as detuning due to variations of C_{GS} is concerned, the effect can be minimised by choosing the effective shunt capacitance of the resonant circuit connected to the FET gate to be large compared with the expected variation in C_{GS}. The conditions for a satisfactory design are thus

$$C_{DG} \ll \frac{C + C_{GS}}{1 - A_V} \tag{12.26}$$

and

$$C_{GS}(\text{max}) - C_{GS}(\text{min}) \ll \frac{C + C_{GS}}{Q} \tag{12.27}$$

where C is the capacitance and Q the in-circuit quality factor of the resonant circuit connected to the gate, A_V is the voltage gain of the stage, and C_{GS} and C_{DG} are the gate–source and drain–gate FET capacitance values.

MOS devices have the advantage in this connection that their inter-electrode capacitances are less voltage dependent than those of a junction device. The difference is only significant at the highest frequencies, however, since at lower frequencies C can be increased to a large enough value to ensure that inequalities (12.26) and (12.27) are valid.

Note that when the FET is completely cut-off the input and output resonant circuits are coupled only by C_{DG}. If the value of this element is too large, the input and output resonant circuits will form an overcoupled band-pass circuit, and a double humped response will occur. Provided that both circuits have the same Q-factor (Q) and total capacitance (C), this situation will be avoided provided that

$$C_{DG} < \frac{C}{Q} \tag{12.28}$$

for large $Q(>10)$, and assuming no neutralisation.

12.7. Common gate amplifiers

All the foregoing describes applications of common-source amplifiers, or the cascode circuit, which may be regarded as a modified common source stage. The common-gate amplifier has the disadvantage of limited power gain and low input impedance. However, at high frequencies, particularly in the UHF band the limited power gain is also a feature of the common-source configuration, due to rising input conductance. Thus the CG configuration may appear less at a disadvantage.

In a well-constructed FET, the drain–source capacitance can be made very small indeed. Thus, even at high frequencies the reverse transfer admittance of a FET in the CG configuration is almost always negligible and the device can be considered as unilateral and unconditionally stable in the Linville sense. For virtually all FETs operated in this mode, therefore, stable and non-regenerative operation is assured, provided only that the circuit layout is good enough to ensure that no unwanted feedback is introduced external to the device. The use of neutralisation, with all its attendant inconveniences, is avoided. It is for this reason that FETs are often operated in the CG configuration in the UHF band. The power gain is easily calculated, and is equal to A_P, where, to good approximation

$$A_P = \frac{g_{fg}}{4g_{og}}$$

Note that, very nearly,

$$g_{fg} = -g_{fs}$$

The input admittance is very nearly given by a conductance equal to g_{fg} in shunt with C_{GS}. The figure of power gain quoted assumes that the signal source is matched to this input admittance and the load is matched to g_{og} in parallel with C_{DG}. This gives the highest attainable power gain, but unfortunately not the lowest noise figure, for which the optimum signal source impedance should be higher, the exact value depending in rather a complicated way on the device working point. Usually the source impedance is adjusted empirically with a given amplifier to give minimum noise, but at a penalty of the loss of one or two dB of gain. A junction FET operating in the CG mode with a suitably adjusted source impedance is the lowest noise transistor amplifier available in the UHF region at present, although MOS devices are only a little inferior.

The application of AGC to CG stages is straightforward and very successful, since the very small internal feedback prevents Miller-effect detuning. The relatively low input impedance of this type of amplifier means that relatively low L/C ratio resonant circuits can be connected to the input terminal and hence variations of C_{GS} are more easily swamped.

12.8. Practical VHF and UHF amplifiers

Actual amplifier designs based on the principles described above are very straightforward indeed at frequencies from AF through to UHF. Designs for the higher end of the frequency range will be described here: those for lower frequencies are, if anything rather simpler since unwanted feedback effects are less evident.

A simple CG amplifier, for use at 100 MHz (using a TIS 58 junction FET) is shown in Fig. 12.12. The circuit is completely stable and requires no complex setting up, but gives a power gain of only

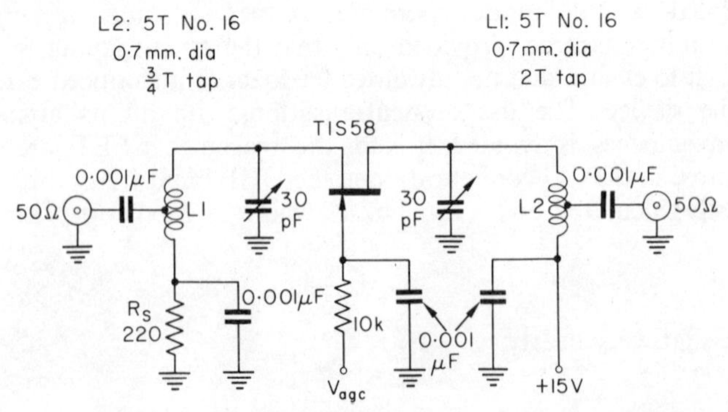

Fig. 12.12. A 100 MHz common gate amplifier using transistor Type TIS 58 (Courtesy Texas Instruments Ltd.)

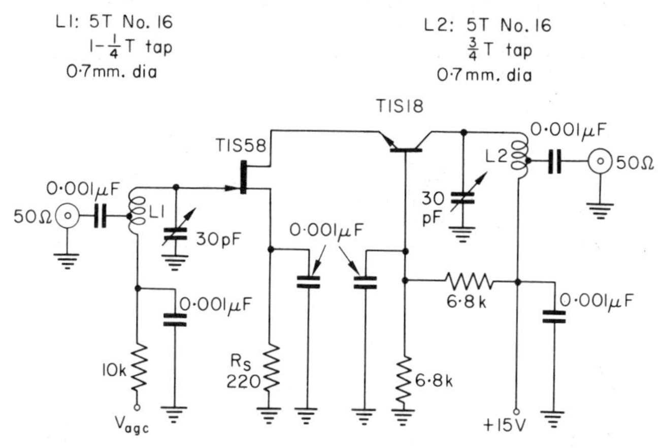

Fig. 12.13. A cascode 100 MHz amplifier. The bipolar transistor Type T1S 18 can be replaced by an FET (T1S 58) with negligible change in performance (Courtesy Texas Instruments Ltd.)

+6 dB, falling to −17 dB with about 3 V of reverse AGC applied. Detuning on application of AGC is negligible. By contrast, at the same frequency and using the same FETs a cascode circuit (Fig. 12.13) will give a maximum gain well in excess of 20 dB, falling to −13 dB with 2·5 V of reverse AGC. Here too no neutralisation is required and detuning by application of AGC is negligible. These examples are based on designs from the paper by Farell[2], to which reference has already been made, who describes a variant of the

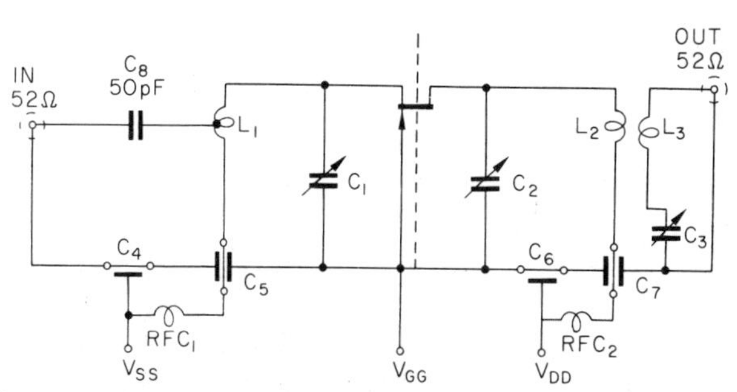

Fig. 12.14. A 450 MHz CG amplifier using FET Type 2N5397 (Courtesy Siliconix)

$C_{1,2,3.}$ = *0·8–12 pf Johanson type 2950*
$C_{4,5,6,7}$ = *1000 pf Allen-Bradley type SS5D*
$RFC_{1,2}$ = *0·15 μF Delevan type 1537–00*
DL_1 = *1·5 in long; # 16 copper*
L_2 = *1·2 in long; # 16 copper*
L_3 = *2 in long; # 22 copper enamel, loosely coupled to* L_2 *0·75 in spacing*

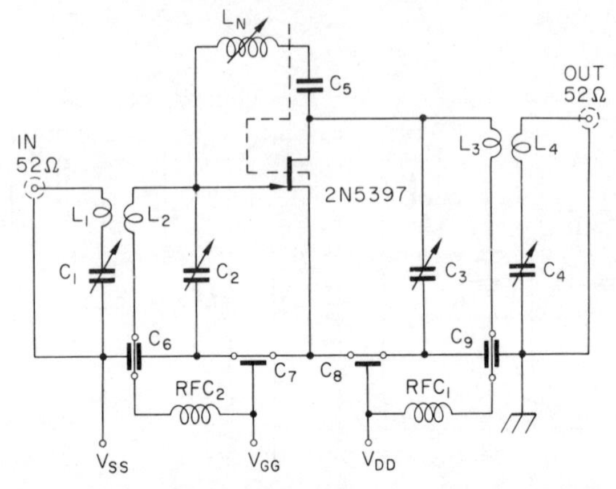

Fig. 12.15. A 450 MHz neutralised CS amplifier using FET Type 2N5397. (Courtesy Siliconix)

C_{1-4} = *0·8–12 pF Johanson type 2950*
C_5 = *40 pF DMS silver micon*
C_{6-9} = *1000 pF Allen-Bradley Type FA5C*
L_1 = *1·4 in long; 22 enamel spaced 0.1 in from* L_2
L_2 = *1·1 in long; # 16 solid copper*
L_3 = *1·3 in long; # 16 solid copper*
L_4 = *1·4 in long; # 22 enamel spaced 0·3 in from* L_3
$RFC_{1,2}$ = *0·15 μH Delevan type 1537–00*
L_W = *3T, # 22 enamel; 0·25 in dia. ceramic form; aluminium skg.*

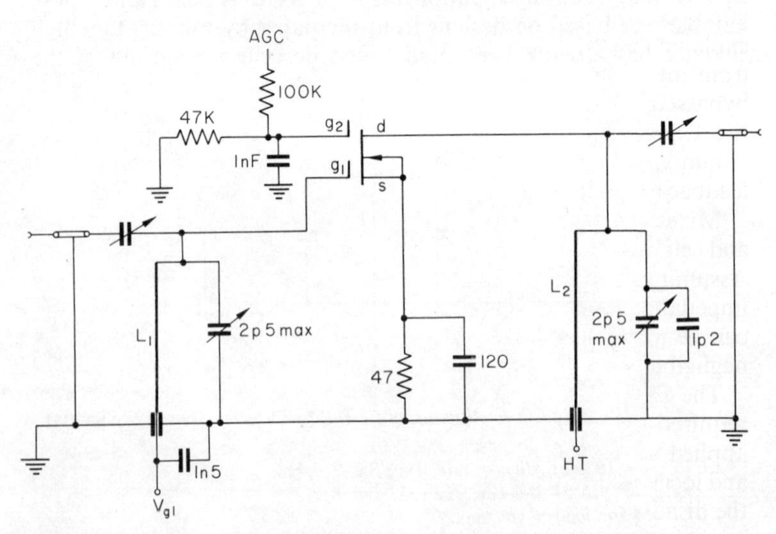

Fig. 12.16. A 600 MHz dual-gate MOS *amplifier* $L_1 = L_2 = 1$ *in length* $\frac{3}{16}$ *in dia. copper rod.* (Courtesy Mullard Ltd.)

cascode circuit in which the CG FET is replaced by a bipolar transistor (TIS 18) in common base configuration with negligible change in performance.

At higher frequencies a device with smaller intelectrode capacitances must be used. For example, the CG amplifier of Fig. 12.14 uses a type 2N5397 FET to achieve a maximum gain of 12 dB at 450 MHz with a noise figure of 4 dB, however, if the source impedance is increased slightly above its value for optimum power gain, by moving the tap on L_1 slightly towards the 'cold' end, the noise figure drops to 3·3 dB, with a gain of slightly over 9 dB. By contrast the CS amplifier of Fig. 12.15, can give a similar noise figure and some 15 dB of gain at the same frequency, but at the cost of introducing a critical neutralisation adjustment. These two circuits were developed by Compton[3].

Fig. 12.16 shows an amplifier designed to operate at 600 MHz using a dual gate MOST, which achieves a gain of 9·5 dB. The circuit is an integrated equivalent of a cascode amplifier. Probably much future development will centre on MOS tetrodes and pentodes, in view of their superior properties.

12.9. Practical mixers

Successful mixer operation, with minimal spurious responses and good conversion conductance, depends, as has already been explained, upon as nearly as possible a perfect square-law response from the FET. For this to be achieved it is important that the unbypassed signal impedance in the source lead of the mixer (assuming the CS configuration) should be minimised. Such impedance is common to input and output circuits and thus results in current feedback which tends to modify the transfer law of the device.

Mixers have two inputs, from the signal and the local oscillator, and can therefore be classified in terms of the way these are combined. Assuming that the IF and signal frequencies are well separated, the impedance to ground at signal frequencies at the drain of the FET can be made very small, thus signal frequency feedback effects are negligible.

The CS configuration is thus widely used and neutralisation is not required. This being so, both signal and local oscillator may be applied to the gate of the FET, or signal may be applied to the gate and local oscillator output to the source. The former arrangement has the disadvantage that local oscillator radiation from the input of the mixer is more likely and hence an RF amplifier stage is almost always used before the mixer. With source injection this problem is less severe, but the internal impedance of the local oscillation generator is connected in series with the source, and there is thus some inevitable deterioration in intermodulation performance. Stevens[4] has described mixers in which the local oscillator drives the gate and signals are applied to the source.

Although good designs are possible on this basis, again care is necessary to ensure that the effective impedance in series with the source is not sufficient to degrade the square transfer law. The

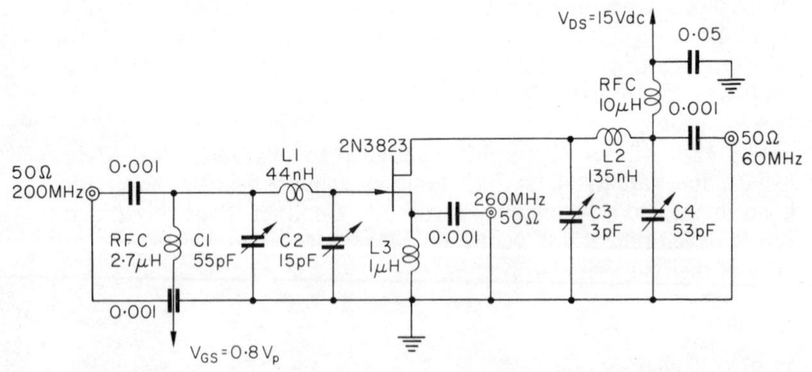

Fig. 12.17. A source injection mixer for use up to 250 MHz L_1 = 5 turns tinned #20 wire; i.d. 0·15 in length 0·6 in. All other inductors are standard moulded chokes. (Courtesy Motorola)

requirements for minimising oscillator radiation and also avoiding the introduction of significant impedance into the source lead is, of course, most easily met if the local oscillator and signal frequencies are widely different, in which case the choice of mixer configuration is not too important.

Typical mixer designs have been described in some detail by Kwok[5], who indicates the conflict between good conversion gain and low intermodulation. For the former (considering depletion mode FETs) the bias point is typically 0·8 of the pinch voltage, with an injected local oscillator peak voltage of similar peak amplitude. The theory developed in section 12.5, above, suggests that a reduction in both cases to 0·5 V_p would reduce intermodulation distortion, however, this is not borne out experimentally due to the transistor deviations from square law and to the linearising effect of source impedance, also the smaller L.O. injection causes a severe loss of conversion gain. Thus mixers are usually designed either empirically or using numerical analysis techniques (such as ECAP) if intermodulation performance is to be optimised. Best noise performance is achieved, to good approximation, when the mixer is adjusted for optimum conversion gain.

A typical circuit (due to Kwok) is shown in Fig. 12.17. Source injection is used and at a centre frequency of 50 MHz a conversion gain of about 10 dB is achieved with a bias voltage of 0·8 V_p and L.O. injection of similar amplitude, this figure of gain being maintained up to about 250 MHz. At 50 MHz noise figures down to a little over 5 dB are obtained, and an interfering signal at 52 MHz of some 250 mV is required to produce 1% intermodulation with a wanted signal of 1 mV.

Dual gate MOS devices can also be used in similar circuits, with source or gate injection, but in this case there is also the interesting possibility of applying the L.O. voltage to the shield gate. In this case the mixer is operating in a fundamentally different way. Since the voltage applied to the upper of the two gates is varied, the effect is that of varying the drain voltage on the lower of the two triode FETs forming the cascode configuration to which the dual gate device is equivalent. When the drain voltage of the lower equivalent FET falls sufficiently for the device to be no longer pinched-off (i.e. to be operating in the non-saturated 'triode' region) the transfer conductance falls sharply, however the transfer law remains substantially square-law.

Mixers of this type are characterised by favourable electrical properties including excellent intermodulation performance and little coupling between the local oscillator and signal input circuits, which last is particularly important when the frequency converter is used without a preceding RF stage, since it reduces oscillator radiation.

REFERENCES

1. Linvill, J. C. and Gibbons, J. F. *Transistors and active circuits*, McGraw-Hill, New York (1961).
2. Farell, C. L. *AGC characteristics of FET amplifiers*, Texas Instruments Application Report CA104 (1968).
3. Compton, J. B. *Characteristics and use of high frequency junction FETs*, Siliconix Inc. (1969).
4. Stevens, M. *FET Mixers*, Texas Instruments Application Report B42 (1968).
5. Kwok, Siang-Ping. *Field effect transistor RF mixer design techniques*, WESCON Convention Record, I.E.E.E. (1967).

13

INTEGRATED CIRCUITS

13.1. Economics of electronic fabrication

A significant proportion of the electronic equipment currently being manufactured uses one or more of the integrated circuit technologies. For the future, this approach can be expected to contribute an increasing proportion of the whole, indeed it is not possible at this juncture to foresee a definite limit to the trend. Three different integrated circuit processes are in current use, differing widely in their technical details and having in common only that the electronic circuit is fabricated in a single sequence of processes, with few or no additional components added later. The prehistory of the integrated circuit can be traced back a long way, at least to the Loewe amplifiers of the early 1930s and to Sargrove's[1] pioneering efforts at automated production of radio receivers in the late 1940s, but its rapid recent development is a consequence of the wastefulness, in economic terms, of the older approach to electronic manufacture, using discrete components.

The constructional techniques in common use in the electronics industry developed out of the methods used at the turn of the century for electrical scientific instruments. Component parts, the most complex of which were thermionic valves and later transistors, were manufactured and tested individually, and were then assembled together on a metal chassis or some other mechanical supporting medium. The pattern of interconnection was established by hand wiring, using soldered connections.

The radio and television industries achieved very large volume production on this basis, adapting well-established techniques of flow production on an assembly line basis to minimise the labour content of the finished equipment. This was never an entirely acceptable production technique, however, and its wastefulness in space, labour and materials became particularly obvious with the introduction of large scale batch fabrication methods for the manufacture of transistors in the early 1950s. The first of these methods was the mesa process, and this in turn was soon superceded by the technically superior planar process which came into use at the end of the decade.

13.1.1 THE PLANAR PROCESS

The planar process, which permits the production of virtually any solid state device fabricated in the form of plane parallel *pn* junctions, has already been described in chapter 2 in connection with the structure of junction gate FETs designed as discrete components.

In essence it depends on the selective diffusion of impurities into the silicon surface through a mask formed by cutting 'windows' in a film of silicon dioxide, grown on the silicon surface by oxidation at an elevated temperature. The 'windows' (which determine the dimensions of the resulting structure) can be made by a process akin to photo-engraving, and consequently extreme resolution is possible by exploiting all the powerful techniques of microphotography. Thus an electronic component, even one as complex as a bipolar transistor, can be fabricated on a surface area only a hundred microns square.

Since silicon slices are readily available several centimetres in diameter it is obvious that many thousands of devices can be formed in the surface of a single slice. In conventional discrete transistor manufacture these individual devices would be cut apart, mounted in suitable encapsulations, bonded to connecting leads, tested (perhaps life tested) and packed for shipment. The paradox arises when it is realised that the system manufacturer who buys them will, in all probability, associate just such a group of transistors together again and interconnect them as part of the process of building his equipment.

13.1.2 MONOLITHIC PROCESS OF INTEGRATED CIRCUIT MANUFACTURE

The obvious path of advance leads to the suggestion that instead of separating the semiconductor devices (only in order to bring them together again later) they should be interconnected *in situ* to form a complete electronic system of some complexity, actually in the silicon chip. This was not possible with the mesa and earlier processes for making transistors, except by tedious hand wire-bonding techniques, because the devices were not fabricated in coplanar form.

The planar process, as its name implies however, does produce arrays of devices with a near flat surface. Thus, interconnection patterns can be applied in the form of thin evaporated metal films, usually aluminium, which forms a very strong bond to silicon oxide. This is the basis of the *monolithic* silicon process for integrated circuit manufacture (see chapter 14), which is the most promising of all from the point of view of very low cost circuit fabrication, particularly when the circuit contains a relatively high proportion of active devices.

13.1.3 HYBRID PROCESSES OF INTEGRATED CIRCUIT MANUFACTURE

The other two integrated circuit processes in current use are the so-called *hybrid processes.* Whereas the monolithic circuit is fabricated on a semiconducting supporting substrate, namely the silicon slice, the hybrid systems use an insulator on which to build the circuit. This may be glass or, more commonly at the present time, a high alumina ceramic. Historically the first of these processes to be fully developed was the *thin film* technique[2]. In this approach, films of metal, semiconductor or insulator are deposited on the substrate by evaporation, or, in a less common alternative form, by cathodic sputtering. The deposition pattern is controlled by masks so that by means of a series of depositions the required circuit configuration can be achieved. Strictly, the evaporated aluminium layer used for interconnecting devices on the surface of a monolithic silicon integrated circuit could be regarded as a very simple thin film circuit.

The alternative hybrid technology is the *thick film* process. In this variant[3] the insulating, conducting or resistive regions are formed by a process descended from that used for applying decorative patterns to porcelain, and ceramic substrates are invariably used. The required circuit configuration is printed, using the silk screen process, direct on to the substrate. The ink employed consists, in the case of a conductor, of a mixture of certain metals (commonly either gold or a mixture of silver and palladium), together with finely powdered glass ('frit') in an organic binder, the latter chosen simply to give satisfactory rheological properties to the resulting ink.

After the printed pattern has dried, the substrate is transferred to a furnace where it is subjected to a firing process at a temperature somewhere between 400 and 1000°C, depending on the formulation of the particular ink concerned. This process burns away the organic binder and melts the glass in the ink, which unites with glasses in the ceramic substrate to give a strong mechanical bond. The ink has been converted into a complex of metals and oxides forming a highly stable conducting or resistive track, according to formulation. By omitting metals from the ink a glass layer may be formed on firing, which can be used to insulate crossovers in the wiring or to form capacitors; although in the latter case it would be normal to add a ferroelectric ceramic to the ink to increase the working dielectric.

Both hybrid techniques lend themselves very well to the formation of highly conducting interconnections and resistors. With only marginally greater difficulty, capacitors and wiring crossovers can be formed. The limitations of these methods arise in respect of active devices. The formation of transistors depends upon the availability of crystalline semiconductors having superior properties, and, particularly for bipolar transistors, the semiconductor must be monocrystalline. This is because the satisfactory operation of a bipolar device depends on the passage of minority charge carriers through the base region of the device. Thus, the mean lifetime of minority carriers must be at least of the order of the transit time of those carriers through the base, otherwise the current gain of the device will be very seriously reduced. Since grain boundaries in polycrystalline materials form ideal sites for electron-hole recombination it proves essential to use monocrystalline material to achieve acceptable minority carrier lifetime.

By contrast, FETs are majority carrier devices so that minority lifetime considerations are unimportant. Successful field-effect devices have indeed been fabricated in polycrystalline materials, and furthermore, in materials deposited by the thin film hybrid circuit technique (see chapter 3). However, at the present time, understanding of semiconductors other than monocrystalline silicon is inadequate, and it is not possible to fabricate thin film transistors having electrical properties comparable with those readily attainable in silicon using the planar process. Thus the thin film transistor (TFT) is likely to be restricted to special applications where its particular properties seem especially suitable: some of these will be reviewed later in this chapter. So far as a transistor fabricated by the thick film process is concerned, although such a 'graphic-arts active device' (GAD) has been proposed, present indications are that its electrical properties will be very inferior, and its possible applications seem limited.

13.1.4 APPLICATIONS

Because of the inadequacy of the active devices, both the thick and thin film processes are used at the moment almost excusively for the fabrication of passive circuit components, the active elements being added subsequently as discrete components, fabricated in the conventional way. It is because a circuit formed in this way combines an integrated approach to the passive elements and the interconnecting pattern with discrete active devices, that these two processes are referred to as producing hybrid circuits.

The chief advantage of hybrid circuits over monolithic integrated circuits is in their ability to produce circuits in which the passive components have superior electrical specifications, but at an initial tooling cost far lower than that for monolithic circuits. They therefore commend themselves particularly when small to medium volumes of circuits are required, but the monolithic process has the cost advantage when the volume of production is really large, so that the per item tooling charge is small. Probably for a long time to come both the monolithic and hybrid technologies are going to co-exist and, indeed, particularly interesting results can be achieved when monolithic integrated circuits are used as the active devices in, say, thick film integrated systems.

However, since in hybrid integrated circuits the active elements are incorporated in discrete component form, circuit techniques are not so very different from those applicable to discrete component circuits. This is far from being the case for monolithic integrated circuits and these are considered further in more detail in chapter 14. In the remainder of the present chapter some of the developments that have taken place using film circuits will be briefly reviewed.

13.2. Thin film integrated circuits

Early work by Weimer[4, 5] and his associates at R.C.A. sought to use TFTs in amplifier and simple logic circuits. Examples are given in Figs. 13.1 and 13.2 of a three-stage DC amplifier and a NOR logic circuit. In both cases the thin film layers are all deposited on glass substrates by successive evaporation during a single pump down of the vacuum system. Events and progress in the silicon integrated circuit field have now made this type of simple thin film circuit uneconomic even if high yields and good stability for the TFTs could be achieved. As a result the current emphasis of the thin film circuit designers is concentrated on image scanning devices where very much larger substrates involving thousands of individual circuit components are needed[6, 7]. This line of work will now be considered further.

Present day image scanning systems tend to use electron beam scanning techniques both in the camera and for the display. An

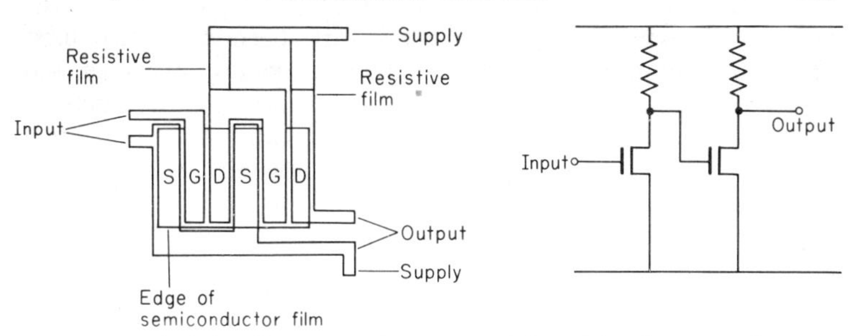

Fig. 13.1. A thin-film version of a two-stage DC amplifier shown at right

Note. *The illustrations in this chapter use a nonstandard symbol to represent* n-*channel or* p-*channel TFTs.*

This symbol has been used in the majority of the papers referenced in this chapter.

alternative arrangement would be to use a matrix array of light sensitive or light emitting elements which are electrically scanned by means of a pulse generator connected to the X and Y co-ordinates connecting strips. The coincidence of pulses from the horizontal and vertical scan generators selects individual elements of the array in sequence in a manner similar to that of a scanning electron beam. In order to obtain a picture having good resolution an array of elements $\simeq 500 \times 500$ would be necessary together with the accompanying scanning circuitry. Fig. 13.3 shows a diagram of part of a solid state image panel where the elements may be either light emitting or light sensitive. The TFT provides a suitable active device for use in the scan generators enabling the entire panel to be made by vacuum deposition of thin films.

Fig. 13.4 shows a scan generator circuit[6] suitable for driving the address strips of the image panel utilising TFTs. Fig. 13.5 is a thin film version of a 180 stage thin film scan generator which has a centre–centre spacing between stages of 50 μ. It is evaporated

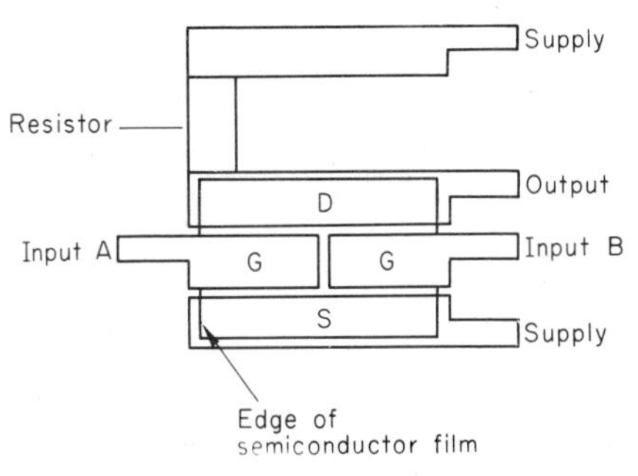

Fig. 13.2. A thin-film NOR *circuit*

on a glass slide. This technique has now been developed using superior masking techniques so that stage spacings of 50 μ can be achieved in a 264 stage thin film shift register suitable for scanning a 256 × 256 element image sensor array[7]. The shift register is based on the circuit of Fig. 13.6 which consists essentially of two sets of complementary inverters connected across the bus bars $GN_1 - V_A$ and $GN_2 - V_B$ with an *n*-channel transistor acting as the transmission gate from the output of one inverter to the input of the

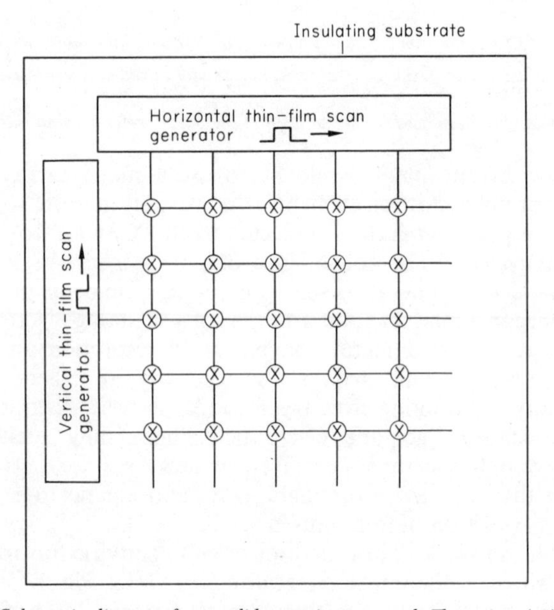

Fig. 13.3. Schematic diagram for a solid state image panel. The points (×) represent light detectors or emitters

next. Fig. 13.7 shows the compact thin film layout used for making the inverters using tellurium for the *p*-type transistors and CdSe for the *n*-type. It was found that the fabrication techniques for both types were entirely compatible and could all be carried out during one pump down of the vacuum system.

An alternative to the shift register approach is to use multistage decoders for scanning the photo-sensitive arrays. Fig. 13.8 shows two types of thin film decoder used by Sadasiv *et al*[7], and Fig. 13.9 shows a photomicrograph of a 256 stage TFT decoder for horizontal scanning. This incorporates also a single line 256-element photosensor and video coupling circuit to facilitate the testing of the unit.

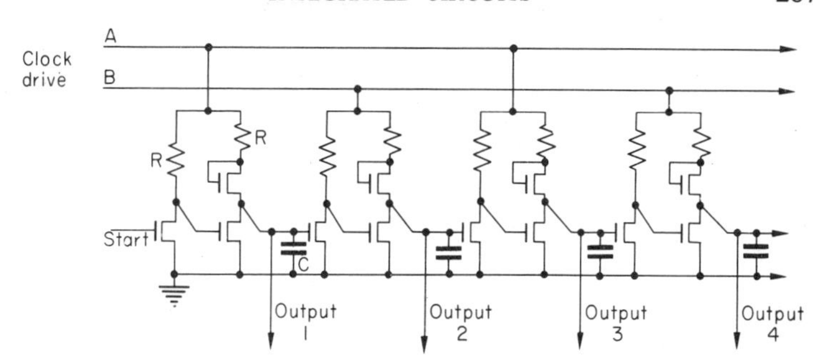

Fig. 13.4. A scan generator circuit based on TFTs suitable for driving 30 and 180 stage image sensor arrays for the type shown schematically in Fig. 13.3[6]

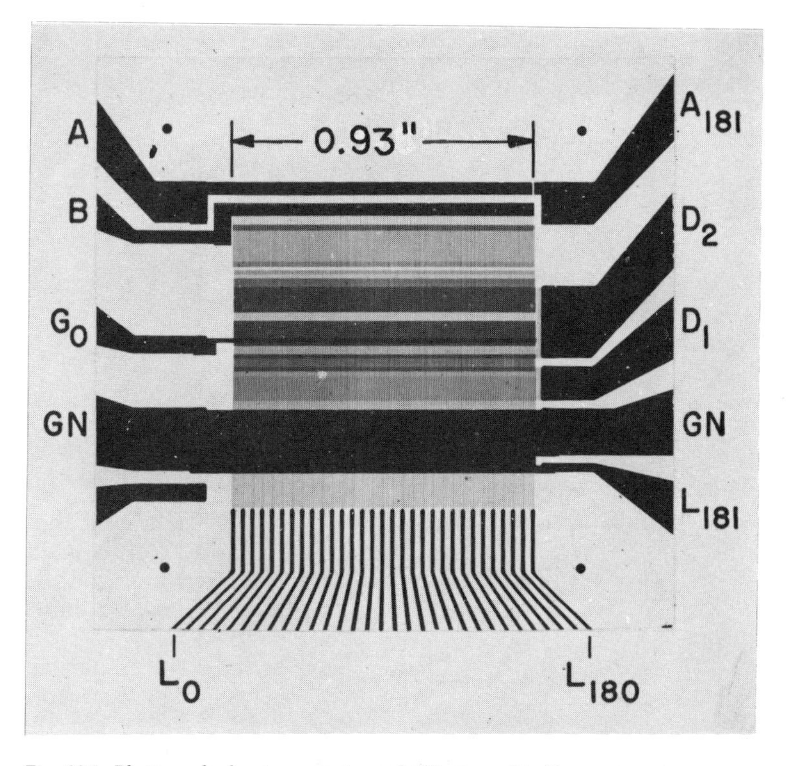

Fig. 13.5. Photograph showing experimental 180 stage thin-film scan generator containing only evaporated thin film components on a glass substrate. The output stages are spaced 50 μ apart[6]

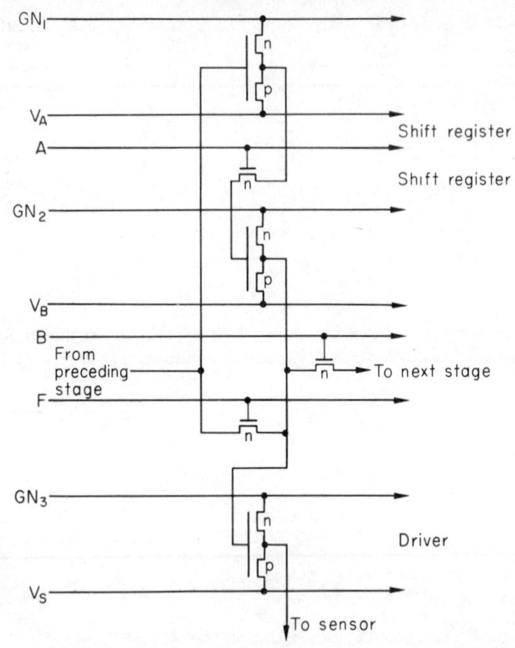

Fig. 13.6. Circuit diagram of one stage of the 264 stage complementary shift register designed by Sadasiv et al.[7]

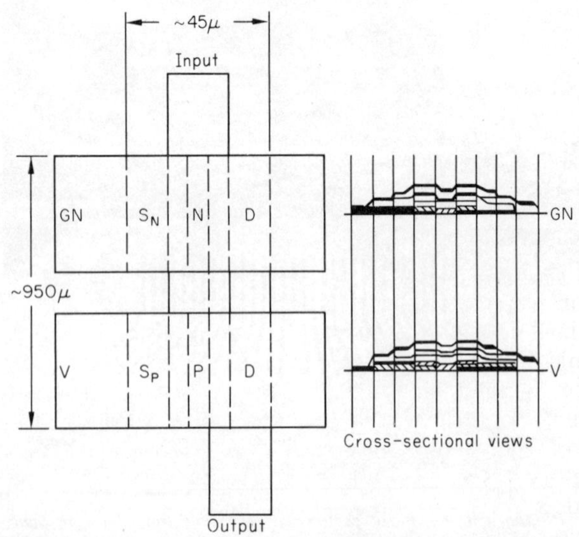

Fig. 13.7. Integrated film layout for the complementary inverter used in the shift register shown in Fig. 13.6 (Sadasiv et al.[7])

All the transistors in this unit are CdSe TFTs and the photoconductor diodes are In-CdS-Te units similar to those used for the 256 × 256 element array. Successful pictures have been reproduced with the integrated decoder line sensor unit though at the present time picture quality and detector sensitivity leave a lot to be desired. At

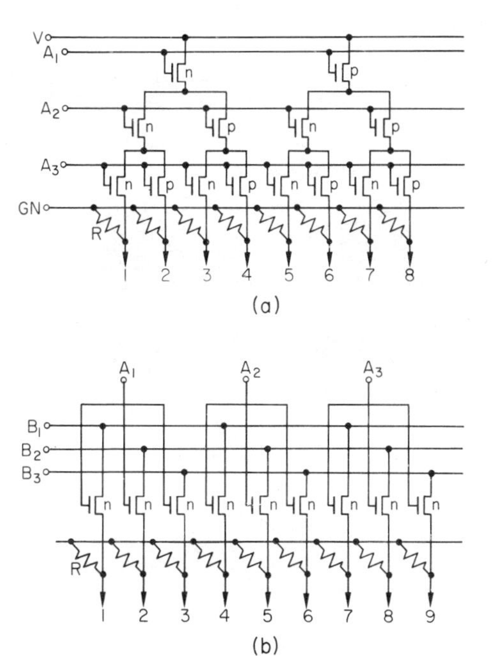

Fig. 13.8. Two types of decoder circuits using TFTs used by Sadasiv et al.[7] *for driving scanning arrays.*

(a) *a three level eight output decoder.*
(b) *a single level nine output decoder*

present it is being claimed that a complete solid state camera unit based on a 500 × 500 sensing array could be housed in a volume approximately 50 mm cube. This would include all necessary electronics but exclude the lens system.

A recent development in the field of thin film active circuits is the replacement of the rigid ceramic or quartz substrate by a wide variety of flexible substrates such as mylar tapes, anodised metal foils, cellulose acetate films, etc.[8]. The advantages of using these materials are the possibility of an almost continuous production process rather than the present small batch process thereby greatly reducing cost, and the possibility of considerably increasing the

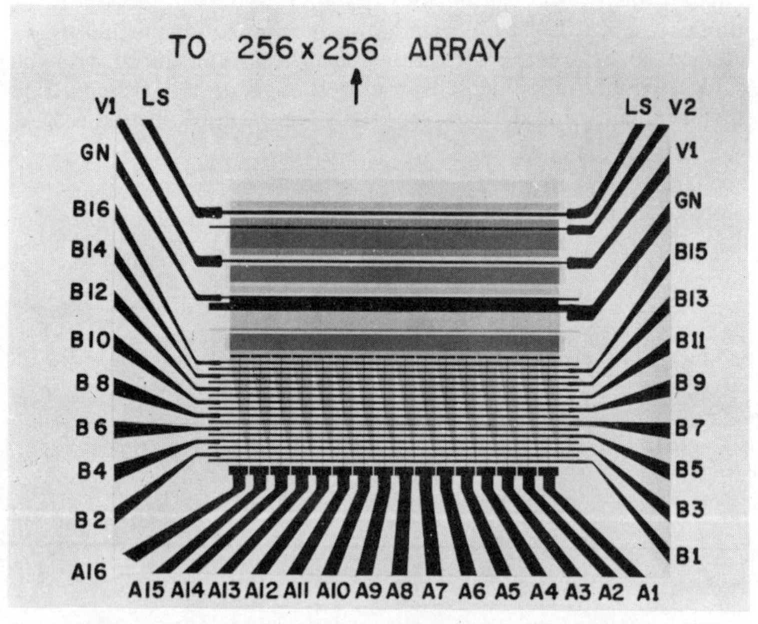

Fig. 13.9. Photomicrograph of a 256 stage transistor decoder. This circuit incorporates an integrated single line 256 element photosensor array and video coupling circuits for test purposes (Sadasiv et al.[7])

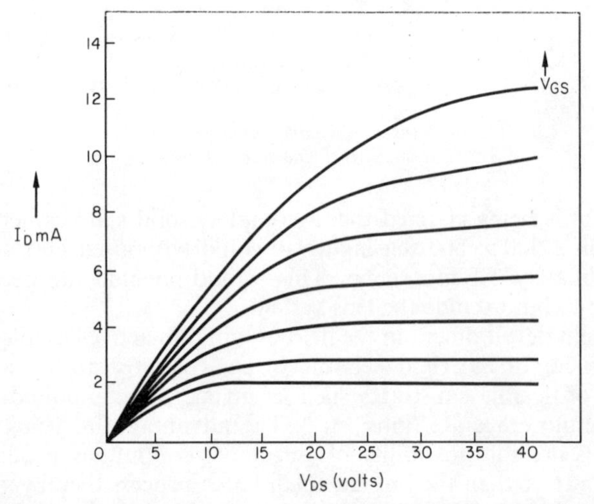

Fig. 13.10. Source–drain characteristics of a half watt flexible thin-film transistor[8]

power dissipating performance of the TRTs. Brody and Page[8] have reported arrays of stable TFTs using tellurium to form enhancement mode devices with various substrates including thin card. Gain bandwidth products of 60 MHz have been achieved with the prediction that 200–250 MHz will soon be possible. The best transconductance quoted is 6000 micromhos at 4 mA drain current. Similar transistors with channel dimensions 12×750 μ on anodised aluminium foil substrates have dissipated a watt of power and others have been operated at source drain voltages >200 V. Fig. 13.10 shows a set of characteristics for a $\frac{1}{2}$ watt device.

Applications for these relatively primitive thin film active circuits on cheap substrates are thought of to lie in various types of mass produced consumer goods where electronics is now not found. Circuits such as cascade amplifiers and oscillators formed on paper, for example, could be used in toys, novelties, etc., when high powers, frequencies and temperatures are not required but where cost is all important.

13.3. Thick film circuits

The possibilities of evaporating circuits by the yard in a semi-continuous form leads naturally to consider the formation of circuits by printing techniques. These are usually referred to as thick film circuits and up to the present time consist entirely of passive components with active devices added either in the form of flip chips, beam lead packages or some form of wire bonded package. Thick film transistors have been reported by Wilkin *et al.*[9] having gain bandwidth product <1 MHz and a transconductance $\sim 0{\cdot}5$ mA/V. Hopes of a large improvement in these values are not high and perhaps the most that can be expected is an order of magnitude increase. One of the principal difficulties lies in forming a satisfactory semiconductor layer using compatible printing and firing techniques. Wilkin *et al.* have used CdS, and CdSe and mixtures of the two and other workers[10, 11] have used CdS, CdSe, ZnS and PbS. It is obvious that active devices of thick film type will be unlikely to supersede high-speed, high-performance bipolar and field-effect transistors now available. There may, however, be specialised applications where the poor performance and low frequency of operation can be tolerated and where the reduction in cost of this type of the circuit is of paramount importance.

REFERENCES

1. Sargrove, J. A. 'New methods of radio production', *J. Brit. I.R.E.*, **7**, No. 1, 2–33 (January–February 1947).

2. Holland, L. (editor). *Thin film microelectronics*, Chapman and Hall (1965).
3. Holland, L. Proceedings of the Joint Conference on Thick Film Technology, *I.E.R.E.* (London 1968).
4. Weimer, P. K. 'The TFT—A new thin film transistor', *Proc. I.R.E.*, **50**, 1462–9 (June 1962).
5. Weimer, P. K. 'Evaporated circuits incorporating a thin film transistor', *Digest International Solid State Circuits, Philadelphia* (February 1962).
6. Weimer, P. K., Sadasiv, L., Meray-Horvath and Homa, W. S. 'A 180-stage integrated thin-film scan generator', *Proc. I.E.E.E.*, **54**, 354–360 (March 1966).
7. Sadasiv, G., Weimer, P. K. and Pike, W. S. 'Thin film circuits for scanning image sensor arrays', *IEEE Trans. on Electron Devices,* ED–15, 215–9, (April 1968).
8. Brody, P. and Page, D. 'Flexible thin-film transistors stretch performance, shrink cost', *Electronics* **41**, 100–3 (August 1968).
9. Wilkin, G. A., Mytton, R. J. and Olsen, G. H. 'Thick film transistors', *Proceedings of the Joint Conference on Thick Film Technology*, 93–8, London (April 1968).
10. Sihvonen, Y. T., Parker, S. G. and Boyd, D. R. 'Printable insulated gate field-effect transistors', *J. Electrochem. Soc.*, **114**, 96–102 (January 1967).
11. Witt, W., Huber, F. and Laznovsky, W. 'Field effect transistors based on silk screened CdS layers', *Proc. IEEE,* **54**, 897–98 (June 1966).

14

THE SILICON MONOLITHIC INTEGRATED CIRCUIT

14.1. Introduction

As indicated in the previous chapter, the basis of monolithic circuit production is the formation of all the circuit components simultaneously by means of a series of processes which include thermal growth of oxide, photo-etching of mask patterns in the oxide, and impurity diffusion through the mask apertures so formed. Epitaxial deposition of an additional layer of silicon is a normal preliminary to the process.

No attempt will be made here to give a detailed account of the manufacturing processes involved in monolithic technology, as excellent detailed descriptions are already available elsewhere[1,2] It will be sufficient for present purposes to note the general ground rules which dictate the overall economics of the system.

14.1.1 RELATIVE COSTS PER AREA

Since all the circuit components are fabricated in a single sequence of processes, the whole of the surface of every silicon slice is subjected to the same sequence of epitaxy, oxidation, photoengraving, and diffusion processes. Thus the cost per unit area of processed slice is constant, regardless of what particular component is being fabricated. It follows that the cost of individual components is just proportional to the surface area that they occupy.

Thus the most expensive components of all are inductances, which would have to be fabricated in the form of flat spirals and would occupy a very large area, even for very small inductance values. Indeed, so expensive is inductance that it is not used at all in integrated circuits at present in production, although its use is being considered in certain experimental monolithic circuits now being designed for the microwave frequency spectrum. Here, the values required are only of the order of a nanohenry, and consequently the cost is not excessive.

Capacitance is also expensive, although not as much so as inductance, in the values commonly encountered. As with inductance, the area occupied by a capacitor is proportional to its value, so use is restricted, so far as possible, to small values. Resistors are rather cheaper than capacitors; provided that values in excess of a few kilohms are not required they can be specified freely in monolithic circuits, but again the area usually increases with value specified, so large values are avoided. Bipolar transistors are cheaper still, being equivalent in area to resistors of about two kilohms in one widely used monolithic process. Thus this component, which is one of the

Table 14.1

Component	*Value*	*Relative area*
Inductor	1 nanohenry	1
Capacitor	10 picofarads	1
Resistor	2 kilohms	1
Bipolar transistor	—	1
MOST	—	$\frac{1}{3}$

more expensive in discrete component technology, would be specified in preference to a passive component, if circuit considerations permitted, in the integrated circuit.

Diodes are normally formed as transistors and the base and collector are subsequently shorted together by the interconnecting metallisation. (Although this is the most common way of reducing a triode to a diode function there are other possibilities which may be advantageous in particular circumstances).

The junction FET is about the same in area as a bipolar device, and will consequently cost about the same, although processing cycles adopted for bipolar device fabrication are not well suited to junction FET production, since the optimum resistivity in the channel is higher than that normally considered suitable as the base diffusion of a bipolar device. The cheapest component of all, however, in the silicon monolithic integrated circuit is the MOST insulated-gate field-effect triode. Due to its great structural simplicity it requires only one third the area corresponding to that needed for a bipolar or junction gate device.

These relative area requirements are summarised in Table 14.1, from which it will be seen that the most economical circuit of all would be one fabricated entirely from MOSTs. Fantastic though such an idea might seem to the circuit designer accustomed only to discrete-component technology, its practicability has now been demonstrated by the commercial success of several integrated circuits based on just such a design procedure.

14.1.2 CHOICE OF THE ENHANCEMENT MODE DEVICE

MOSTs can be fabricated in either enhancement or depletion mode form, as explained in chapter 3. For integrated circuit applications attention has been concentrated principally on the enhancement mode device, for two reasons. The first reason is that overwhelmingly the devices will be used in the common source configuration in both digital and analogue applications. This must necessarily be so, since this configuration gives the desirable high input impedance with greater than unity voltage gain, which is what the circuit designer will usually require.

The enhancement mode MOST can be designed so that in the common source configuration its gate and drain terminals are at about the same DC level. This is a great advantage since it means that subsequent stages can be coupled together without the need for interstage passive circuitry to effect a change in DC level. With bipolar devices or depletion mode FETs, whether of insulated-gate or junction type, such interstage circuitry would be virtually unavoidable. Thus, simpler circuit layouts are possible with the enhancement mode device and, in particular, DC logic circuits are feasible. This gives MOST circuits using enhancement-mode devices a considerable economic advantage, within their technical limitations, compared with competing systems.

The second reason for choosing the enhancement mode device is that it is structurally simpler. The channel in such a transistor is induced by the potential applied to the gate, as explained in chapter 3, and therefore does not have to be fabricated by a diffusion process, as is the case for depletion MOSTs. Thus, the number of processing steps is reduced. As a result, processing costs are lower and yields on production are higher, often substantially so.

The cost per unit area of a successfully processed slice will be low, perhaps less than half of the cost per unit area for a bipolar process. Therefore, in addition to its advantage of small area, the MOST also scores in having a lower cost per unit area, particularly in enhancement-mode form.

The overall cost advantage of enhancement MOST circuits compared with bipolar counterparts (or junction FET circuits, for that matter) is at least an order of magnitude. For this reason monolithic silicon integrated circuits of this type represent one of the most rapidly growing sectors of the electronics market at the present time and are likely to go on doing so for the foreseeable future.

14.1.3 ADVANTAGES OF HIGH YIELD CIRCUITS

MOS circuits are often simpler and have fewer components than competing counterparts in bipolar technology, the MOST itself being a relatively simple device. Also, the fabrication of the circuits involves fewer steps, and, therefore, the yield on circuits of this kind is likely to be far higher than for bipolar circuits performing comparable functions. High yield can be exploited to make possible the fabrication of circuits a whole order of magnitude more complex than would be feasible in bipolar form. This can have important system advantages.

The cost of assembling sub-systems together to form the final completed system is obviously less if the sub-systems themselves have greater functional complexity and there are consequently fewer of them. It is also a fact that if the integrated circuit designer is free to consider building more complexity onto the silicon chip he can choose the boundaries of the sub-system that he incorporates on the chip in a way which is more nearly optimal in respect of the design of the whole system. Thus, the chip can be chosen to represent a sub-system of some degree of functional integrity so that the number of connections required between it and the remainder of the system is minimized: it being as far as possible self-contained.

In this way the number of connecting leads required on the integrated circuit encapsulation (or, what amounts to the same thing, the number of bonds to be formed in an unencapsulated system fabrication technology) is minimised. This leads to a saving on labour when it is inserted into the system, reduction in cost of the encapsulation, and improvement in reliability, since every connection to the chip is subject to the possibility of failure. The incorporation of relatively complex system functions on a single chip has many advantages only briefly mentioned here: it is one approach to the technique of large scale integration (LSI) and so far it has been carried out with very much more conspicuous success using enhancement MOSTs than with any other monolithic integrated circuit structure.

14.2. Principles of enhancement MOS integrated circuit design

The enhancement MOST is a voltage-controlled device characterised, when operated in the common-source configuration, by extreme current gain and a voltage gain which can easily be substantially greater than unity.

Its physical properties have been discussed very fully in chapter 3 and particular circuit applications have been considered in detail in other chapters. This section will review principally the factors bearing on MOS circuit design in integrated form, insofar as they differ from corresponding procedures in discrete-component counterpart circuits.

14.2.1 POWER, VOLTAGE AND CURRENT LEVELS

So far as power, voltage and current levels are concerned, MOSTs need not differ very much from bipolar devices. In particular current levels can be very similar, and in earlier generations of devices, generally were.

Field-effect devices are different from bipolar devices, however, in that current levels can be reduced without any penalty in either gain or (in integrated circuits) bandwidth. Whereas reduction of collector current from a bipolar device will result in a reduction of current gain with almost no change in inter-electrode capacitances, by reducing the channel width of the MOST, the standing drain current at a given bias level, forward transfer conductance, and gate–drain and gate–source capacitances are all reduced in the same ratio. The current gain is unaffected and the voltage gain can be retained at its original value by a proportionate increase in the value of the resistive load presented to the FET without change in the DC working conditions or in the upper cut-off frequency. Thus the gain-bandwidth product is conserved, as is the speed of the device considered as a switch.

This fact is exploited in MOS integrated circuits to permit the use of devices having minimal channel width, thus saving both power and chip area. Usually only those MOSTs which are directly connected to the output terminals of the circuit, and consequently have to drive external loads which cannot be scaled in this way, are of the dimensions which would be used for MOSTs designed as discrete components. This scaling down of size cannot be used to the same degree in discrete component technology, because the existence of 'parasitic' stray capacitances due to the encapsulation and particularly the circuit wiring prevent the gain-bandwidth product from being conserved.

The total circuit capacitances, comprising those of the device added to the strays, decrease less rapidly than does the forward transfer conductance when the dimensions of the MOST are reduced. In the integrated circuit the dimensions of the interconnections are so small that their stray capacitance is negligible at the sizes to which MOSTs are currently being fabricated.

MOSTs are at a disadvantage relative to bipolar devices in respect of voltages required to switch them, and this has an important consequence for the speed of digital systems fabricated in the two forms. Bipolar devices will typically switch from the OFF to the ON state with a change of potential applied to the base of as little as half a volt. MOSTs at present being fabricated require several volts swing applied to the gate to achieve the same function. This voltage swing is produced by routing a current to the circuit node concerned which charges the total capacitance at the node through the voltage swing required. Thus the switching time is proportional to the product of the required swing and the ratio of node capacitance to current flowing to the node. This last ratio is characteristic of the active devices being used. With comparable fabrication techniques the ratio is about the same for *npn* bipolar devices and *n*-channel FETs, whilst for *p*-channel FETs and *pnp* bipolars it is larger by a factor of almost two. Thus the result is that as compared with *npn* bipolar devices, *n*-channel MOSTs have a longer switching time roughly in the ratio of the voltage swings needed for switching, whilst *p*-channel devices are substantially worse.

In practice, *n*-channel MOST digital circuits are designed with clock rates slower than those for *npn* bipolar circuits by a factor of at least ten, whilst for *p*-channel devices the ratio is nearer twenty, a clock rate of 10 MHz being about the best currently feasible in the latter case. This is the principal, and some would say the only, disadvantage of MOS integrated circuits for digital applications. It must be appreciated that there is a wide range of digital systems outside the central processors of large computers where this speed disadvantage is of little or no importance. In industrial controls, automatic instrumentation, electronic telephone switching, and many optoelectronic applications, the speed of the system is determined by other factors than the speed of the electronics. In these and similar cases the cost advantage of MOSTs is very much more important than their speed deficiency.

In linear applications the voltage swing required is often much less than that required to switch the device, so that a direct comparison of the frequency limitations of MOST and bipolar transistors is less easy. In many typical applications there appears to be little difference between what can be achieved in this respect with the two families of devices. However, the relatively large pinch-off voltage of MOSTs is a disadvantage in another respect, since this represents the lowest value which the drain–source voltage can be allowed to assume if reasonably linear operation is to be retained. It represents the 'bottoming' voltage of the device.

When the device is used as a power amplifier this relatively large minimum voltage across the transistor may be a substantial fraction of the supply voltage, and thus the power efficiency of the amplifier is poor. Since the total power that the integrated circuit can dissipate is anyway likely to be quite small, this poor output stage efficiency severely limits the power which a MOST amplifier can deliver to an external load. A typical contemporary figure is a few tens of milliwatts.

14.2.2 NOISE AND DRIFT

Another limitation in the use of MOST circuits in linear (analogue) systems, arises at the other extreme of signal amplitude where noise and drift become important. As mentioned in chapter 4, the MOST is a relatively noisy device in that part of the frequency spectrum where the $1/f$ noise is significant. To put the matter in another way, the corner frequency at which $1/f$ noise becomes significant relative to white noise is relatively high for these devices, of the order of megahertz, compared with a few hundred hertz for good bipolar devices and less than one hundred for junction FETs. Thus, signals in the audio and low radio-frequency range and of less than a millivolt in magnitude are likely to be seriously degraded. The problem has already been considered quantitatively in chapter 4. In circuit terms its effect has been to inhibit the development of high gain operational amplifiers using MOS technology.

The relatively poor long term stability of device characteristics which was typical of early MOSTs had a similar effect, since it was impossible to guarantee the very low levels of long term drift required in DC amplifiers designed to operate with low-level input signals. Great improvements in characteristic stability have resulted from a growing understanding of the physical principles governing surface devices like the MOST in recent years, and the point is now being reached where the drift levels achieved are acceptable for the less critical applications. It is perhaps as well to emphasise that it is not temperature dependent drift which is in question here: this is no worse in insulated-gate than in junction FETs, and can be significantly less than in bipolar devices. Rather it is a longer-term change in device properties, particularly threshold voltage, due to physical changes which may be permanent or may depend on the immediate past history of the device.

For example, applying a large voltage to the gate will produce a change in threshold voltage which may however, be reversed in part or wholly by applying a voltage to the gate opposite in polarity to that originally applied. This effect has already been discussed in chapter 3. It is not large enough, certainly with modern MOSTs, to effect digital circuits, but is obviously important in small-signal DC amplifiers. In AC amplifiers, even of high gain, the effect is not important since the working point can be stabilised by the application of overall DC negative feedback.

14.2.3 SUMMARY

From what has been indicated here of the properties of MOSTs in integrated circuit applications, it will be appreciated that they are complementary to bipolar devices, and by no means wholly replace them. MOST circuits can be produced at lower cost and with better yield, but cannot achieve the same speed as switches and are less suitable for use as power amplifiers.

In small-signal applications, bipolar devices or junction FETs are superior both in respect of drift and $1/f$ noise, although in respect of the last it is worth observing that the $1/f$ noise generator associated with the MOST can be represented almost perfectly by a noise voltage generator in series with the gate. Its effect is thus least when the signal source impedance is very high. At signal source impedances of the order of several megohms the $1/f$ noise ceases to be significant and the MOST would then be at least as good in respect of noise as any other contemporary active device.

This is of practical significance in some of the optoelectronic devices, which will be discussed subsequently, where the signal source, being a reverse-biased silicon junction photoconductive cell, has a very high internal impedance indeed.

14.3. MOST logic circuits

The system of logic invariably used in MOST circuits is a direct-coupled system closely related to bipolar DCTL. Considering p-channel devices operated with negative logic (that is with a negative voltage representing a logic ONE-state and a zero or positive voltage representing the ZERO-state) the most widely used circuits are the NOR and NAND of Figs. 14.1(*a*) and (*b*). Although circuits are built with a resistor load as shown, if it is desired to use MOSTs of small channel width at low current levels the resistor values become inconveniently large, with consequent wasteful use of chip area. In this case the load is replaced by another MOST, as shown in Fig. 14.2 for a simple inverter (NOT).

In this and all subsequent integrated MOST circuits the substrate connections of the MOSTs are not shown. They are, in fact,

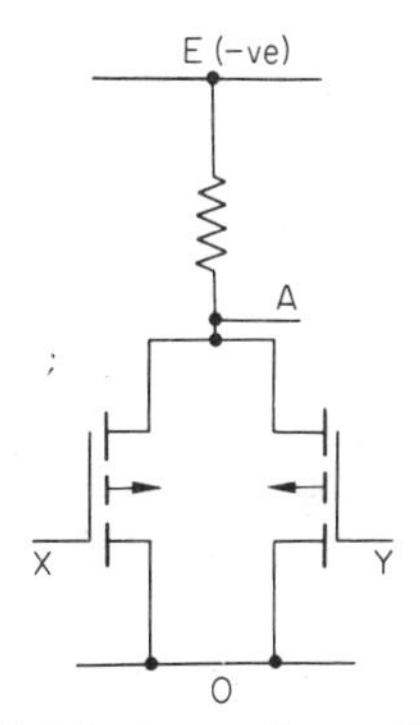

Fig. 14.1(a) A two input NOR for negative logic using p-*MOSTs. For logic signals in which '1' is represented by a negative voltage greater than* V_p *this circuit has the logical property* $A = (x + y)'$

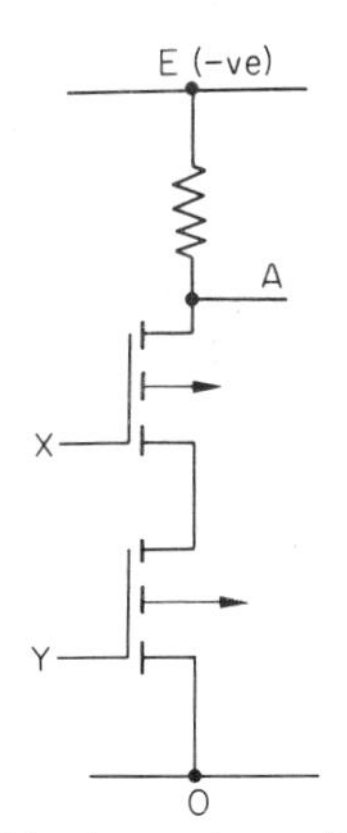

Fig. 14.1(b). A two input NAND for negative logic. $A = (X . Y)'$. *Subject to the conditions for Fig. 14.1(a)*

connected to a point in the circuit which will ensure that the substrate-channel junction is reverse-biased. Thus with *p*-channel devices the substrate is usually connected to the most positive of the supply rails and for *n*-channel circuits the corresponding connection is to the most negative of the supply rails. This is all that is necessary to ensure isolation of the FET from the rest of the circuit, and contrasts favourably with the complex isolation arrangements in bipolar integrated circuits.

At first sight this circuit looks unsatisfactory, since the load MOST has its gate and drain tied together to a constant potential while the source terminal is used as the load point. The impedance seen at this point is just the output impedance of a common drain amplifier and is thus not far from $1/g_{fs}$, which is a relatively low value. However, the integrated circuit designer has complete freedom to choose the g_{fs} of the device, within wide limits, by choosing the width of the channel. Thus provided that the width of the channel of the device used as a load is considerably less than that of the two input devices the voltage gain can be made greater than unity, as required to ensure complete switching of all gates in the system. The ratio of widths will also determine the voltage drop

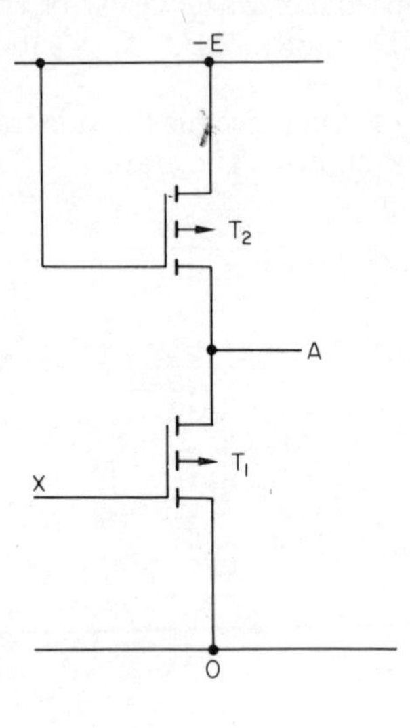

Fig. 14.2. An inverter using a MOST load, for negative logic A = X′

This circuit is also that of a common source ALMOST amplifier

across the active MOST in the ON state. Thus from equation (3.14) (section 3.2.2):

$$I_D = -\frac{\beta}{2}(V_{GS} - V_{th})^2$$

$$g_{fs} = \frac{\partial I_D}{\partial V_{GS}} = -\beta(V_{GS} - V_{th})$$

Here β and V_{th} are characteristic of the transistor concerned, but in general, provided that the same thickness of insulating oxide is used under the gate electrode the value of V_{th} will be almost identical for MOSTs within the same monolithic circuit. Letting β_1 be the value of β for the active MOST whilst β_2 refers to the load device, the voltage gain is just

$$A_V = -\frac{g_{fs,1}}{g_{fs,2}} = -\frac{\beta_1(V_{GS,1} - V_{th,1})}{\beta_2(V_{GS,2} - V_{th,2})}$$

where, again, $g_{fs,1}$ and $V_{GS,1}$ refer to the active MOST; $g_{fs,2}$ and $V_{GS,2}$ to the load. The two devices are in series and hence must conduct the same drain current, so, substituting into the expression relating drain current to V_{GS},

$$\frac{\beta_1}{2}(V_{GS,1} - V_{\text{th},1})^2 = \frac{\beta_2}{2}(V_{GS,2} - V_{\text{th},2})^2$$

Hence it follows that

$$A_V = -\left(\frac{\beta_1}{\beta_2}\right)^{\frac{1}{2}} \tag{14.1}$$

This expression for voltage gain can be related more directly to device parameters by noting that

$$\beta \propto \frac{W}{L}$$

where W is the width and L the length of the MOST channel. For devices within the same monolithic chip the constant of proportionality will be the same, thus

$$\frac{\beta_1}{\beta_2} = \frac{W_1}{W_2} \cdot \frac{L_2}{L_1} \tag{14.2}$$

However it is usual to make L as small as the process accuracy will permit, since in this way the shortest possible switching time is attained. Hence L will usually have the same value for all devices on the chip. In this event the right-hand side of equation (14.2) becomes just the ratio of widths and hence

$$A_V = -\left\{\frac{W_1}{W_2}\right\}^{\frac{1}{2}} \tag{14.3}$$

This is an extremely important equation, since it is clear that values of A_V greater than unity are possible. In considering linear circuits, subsequently, further important implications will be drawn. However, for digital circuits, equation (14.3) is not sufficient to determine the actual value of the ratio W_1/W_2. This is set by the magnitude of the voltage across T_1 when ON. Since under these conditions T_1 will be bottomed, its drain current is given not by equation (3.14) but by equation (3.7), namely

$$I_D = -\beta\{(V_{GS} - V_{\text{th}})V_{DS} - \tfrac{1}{2}V_{DS}{}^2\}$$

However, before the value of V_{DS} in the ON state can be calculated it is necessary to investigate the gate voltage likely to be applied in a typical logic network.

Consider first the magnitude of the supply voltage. When a gate is in the OFF state, the active transistor, T_1 is not conducting and hence the voltage at its drain terminal, which is also the source terminal of T_2, will go negative. However, if the supply voltage is E, the output voltage cannot go more negative than $(E - V_{th})$ since the gate-source voltage of T_2 must be at least equal to V_{th} for conduction. This voltage is applied to the succeeding MOST, which conducts a current determined by the extent to which the magnitude of the gate–source voltage applied exceeds V_{th}. Thus the effective voltage $|(V_{GS} - V_{th})|$ for this transistor is $|(E - 2V_{th})|$. This must be greater than zero for conduction, hence

$$E = mV_{th} \tag{14.4}$$

where $m \geqslant 2$.

The equality in the above condition would correspond to the following transistor only just going into the ON state, thus the drain current available to charge circuit capacitances would be vanishingly small, and the switching time would tend to infinity. Usually $m = 3$ or more is chosen to give rapid switching. Since V_{th} is typically -5 V, supply rails of -15 to -20 V are common. However, for more recently available MOSTs the value of V_{th} is only some -1 to -2 V, and logic circuits will thus work satisfactorily with a supply of -5 V or so.

Substituting $(E - V_{th})$ for $V_{GS,1}$

$$I_{D,1} = -\beta_1\{(E - 2V_{th})V_{DS,1} - \tfrac{1}{2}V_{DS,1}{}^2\}$$

$$= -\beta_1\{(m - 2)V_{th}V_{DS,1} - \tfrac{1}{2}V_{DS,1}{}^2\}$$

If it can further be assumed that E is more negative than $V_{DS,1}$ when T_1 is in the ON state, as is reasonable, then both the gate–source and drain–source voltages for T_2 are nearly equal to E, hence T_2 is saturated and its drain current is

$$I_{D,2} = -\frac{\beta_2}{2}\{E - V_{th}\}^2$$

$$= -\frac{\beta_2}{2}\{(m - 1)V_{th}\}^2$$

Equating $I_{D,1}$ and $I_{D,2}$

$$\beta_1\{(m - 2)V_{th}V_{DS,1} - \tfrac{1}{2}V_{DS,1}^2\} = \frac{\beta_2}{2}\{(m - 1)V_{th}\}^2$$

Since $(m - 2)V_{th}$ is much more negative than $\tfrac{1}{2}V_{DS,1}$ in all real cases, to good approximation

$$\beta_1\{(m - 2)V_{DS,1}\} = \frac{\beta_2}{2}(m - 1)^2 V_{th}$$

or

$$V_{DS,1} = \frac{\beta_2}{2\beta_1} \frac{(m-1)^2}{(m-2)} V_{th} \tag{14.5}$$

For example if $m = 3$, $(\beta_2/\beta_1) = 0{\cdot}1$,

$$V_{DS,1} = 2\frac{\beta_2}{\beta_1} V_{th}$$

$$= 0{\cdot}2\, V_{th}$$

For $V_{th} = -5$ V the drain–source voltage of T_1 would thus equal -1 V.

The actual value of drain–source voltage in the ON state which can be accepted depends on the circuit details. Consider a NAND circuit (Fig. 14.3). There are n-MOSTS in series to give an n-input

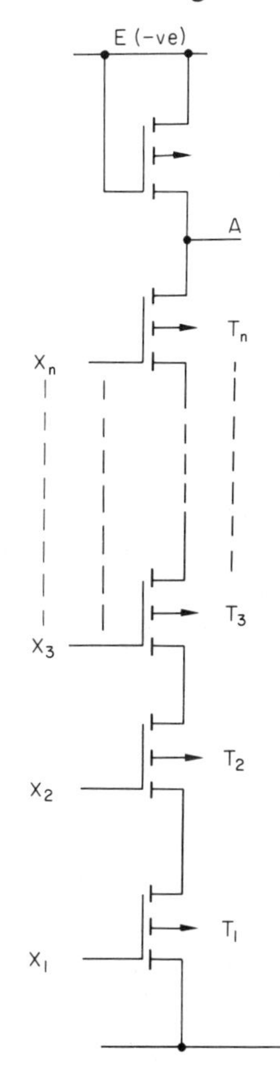

Fig. 14.3. An n*-input NAND for negative logic* $A = (X_1 . X_2 . X_3 + \ldots X_n)'$

NAND gate with MOST load. Provided that V_{DS} in the ON state is much less negative than E (so that the fact that T_2 is forward-biased by a voltage V_{DS} less negative than that applied to the gate of T_1, T_1, T_3 by a voltage $2V_{DS}$ less negative than that applied to $T_1 T_1$, and so on, can be neglected) the maximum value of n for a

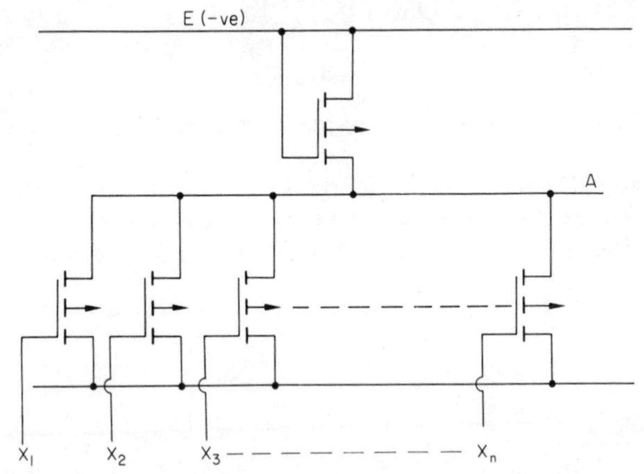

Fig. 14.4. An n-*input NOR for negative logic* $A = (X_1 + X_2 + X_3 + \ldots X_n)'$

given noise immunity can be calculated. If the following gate is to remain in the OFF state with $T_1 \ldots T_n$ in the ON state and an additional noise signal V_N, then

$$|nV_{DS} + V_n| < |V_{th}|$$

or

$$n < \left|\frac{V_{th} - V_N}{V_{DS}}\right|$$

Hence, substituting the value of V_{DS} from equation (14.5),

$$n < \left|\frac{2\beta_1}{\beta_2} \cdot \frac{(m-2)\{1 - V_N/V_P\}}{(m-1)^2}\right| \qquad (14.6)$$

This equation defines the maximum permissible fan-in for a NAND gate. With noise of the order of one volt, a fan-in of three or four is quite easily achieved. When high fan-in is desired, however, the NOR configuration is preferable (Fig. 14.4). In this case when one or more of the lower transistors is conducting, the voltage at the output is V_{DS} or less, and no fan-in limitation results from output voltage considerations. The condition for noise immunity is just

$$|V_N| < |V_P - V_{DS}|$$

or

$$\left|\frac{V_N}{V_C}\right| < 1 - \left|\frac{\beta_2}{2\beta_1} \cdot \frac{(m-1)^2}{(m-2)}\right| \tag{14.7}$$

and circuits can easily be made immune to noise of several volts. In this case the fan-in ratio of the NOR circuit, like its fan-out ratio and the fan-out ratio of the NAND, is determined solely by the degradation of gate switching time due to capacitance loading; an effect to be considered further subsequently.

14.4. Logic circuit speed

In calculating the waveforms encountered in MOST logic circuits, it is not necessary to consider transit time effects within the devices, which are negligible compared with the slowing down effect of circuit capacitances. Thus the switching time of a MOS gate can be found by the solution of a capacitor charging problem; complicated, however, by the fact that the charging current is neither constant nor derived from an ohmic source, and the capacitors themselves will to some extent be voltage dependent. Since this last effect is much smaller than in corresponding junction devices it is usually neglected in approximate treatments of the MOST case.

If the initial conditions and the gate voltage waveform applied to the active MOST are known, as is the voltage–current characteristic of the MOST connected as a load, the difference between the current supplied by the active device and that taken by the load is available to charge the capacitance at the node concerned. Because of the non-linear current–voltage relationship, analytical solutions are difficult. Results are thus usually obtained numerically. The procedure is to assume that the charging rate remains constant over a very short interval, say 10^{-3} of the total switching time or less. Using this charging rate, the voltage change at the node can easily be determined using the relationship

$$C_n \Delta V = I_{CH} \Delta t \tag{14.8}$$

where ΔV is the voltage change in time Δt at a node having a total capacitance C_n to which a charging current I_{CH} flows. A typical algorithm would be as follows:

(i) Assume that at time t_0 the voltage at the node is known (e.g. that at $t_0 = 0$ the voltage $V_n(t_0)$ at the node is E volts).
(ii) Using this node voltage calculate $I_{CH}(t_0)$ from available data on MOST DC characteristics.
(iii) Using equation (14.8) calculate $\Delta V(t_0)$.

(iv) The value of $V_n(t_1)$, where $t_1 = t_0 + \Delta_t$, is then $V_n(t_0) + \Delta V(t_0)$ to good approximation, and this value may be printed out.

(v) To calculate the next voltage increment a new value of charging current is required. This is the value of charging current averaged over the next Δt, or, very nearly, the value of charging current halfway through the next Δt. To approximate this it is assumed that the voltage will keep on changing very nearly at the same rate so that the node voltage at time $t_{1\frac{1}{2}}$ (halfway between t_1 and $t_2 = t_1 + \Delta t$) is $V_n(t_{1\frac{1}{2}})$ where

$$V_n(t_{1\frac{1}{2}}) = V_n(t_1) + \tfrac{1}{2}\{V_n(t_1) - V_n(t_0)\}$$
$$= V_n(t_0) + \tfrac{3}{2}\{\Delta V(t_0)\}$$

(vi) The mean charging current $I_{CH}(t_{1\frac{1}{2}})$ is then calculated from $V_n(t_{1\frac{1}{2}})$

(vii) Using equation (14.8) calculate $\Delta V(t_{1\frac{1}{2}})$ and hence $V_n(t_2) = V_n(t_1) + \Delta V(t_{1\frac{1}{2}})$, which print out

(viii) Repeat instructions (v) through to (vii) for successive intervals. In the general case

$$V_n(t_{r+\frac{1}{2}}) = V_n(t_r) + \tfrac{1}{2}\{V_n(t_r) - V_n(t_{r-1})\}$$
$$= V_n(t_r) + \tfrac{1}{2}\,\Delta V(t_{r-1})$$

and $I_{CH}(t_{r+\frac{1}{2}})$ is calculated from the MOST DC characteristics and $V_n(t_{r+\frac{1}{2}})$.

(ix) The process is continued until the value of V_n has got reasonably close to its terminal value, as calculated from DC considerations.

(x) If the fact that C_n is voltage dependent cannot be neglected, the constant value of C_n assumed above is replaced by $C_n(V_n)$, a value which depends on the node voltage in a predetermined way. The algorithm is modified in that every time a new $V_n(t_{r+\frac{1}{2}})$ is determined the corresponding value of C_n is calculated, and it is this value which is subsequently used in equation (14.8) to calculate $\Delta V(t_r)$

Generally in *p*-channel MOST inverters of typical contemporary design, delays are of the order of a few tens of nanoseconds. It is reasonable to expect this to get shorter in future, and with the complementary technique (to be described subsequently) a speed increase of half an order of magnitude can be looked for. Where the circuit is more complex than a simple inverter, effects arise due to interconnection which limit fan-in and fan-out if excessive speed penalties are not to result.

Consider a NOR coupled to other similar circuits, as in Fig. 14.5. If the fan-in ratio is p and the fan-out ratio is q, attached to the output terminal will be q connections to other NOR circuits each of which has an input capacitance, say C_{in}. The capacitance load due to this cause is thus $q\ C_{in}$. Also connected to this load are p drain connections, and if associated with each is a MOST output capacitance C_{out}, a capacitance $p\ C_{out}$ will be attached to the node for this reason. In addition there will be a capacitance C_L across the load device, so that the total node capacitance will be

$$C_n = p\,C_{out} + q\,C_{in} + C_L \tag{14.9}$$

All three terms will be voltage dependent although not very strongly so. Since, from equation (14.8), the product $C_n \Delta V$ is a

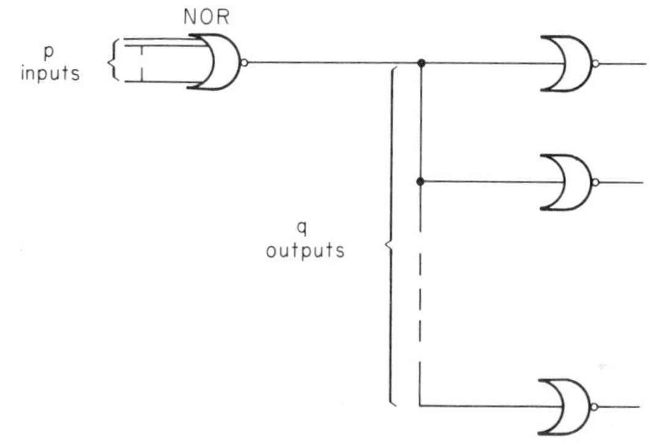

Fig. 14.5. A p-*input NOR feeds* q *following circuits of a similar kind*

constant for a given charging current and time interval, evidently to maximise ΔV, giving rapid switching, C_n must be minimised. This implies restriction of fan-in and fan-out ratios.

It is very difficult to give specific numerical examples in this case because the results are so very dependent on circuit details. It may, however, be instructive to consider the effect of β ratio between active and load devices on circuit speed. For simplicity the discussion will be limited to a simple inverter.

Consider the case where the active device is suddenly cut off by a step voltage applied at the gate. The capacitance at the node is the sum of the gate–drain capacitance of the active device and the capacitance of the load, which will be very nearly C_{GS}. The current available to charge the total capacitance is simply provided by the load MOST and is thus proportional to its width. So also is its C_{GS}, thus increasing the width of the channel of the load MOST will increase the charging current and C_{GS} in the same ratio, but since the C_{GD} of the active MOST is unaltered the total node capacitance will increase less than the charging current and the inverter will switch faster. Similarly, reducing the width of the active MOST will reduce node capacitance leaving the charging current unchanged and will thus also increase speed. Thus, if β_1 refers to the active device and β_2 refers to the load, reduction in the ratio β_1/β_2 will give a faster circuit. Although proved here for only one case this proposition is generally true.

There are, however, two disadvantages in increasing speed by reducing the β ratio. The first is that the drain–source voltage of the active MOST when in the ON state will thereby be increased, in accordance with equation (14.5). Since in order that following circuits shall not incorrectly turn on, it is necessary that

$$|V_{DS,1}| < |V_{th}|$$

it follows from equation (14.5) that this condition is equivalent to

$$\frac{\beta_1}{\beta_2} > \frac{(m-1)^2}{2(m-2)} \tag{14.10}$$

This sets a limit to the minimum β-ratio for a given $m(=$ supply voltage/pinch voltage). Thus for a typical $m = 3$, the β-ratio must exceed two.

The other disadvantage of reduction of β-ratio as a means of increasing speed is that the total current drawn by the gate in the ON state is increased if the width of the load MOST is increased, and in many applications minimisation of power consumption is a design criterion.

14.5. Complementary MOST logic circuits

So far, the logic circuits described use p-channel enhancement MOSTs exclusively. Identical circuits can be designed using n-channel enhancement devices, with the advantage of up to two to one speed increase due to the higher mobility of electrons than holes. Unfortunately the manufacture of n-channel enhancement devices presents problems not yet wholly resolved. Both n- and p-channel depletion devices are possible (although fabrication of the latter is not entirely straightforward either) but are unattractive for digital circuits since they must be operated with a polarity of gate voltage opposite to that at the drain. There would thus be DC level shifting problems between stages which would unnecessarily complicate circuit design.

It is also possible to design circuits utilising more than one type of MOST. For the four types of MOST there are six different ways, for example, in which circuits using two types can be designed. Of these only one is of substantial interest, namely so called complementary MOST logic, which combines *p*- and *n*-channel enhancement devices. This logic system has two main advantages, namely a three-fold increase of speed over *p*-channel systems and the possibility of designing logic circuits which only consume power when changing

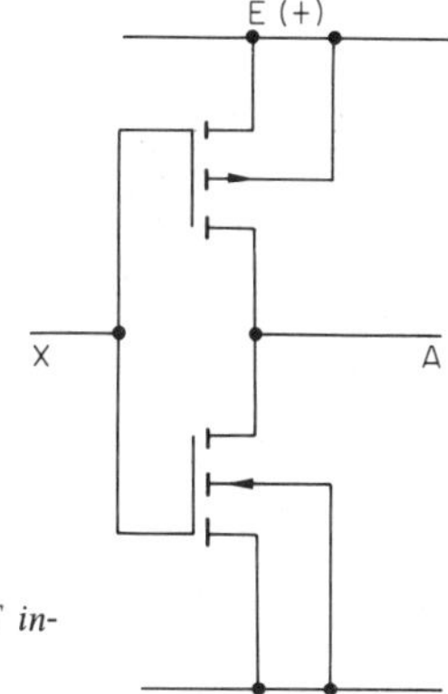

Fig. 14.6. A complementary MOST inverter. A = X′

state, and which consume no power at all in any stable state. Its disadvantage is increased processing complexity, with consequently lower yields and higher cost.

A typical complementary MOST inverter is shown in Fig. 14.6. The collector and base terminals of the FETs are connected together, while the source and substrate terminals are connected to the supply rails as shown. This illustrates the philosophy of connection of the substrates as described previously: the *p*-channel substrate is connected to the positive supply rail and the *n*-channel to the negative. In subsequent diagrams these connections will be omitted for clarity.

Note that the two devices cannot now be classified as active or load (passive), since both are driven by the input signal and each acts as a load to the other. Also the devices each see as a load the drain impedance of a common source amplifier, which is far higher than the source impedance of a common drain amplifier, as in the non-complementary case of Fig. 14.3. The circuit thus has a very high voltage gain indeed, in between its two stable states and, more important for digital circuits, gives a voltage swing at the output terminal which, provided that the circuit is properly designed, is almost equal to the total supply voltage.

If the input voltage to the circuit of Fig. 14.6 is V_X then T_2 is OFF and T_1 is ON provided that

$$V_X > E - |V_{th,2}| \qquad (14.11)$$

and

$$V_X > V_{th,1}$$

where $V_{th,1}$ is the threshold voltage of T_1, the *n*-channel device, and $V_{th,2}$ is the threshold voltage of the *p*-channel, T_2. The alternate state, with

$$V_X < E - |V_{th,2}| \qquad (14.12)$$

$$V_X < V_{th,1}$$

is that in which T_2 is ON and T_1 OFF.

The required values of supply voltage are determined by the need to operate similar circuits which follow this one, thus the output swing must be sufficient for this purpose. Since the OFF transistor passes almost no current regardless of the voltage applied to it (up to its breakdown point), very little current flows through the ON transistor in series with it, provided that current is not drawn from the output terminal. Thus the voltage drop across the ON transistor is negligible, and the output voltage is very nearly O and E in the two extreme states. Evidently the conditions of equations (14.11) and (14.12) will be met provided that E is greater than the larger of $|V_{th,1}|$ and $|V_{th,2}|$. This maximum of the two threshold voltages will be referred to as $|V_{th,max}|$. Then the condition is

$$E = m|V_{th,max}| \qquad (14.13)$$

$$m \geqslant 1$$

As in the non-complementary case, the minimum value of m (unity) would give rise to infinitely slow switching, since the ON transistor would only just be conducting and hence the current available to charge circuit capacitors would be very small. Typically m is chosen equal to two or more, to give fast switching. The procedure for calculating switching transients is exactly comparable with that in section 14.4, and will not, therefore, be repeated here.

Since one of the two transistors in series is always non-conducting, the circuit draws virtually no current in the steady states. However, during the switching process the ON transistor will pass a substantial current in order to charge or discharge the effective capacitance at the output node. This current is computed directly by the procedure of section 14.4 and can be printed out during the computer programme if the transient current demanded by the inverter is of interest.

The circuit consumes almost no power except during the change of state transient. It is therefore suitable for micropower applications, but the power supply must retain a low impedance for transient demand, otherwise defective operation is likely. Due to the low power consumption, extremely high equipment packaging densities are theoretically possible without encountering thermal dissipation problems. This advantage is not achieved in practice, however. Due to isolation diffusions the complementary circuits occupy more area on the monolithic chip than do non-complementary equivalents, so that higher packing densities are not actually achieved.

Complementary circuits do have a speed advantage over ordinary MOST logic using *p*-channel devices. Partly, this is because they embody *n*-channel MOSTs which, due to the superior mobility of electrons compared with holes, are faster than *p*-channel types. More important, however, is the fact that the output node is driven hard in both the turn-on and turn-off transients, whereas this is not so for non-complementary logic.

Consider the simple non-complementary inverter of Fig. 14.2. When the active MOST (T_1) turns on, a large current flows through it to charge the output node capacitance, and, depending on the voltage applied to the gate this current may be much larger than the final value of drain current once the circuit has settled: indeed the gate voltage applied must be sufficient for this to be so, otherwise the active MOST will not bottom satisfactorily in the ON state. The turn-on is thus quite fast. However, at the other extreme of gate potential, the active MOST will turn off and the current available to alter the potential of the output node, by charging or discharging the effective capacitance, will only be the relatively smaller equilibrium current flowing through the load MOST. Thus the voltage change at the drain of T_1 when it turns off is rather slow. This type of switching characteristic is sometimes termed 'pull-on, sag-off switching'

The complementary circuit, by contrast, is a 'pull-on, pull-off' switch. For a transition in either direction one of the MOSTs will be conducting hard. In Fig. 14.6, T_1 will be conducting hard when A is going positive, and T_2 when A is going more positive, the other MOST being cut-off in each case. Thus, in both cases, large currents are available to charge the node capacitances. The speed advantage from this effect is, however, partly offset by the fact that the node capacitance for complementary circuits is typically twice that for non-complementary equivalents. This arises from the greater complexity of complementary circuits, for the usual logic functions, which will be described subsequently.

Typically the overall speed advantage of complementary circuits is above $2\frac{1}{2}$ to 1 compared with *p*-channel non-complementary logic.

14.6. More complicated complementary circuits

The increased complexity of complementary circuits, and hence increased chip area, arises because each input terminal to a logic block drives two MOSTs, turning one on and the other off. Thus, for example, a three input NAND for positive logic will be as in Fig. 14.7.

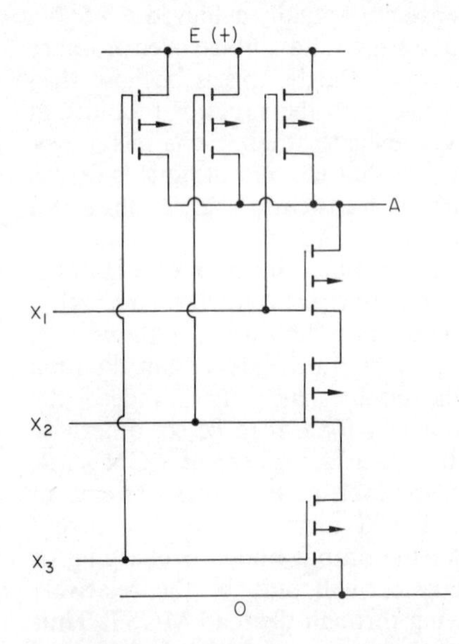

Fig. 14.7. A complementary three-input NAND for positive logic $A = (X_1 . X_2 . X_3)'$

The same circuit will also function as a NOR for negative logic, and if *p*- and *n*-channel MOSTs are interchanged will act as a positive logic NOR or a negative logic NAND. In all cases at least one string of MOSTs are connected in series. However, the voltage drop across ON devices is so small that this does not lead to fan-in restrictions for the reasons which apply in the case of the *p*-channel NAND, described previously. However, the effect on having three devices in series is to increase (by a factor of three) the resistance through which the output node capacitance is charged. Thus there is a speed penalty for too large a fan-in, additional to that already discussed, which results from increasing node capacitance.

Other complicated logical function, combining both AND and OR conjunctions can be realised in an entirely straightforward way.

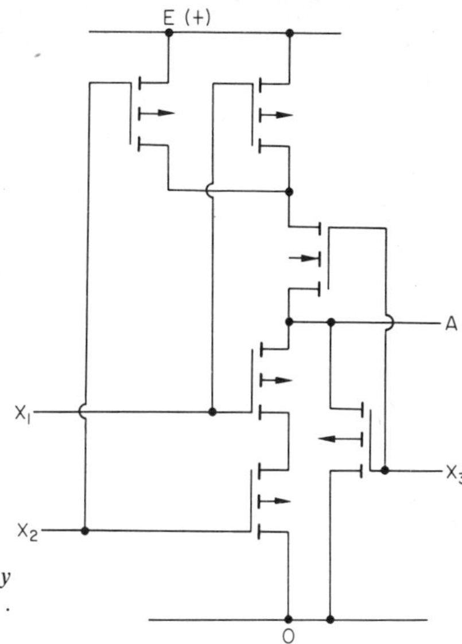

Fig. 14.8. A positive logic complementary circuit having the property $A = (X_1 . X_2 . X_3)'$

For example, Fig. 14.8 shows a circuit for which the output A (assuming positive logic) is related to the inputs X_1, X_2, X_3 by

$$A = (X_1 . X_2 + X_3)$$

If p- and n-channel devices were interchanged (together with the supply rails) the function would become

$$A = \{(X_1 + X_2) . X_3\}'$$

With the output inverted and with a two input NAND circuit to generate X_3 as

$$X_3 = (X_1 . X_2)'$$

the circuit of Fig. 14.8 yields a half-adder or exclusive OR, one of the most useful of logic blocks.

14.7. Dynamic storage

Used as a digital circuit element the MOST has an important property, not so far considered, which is not shared by bipolar devices. This is the possibility of dynamic storage, so called.

Since the gate leakage current of a MOST is so very small, typically less than 10^{-15} A, a charge applied to the capacitor formed by the gate and channel will decay only very slowly. For example, if the gate capacitance were 1 pfd and the leakage 10^{-15} A, any charge applied to the gate would decay at a rate corresponding to only one millivolt per second, or one volt in about sixteen minutes. Thus, although this property obviously could not be used for permanent storage, it does give a temporary storage facility over a period which is very long on the time scale normally encountered in digital circuits.

An example occurs in a shift register. In any shift register design, two storage elements must be present for every bit stored. The main array of storage elements, which must have permanent storage

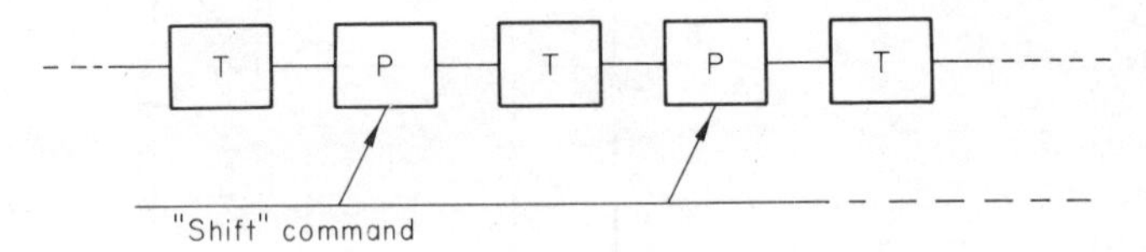

Fig. 14.9. Simplified block diagram of a right-shifting register
T = temporary store
P = permanent store

properties, actually holds the stored bits but is cleared on the receipt of a shift command. Each element is subsequently reset to the previous state of the adjacent element. In this way a given state can be made to propagate along the array by a succession of shift commands. Fig. 14.9, which is a simplified block diagram of a shift register, makes this clear. Essentially, the clearing process must occur before the storage elements can be re-set, hence temporary storage is required to 'remember' the states of the stores before the shift command is received and clearing of the permanent storage elements takes place.

In a dynamic MOST shift register at least, the temporary storage function is performed by a charge applied to the gate. Reliable operation at any normal shift rate is obtained. A typical circuit is shown in Fig. 14.10. The shift pulses are applied at CP_1, and also reach the line CP_2 after a delay long enough to ensure that the pulse applied to CP_2 does not begin until that at CP_1 has finished. Referring to the circuit, if the input is at, say, -10 V, corresponding to a logic ONE-state, T_1 will turn on and when the pulse CP_1 arrives at the gates of T_2 and T_3 a low voltage, logic ZERO, is communicated to the gate of T_4. After CP_1 falls back to zero, T_3, which functions merely as a switch, turns off, leaving the gate of T_4

isolated. The state of T_4 will thus remain unchanged, and when CP_2 appears, turning T_5 and T_6 on, the gate of T_7 will be taken up to a relatively high voltage, say -10 V, corresponding to a logical ONE-stage. The ON state of T_1 has thus been transferred to T_7 by the application of a suitable pulse to the shift terminal. Here T_4 provided the requisite temporary storage during the shifting process. Had the input voltage been initially low (logic ZERO) the transistor T_1 would have been turned off, with the result that a 10 V level would have been applied to the gate of T_4. This transistor in the ON state will switch T_7 into the OFF state on application of CP_2.

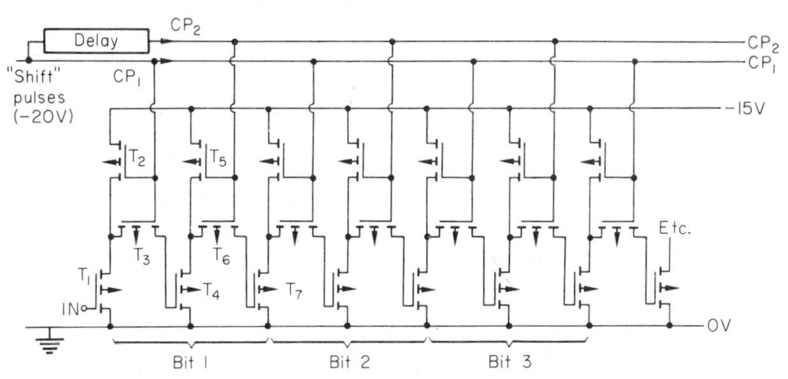

Fig. 14.10. A dynamic storage shift register. The first three bits of storage are shown. Negative logic

Thus the logic ZERO is shifted from the gate of T_1 to that of T_7, just as in the case of a logic ONE.

This particular register uses gate charge retention as both the temporary and 'permanent' bit storage media. This charge decays with time, and at a rate much faster than that calculated above for a disconnected gate. This is because, in addition to normal leakage, the gate capacitors also discharge through the series switching transistors such as T_3 and T_6. Even when turned off, these will pass a significant leakage current, typically up to as much as one nanoamp, depending on ambient temperature and the cross-sectional area of the transistor. The calculation is quite straightforward, involving only consideration of the total leakage current and the effective capacitance to be charged, but leads to typical lower limits on shift rate of about one kilohertz, in circuits, such as that of Fig. 14.10, which use gate charge for both temporary and permanent storage. An alternative family of circuits uses gate dynamic storage as the temporary store, but bistable (flip-flop) circuits for the permanent elements. In this case lower limits on shift rate are eliminated at the cost of greater circuit complexity.

Other digital circuits using dynamic storage are obviously possible, and indeed it is a generally useful technique in circuit design which can often result in substantial reductions in complexity for a given function. Probably even more significant for the future is the application of gate storage techniques in connection with multi-phase logic systems. This approach, which promises substantial advantages in terms of speed of operation and economy of power consumption, is sufficiently significant to justify a section to itself.

14.8. Multi-phase logic

Although all the MOST logic circuits so far described have been illustrated with DC power supplies this is not essential and clocked (pulse) power supplies may be substituted provided that a continuous output from the circuits is not essential. The obvious advantage of such an arrangement is that the mean power consumption is reduced. This is important in itself for many applications, but even where the level of power consumed does not affect system cost significantly it is essential that it should be minimised in order to limit the power dissipated in the integrated circuit chip. For devices of a given packing density, maximum operating temperature is set by the power dissipation in the devices. To put the matter in another way, pulsing the supplies makes it possible to increase the peak dissipation in the devices without affecting the mean, or even perhaps whilst reducing it. Higher peak dissipation implies larger charging currents to circuit nodes and hence faster circuit operation. The gain can be well worthwhile.

Systems of this type use the ordinary circuit configurations of Figs. 14.1 to 14.4, but with the supply, $E(+\text{ve})$, replaced by a negative-going pulsed supply of similar magnitude. However, the same pulsed supply cannot be used for subsequent circuits, to which the first circuit passes its output. This is because, as indicated in section 14.4, the output from the logic circuit does not come instantaneously to its final value. Instead, there is a time delay as the output node capacitance charges. Thus the output reaches its final value some time after the leading edge of the pulsed power supply (the actual value of the delay can be calculated by the numerical approach described in section 14.4). Since this delay is cumulative through successive circuits, if the same pulsed supply were used for all circuits malfunction would occur when the logic pulse delay equalled the width of the power supply pulse.

14.8.1 TWO-PHASE LOGIC

This difficulty is overcome by the use of a two-phase supply, as illustrated in Fig. 14.11 for the case of three cascaded negative logic NOR circuits using *p*-MOSTs. The NOR T_1, T_2, T_3 is fed by a pulsed supply ϕ_1. The output from this circuit settles close to its final value after a time t_s, thus the following NOR has its supply in pulse form commencing at least t_s later. Actually the interval can be slightly

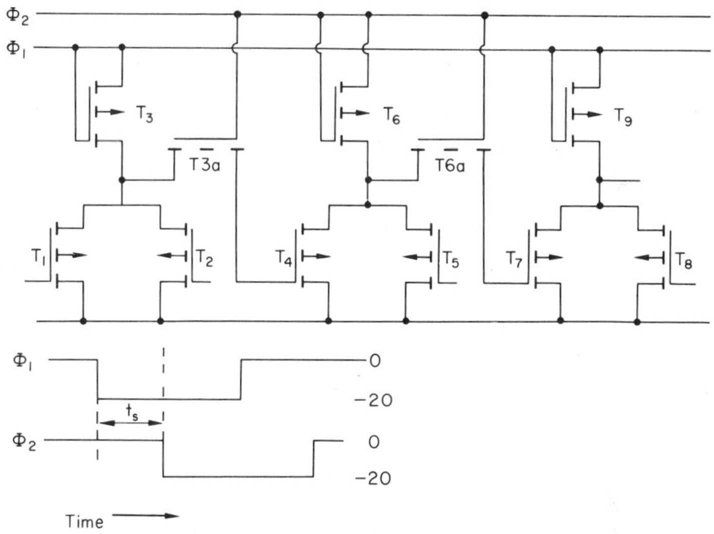

Fig. 14.11. Clocked negative logic

less than t_s (assuming a worst case value) but this is a fine point which will not be pursued here.

The pulse ϕ_1 need not last longer than t_s, since when ϕ_1 returns to zero, T_{3a} will be reversed and biased to zero gate–source voltage, and therefore turns off. Thus, if T_1 and T_2 were previously non-conducting, the voltage on the gate of T_4 will remain high, due to charge storage, whilst if either T_1 or T_2 are in the ON state the zero output will not be disturbed. The clocked supply pulses ϕ_1 and ϕ_2 need not, therefore, necessarily overlap as shown in Fig. 14.11, provided that the interval between them is not so long that there is significant decay of stored gate charge.

14.8.2 FOUR-PHASE LOGIC

Circuits of this kind show a reduction of total power consumed but do not achieve the full potentialities of multi-phase logic systems.

The full sophistication of four or even six phase systems can achieve power consumption levels several orders of magnitude lower, or can give a substantial speed advantage, or both. It will be convenient to describe four-phase logic first, and to do so by building up from a simple two phase inverter.

Consider the circuit of Fig. 14.12. In this version there are no earth (ground) connections and no DC supplies to the circuit. Two clock pulses, ϕ_1 and ϕ_2 are used: in the case of *p*-channel MOSTs, as

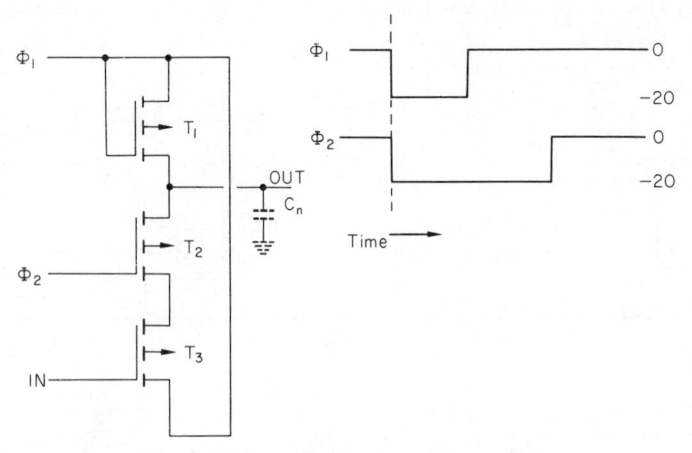

Fig. 14.12. A two-phase inverter. Negative logic

shown, the clock pulses go negative, say to -20 V. The effective load for the circuit is the capacitance to ground at the output node C_n, provided partly by the channel-substrate capacitance of transistors T_1 and partly by the capacitance of the gate or gates of MOSTs in subsequent circuits to which it is connected. One possible timing for the pulsed supplies is as shown in the diagram.

The basis of operation of the circuit is that it operates in two stages: in the first C_n is charged through T_1, and in the second it may or may not be discharged, depending on the state of the input. When the clock pulse ϕ_1 goes negative T_1 turns on, thus C_n is charged to very nearly $(-20 - V_{th})$ volts. It may also charge through T_2 and T_3 if both of these are conducting, but this is unimportant to the functioning of the circuit. When ϕ_1 drops to zero the MOST T_1 is inverted, that is to say that the polarity is such that the source and drain interchange their functions. Thus the gate is now tied to the effective source and T_1 turns off. Since ϕ_2 is negative, however, T_2 is in the ON state, thus if T_3 also conducts the capacitor C_n will discharge and the output voltage will fall to zero. If, on the other hand, T_3 is not conducting the output will remain high.

The output during the period after ϕ_1 returns to zero but whilst ϕ_2 is negative is thus the logical inverse of the input. When ϕ_2 returns to zero T_2 turns off and the output node is then electrically isolated. The output state set during the ϕ_2 output period thus remains at the output node as a stored charge, decaying very slowly in the usual way. Two points about this circuit are particularly important. The first

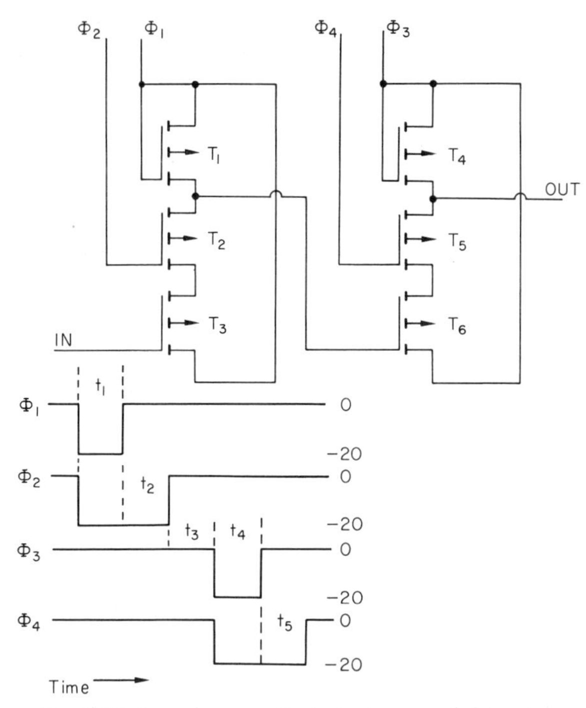

Fig. 14.13. Four-phase negative logic; two cascaded inverters

is that during the ϕ_1 negative period the circuit always charges C_n fully, giving an output corresponding to a logic ONE, regardless of the circuit input. The second is that the inverter action depends on the input sampled for the period after ϕ_1 returns to zero and when ϕ_2 is still negative. At other times the input has no effect at all on the output.

Thus, other circuits following that of Fig. 14.12 must be insensitive during the period when ϕ_1 is negative, but sensitive after it returns to the grounded state. This leads to four-phase circuits as in Fig. 14.13. The two inverters operate as described in the preceding section. When ϕ_1 goes negative the gate capacitance of T_6 charges through T_1. The time t_1 must be long enough to allow the gate of T_6 to be nearly fully charged. After ϕ_1 returns to zero, ϕ_2 remains negative long enough (t_2) for the T_6 gate to discharge if the input to the first inverter corresponds to a logic ONE. After an interval t_3, which may vary from zero to the longest period over which self discharge of T_6 gate may be neglected, ϕ_3 goes negative, allowing the output node to be charged through T_4. A time interval t_4 then elapses, sufficient to permit the completion of charging, after which ϕ_3 returns to zero

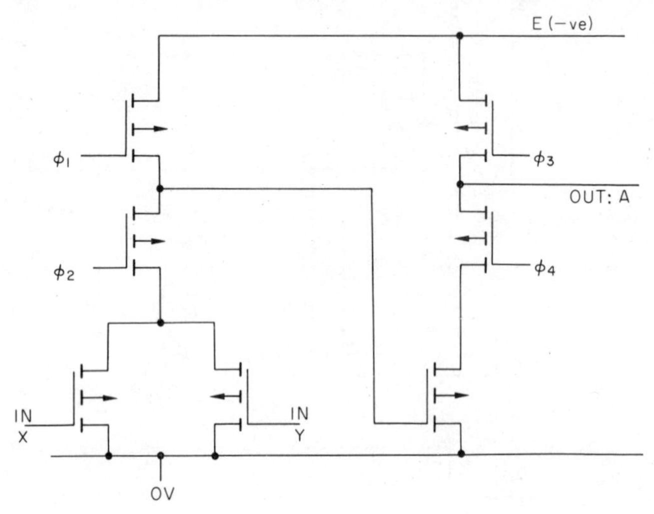

Fig. 14.14. Four-phase OR circuit. A = X + Y. *Timing of* Φ_1, Φ_2, Φ_3, Φ_4 *as for Fig. 14.13. For negative logic*

and the output is set, during a period t_5, to the logical inverse of that at the gate of T_6. The pairs of pulses ϕ_1, ϕ_2 and ϕ_3, ϕ_4 need not start together as shown, since ϕ_2 may follow at any time after ϕ_1, and ϕ_4 after ϕ_3. However, the timing shown (with t_3 reduced to a very small value) gives the fastest possible operation of the circuit. Provided that the pulsed supplies are made large enough to give high currents through transistors in the ON state, fast circuit operation is possible, yet the mean power consumption is much less than for the DC circuits described earlier.

Still further improvements are possible by slight circuit modifications. Referring to the inverter of Fig. 14.12, charging current to the output node is limited by the fact that the gate and drain of T_1 are tied together. Not only does this result in the full drain current of T_1 being drawn from the pulsed supply, but also the drain–source voltage of T_1 is constrained not to be less than the device threshold voltage. Thus, C_n charges toward a final voltage which is less than the supply voltage by an amount V_{th}. For a given negative excursion of ϕ_1 this implies a lower rate of charge for $C_{n'}$. Hence a speed advantage can be gained by connecting the drain of T_1 to a fixed negative supply, instead of to the pulsed supply ϕ_1. The penalty paid for a gain in speed is a loss in layout convenience. Another variant is a circuit in which the source of T_3 is connected permanently to ground instead of to ϕ_1, thus avoiding the need for the pulsed supply line to accept the rather large discharge current of $C_{n'}$. Again, layout may be more difficult.

So far only simple inverters have been considered. Fig. 14.14 shows a two input OR circuit designed on the same principles, and other logic functions can be synthesised in a straightforward manner. At present four-phase circuits of this kind are commercially available for clock rates up to 2 Mhz, but experimental units operating up to 10 Mhz have been reported, and will probably be marketed soon.

14.8.3 SIX-PHASE LOGIC

Six-phase techniques have so far not been exploited commercially: they offer still further reduction of power consumed at the cost of increased complexity. All the multi-phase logic schemes depend upon the charging and discharging of node capacitance. Unfortunately, this is an inefficient process. Consider the case of a voltage $e(t)$ applied to a series combination of a resistor R and a capacitor C. In the particular case of interest, the voltage is a step function, increasing from zero to, say, E volts at a certain time $t = 0$, that is

$$e(t) = Eu_{-1}(t)$$

where $u_{-1}(t)$ is a unit step function. As is well known, the voltage across the capacitor $e_c(t)$ will follow the simple charging curve:

$$e_c(t) = E\left\{1 - \exp\left(-\frac{t}{CR}\right)\right\} u_{-1}(t)$$

and the voltage drop across the resistor is

$$e_R(t) = E \exp\left(\frac{-t}{CR}\right) . u_{-1}(t)$$

Thus the total power dissipated in the resistor is

$$\frac{1}{R}\int_{-\infty}^{+\infty} e_R^2(t)\,dt = \frac{1}{R}\int_{-\infty}^{0^-} 0\,.\,dt + \frac{1}{R}\int_{0^+}^{\infty} E^2 \exp\frac{-2t}{CR}\,dt$$

$$= 0 + \frac{1}{R}\left[\frac{-E^2CR}{2}\exp\left(-\frac{2t}{CR}\right)\right]_0^\infty$$

$$= \frac{E^2C}{2}$$

However, the energy finally stored in the capacitor when the circuit reaches equilibrium is also $\frac{1}{2}E^2C$. The energy lost in the resistor is exactly equal to the energy stored in the capacitor, independent of the value of R. Thus the charging of a capacitor

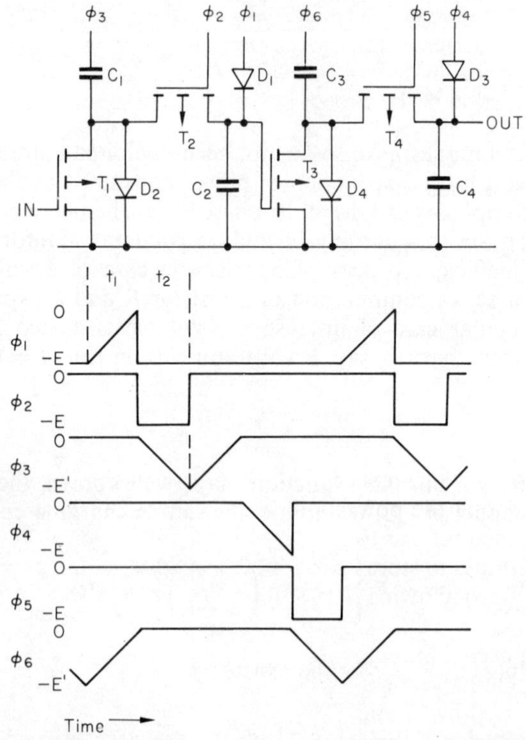

Fig. 14.15. A six-phase one bit delay (two cascaded inverters). Ideal clock waveforms shown assume that the transistor switching time is negligibly short

through a resistor by the application of a voltage step has an intrinsic efficiency of 50% and reducing the value of the resistor does not improve this. In a four-phase logic system, half the total energy drawn from the supply to charge the node capacitances is wasted, and it is this inefficiency which six-phase operation seeks to overcome. If the capacitor is connected to a pulsed supply line at a time at which the voltage is zero and it is subsequently increased relatively slowly it can be shown that the power loss in the series charging resistor is much reduced. This is the principle on which the six phase digital circuits are designed.

Consider Fig. 14.15, which is of a six-phase one bit delay circuit. There are four diodes in this circuit, but D_2 and D_4 are not separate components as they are intrinsically present as the drain-substrate diodes of T_1 and T_3 respectively, and similarly C_2 and C_4 are 'implicit' components, corresponding to the capacitance to substrate at the nodes concerned. However, C_1 and C_3 and D_1 and D_3 must be fabricated, and since they are not MOSTs this inevitably complicates the fabrication technology. Ideal clock waveforms for the circuit are also shown in Fig. 14.15, where it is assumed that the MOSTs turn on or off in negligible time.

The operation of the circuit is as follows. The supply ϕ_1, normally at $-E$ returns slowly to zero, taking a time t_1. If C_2 is negatively charged D_1 will conduct and the capacitor will be discharged to zero, however, provided that t_1 is sufficiently long the discharge current, and hence the power loss, can be made arbitrarily small. As ϕ_1 returns to $-E$, turning off D_1, ϕ_2 goes negative and T_2 turns on. At the same time ϕ_3, previously zero, slowly changes to $-E'$. If T_1 is non-conducting, since D_2 will also be non-conducting, the effect is for C_2 to charge, the gate of T_3 going negative. Provided that

$$E' = E\left(1 + \frac{C_2}{C_1}\right).$$

the gate of T_3 will charge very nearly to $-E$. Also provided that t_2 is sufficiently long the power loss may be made negligibly small. The supply ϕ_2 then returns to zero, turning off T_2 and hence isolating C_2 which continues to store its charge. The return to zero of ϕ_3 is, again, relatively slow, allowing C_1 to discharge through D_1 with negligible loss.

Had T_1 been conducting during the negative excursion of ϕ_3, no charge would have appeared on C_2 which would have, in effect, been short circuited by T_1 and T_2 in series.

To summarise, if the gate of T_1 is initially negatively charged no charge is transferred to the gate of T_3, and conversely. Thus the first half of the circuit (fed by ϕ_1, ϕ_2, and ϕ_3) acts as an inverter. The second half of the circuit behaves identically. Overall, therefore, the sub-system of Fig. 14.15 may be regarded as a one bit delay. Other logic functions are easily synthesised: for example, if T_1 were replaced by two MOSTs with source and drain in parallel the circuit would become a two input OR.

Because the rate of change of ϕ_1, ϕ_3, ϕ_4 and ϕ_6 is not infinitely slow the energy loss in charging and discharging the capacitors is not, in fact, zero, however it can be very small. The total power consumption at 1 MHz clock rates can be one or two orders of magnitude lower than for simple DC logic systems, and at lower clock rates the advantage becomes progressively greater.

14.9. Logic systems compared

All the logic systems described have been used, some to a greater and some to a less extent. Comparisons of *p*-channel DC logic with complementary logic are discussed in considerable detail by Evans[3], whilst Watkins[4] has reviewed the advantages to be gained by multi-phase operation.

Simple *p*-channel DC and four-phase logic have the great advantage that the fabrication processes required to manufacture integrated circuits using this technology are relatively very simple: no more complex in fact than those required to make *p*-channel discrete MOSTs. Thus yield for circuits of this kind is very good. The four phase system can greatly reduce circuit power consumption or increase operating speed, or can give a modest gain in both. Its principal disadvantage is the relative complexity of the power supplies used, and the larger chip area occupied by the slightly more complex circuits. This last factor is not too serious, amounting typically to about a fifty per cent increase in area even in less favourable cases. So far as the complicated pulsed power supplies are concerned, the complexity of the generating circuits is not great by contemporary standards and integrated circuit pulse generating sub-systems are commercially available. Obviously the cost of this additional equipment adds to that of the logic itself, and thus makes four-phase operation uneconomic for very small systems. Four-phase logic, as explained, gives a speed advantage. The extent of this is to some degree a matter of dispute. A tenfold increase in speed over simple *p*-channel DC MOST logic has sometimes been claimed, but published practical results have not so far substantiated this. Certainly, however, it is possible to improve speed by a factor of two to three, which is similar to the improvement obtained with complementary circuits.

Although complementary logic gives a useful speed advantage, it costs considerably more to implement than *p*-channel DC because of the much greater complexity of processing. The component count in complementary systems is marginally greater than in four phase logic thus the latter has a cost advantage and no speed disadvantage. However, the advantage of complementary circuits is that pulsed power supplies are not needed. This is significant only for small systems: large digital systems will inevitably include some arrangement for strobing or clocking logic to improve the definition of logic circuit operating times.

Six-phase logic must at present be regarded as of potential importance only. Since it embodies diodes and capacitors as well as MOSTs, fabrication methods are relatively complex. Capacitors can be fabricated in MOS form at the same time as transistor gates, but D_1 and D_3 of Fig. 14.15 would necessitate a double diffused process. Various circuit modifications are possible to circumvent this difficulty, but only at the sacrifice of some desirable performance parameters. The clock waveforms required for six-phase systems are also relatively complex and need to be more closely controlled than the simple pulsed waveforms used for four phase. Again degraded waveforms can be used, at some performance sacrifice. Generally, however, six-phase logic will only be considered when the ultimate in reduction of power consumption is mandatory, and even then power losses in the pulse generating circuits must be given close attention if the overall reduction in system power consumption is not to be disappointing.

To summarise: simple DC *p*-channel logic will be used wherever cost is the main consideration and speed and power consumption are of secondary importance. A $2\frac{1}{2}:1$ speed advantage or better can be obtained by the use of complementary logic, and a possibly greater advantage using four phase *p*-channel logic. However, the cost penalty is much greater in the former case. Thus, complementary logic is probably only preferable where pulsed power supplies are unacceptable. Six-phase systems present serious fabrication problems, but could be of interest where power economy is of over-riding importance.

14.10 Linear MOS monolithic circuits

The MOS integrated circuit first became of commercial significance in the digital field. Only relatively recently have the advantages of MOS linear circuits come to be fully appreciated. Partly this is because MOSTs have certain parameter deficiencies, particularly in regard to noise and drift, which are more serious in linear applications. Recent advances in fabrication techniques have made these deficiencies less, as already indicated in section 14.2, and for this reason interest in linear MOS integrated circuits is now increasing rapidly. In this section the principles of MOS linear circuit design will be considered, without further allusion to device limitations until experimental results are considered.

In many other places in this book, use of MOSTs in discrete component circuits has been described. Indeed, circuit differences between MOST and junction FET applications can often be quite small, and much circuit design information is applicable to both so far as linear discrete circuits are concerned. However, in monolithic technology this is not so, because economic considerations, outlined in section 14.1, make the so-called ALMOST family of circuits particularly attractive. As the name implies, ALMOST circuits use MOST structures in place of all circuit components, not only as active devices but also in place of resistors and capacitors. The motive for this type of design is initially economic, but it can be shown that circuits of this type have unique advantages not shared by other types of circuit, the most important of which is the possibility of achieving extremely stable and well defined voltage gain without the use of negative feedback.

14.10.1 THE CS ALMOST AMPLIFIER

In section 14.3, above, equation (14.1) is derived, which shows that for a simple enhancement MOST common source amplifier (Fig. 14.2) with a similar device used as a load (its gate and drain being connected together) the voltage gain is precisely

$$A_V = -\left\{\frac{\beta_1}{\beta_2}\right\}^{\frac{1}{2}}$$

or, if the two devices have the same channel length, from equation (14.3),

$$A_V = -\left\{\frac{W_1}{W_2}\right\}^{\frac{1}{2}}$$

It is important to note that these expressions for gain hold good even if the two MOSTs do not have the same threshold voltage. This is because the terms in the expression for gain which contain the threshold voltage are eliminated by applying the condition that both MOSTs pass the same drain current. This point is important, because the threshold voltage for T_2 will in fact be greater than that for T_1, due to the greater reverse bias on its substrate.

Since the gain is defined simply in terms of the ratio of two channel widths, that is to say the ratio of two dimensions on the same mask, it can be held very closely, both in respect of variations between different samples of amplifiers made from the same mask and also with respect to any one amplifier over its range of working supply voltages and temperatures. Typically, a 1% variation in gain over a 100°C temperature range and a 2:1 variation of supply voltage can be achieved without particular care in fabrication.

This rather striking result, however, was obtained in section 14.3 subject to certain conditions, which must be made explicit. They are as follows:

(i) The effect of interelectrode capacitance is ignored, that is to say the results are only valid at low frequencies.
(ii) It is assumed that the drain-source voltage is sufficiently large that both MOSTs operate as constant current devices.
(iii) The square law of equation (3.14), (section 3.2.2) is assumed to be exactly followed.

These assumptions can be dealt with in different ways. That quoted under (iii) cannot conveniently be relaxed; however, the error it contributes is small. Assumption (ii) can be met perfectly provided that the supply voltage $E(-\mathrm{ve})$ is large enough. Just how large this is, is easily determined below. As for assumption (i), a modified analysis which takes into account high frequency effects is not too difficult, and will also be considered subsequently.

First, however, consider the DC circuit conditions for the amplifier of Fig. 14.2. Let the minimum drain–source voltage for constant current operation be V_S. (To give a precise definition: let V_S be the drain–source voltage at which the rate of change of drain current with drain–source voltage is less than one hundredth of the g_{fs} for the device concerned.) For present ALMOST amplifiers using *p*-channel enhancement devices V_S is usually about equal to V_{th}, though this need not necessarily be so.

The voltage gain of ALMOST amplifiers is necessarily quite modest. Partly, this is because the W/L ratio for devices cannot vary by more than about 25:1 without inconvenience in layout, leading to voltage gains of 5:1 or less; and partly because attempts to realise higher gain lead to most unsatisfactory DC conditions and rather poor high frequency response. Thus in many applications, perhaps almost all, it will be necessary to use several stages of amplification in cascade. For this reason it is usual to apply a further condition to the design of such amplifier stages, namely that the input and output terminals shall be at the same DC level, to permit direct coupling of successive stages, without the need for level shifting between stages.

Thus, using the subscript (1) for the active MOST and (2) for the load, as before,

$$V_{GS,1} = V_{DS,1} \tag{14.14}$$

is the relevant condition. Also

$$\begin{aligned} V_{GS,2} = V_{DS,2} &= E - V_{DS,1} \\ &= E - V_{GS,1} \end{aligned} \tag{14.15}$$

Further, since the same drain current flows through both devices

$$\frac{\beta_1}{2}(V_{GS,1} - V_{\text{th},1})^2 = \frac{\beta_2}{2}(V_{GS,2} - V_{\text{th},2})^2$$

so that

$$\frac{V_{GS,1} - V_{\text{th},1}}{V_{GS,2} - V_{\text{th},2}} = \left\{\frac{\beta_2}{\beta_1}\right\}^{\frac{1}{2}} = -\frac{1}{A_V}$$

Although both MOSTs are fabricated in the same process, it is not correct to assume that $V_{\text{th},1}$ and $V_{\text{th},2}$ will be identical. This is because the upper transistor, T_2, has its substrate much more strongly reverse biased than the lower. Indeed the substrate-source voltage for T_2 is just $V_{DS,1}$, whereas the corresponding voltage for T_1 is zero. This increases the apparent threshold voltage. Hence if $V_{\text{th},1}$ has the value V_{th}, $V_{\text{th},2}$ will be $(V_{\text{th}} + \Delta V_{\text{th}})$. The value of ΔV_{th} has been considered in chapter 3. The precise expression for ΔV_{th} is fairly complicated; however, the effect of this term in the present analysis is quite small, so that an approximate expression, due to Williams, may be used namely

$$\Delta V_{\text{th}} = (V_0 V_{DS,1})^{\frac{1}{2}} \tag{14.16}$$

where V_0 is a constant having the dimensions of voltage the value of which is approximately 0·7 V for MOSTs having threshold voltages between 2·5 and 5 V

(Since ΔV_{th} is itself small and varies only as the square root of $V_{DS,1}$ in many small signal amplifiers, the expression may be still further approximated by writing V_{th} for $V_{DS,1}$.)

Making this substitution for $V_{\text{th},1}$ and $V_{\text{th},2}$ an equation may be obtained for $(V_{GS,1} + V_{GS,2})$ thus

$$V_{GS,2} + V_{GS,1} - (2V_{\text{th}} + \Delta V_{\text{th}}) = (1 - A_V)(V_{GS,1} - V_{\text{th}})$$

and so, using equation (14.15)

$$\frac{E - 2V_{th} - \Delta V_{th}}{1 - A_V} = V_{GS,1} - V_{th} \tag{14.17}$$

or

$$I_D = \frac{\beta_1}{2}\left\{\frac{E - 2V_{th} - \Delta V_{th}}{1 - A_V}\right\}^2 \tag{14.18}$$

Fig. 14.16 indicates how very well theory, as represented by equation (14.18), agrees with practice in the case of a commercial ALMOST amplifier (Marconi E6022), assuming $\beta_1 = 72$ µA per volt^2 and $(2V_{th} + \Delta V_{th}) = -8$ V The graph shows the relationship

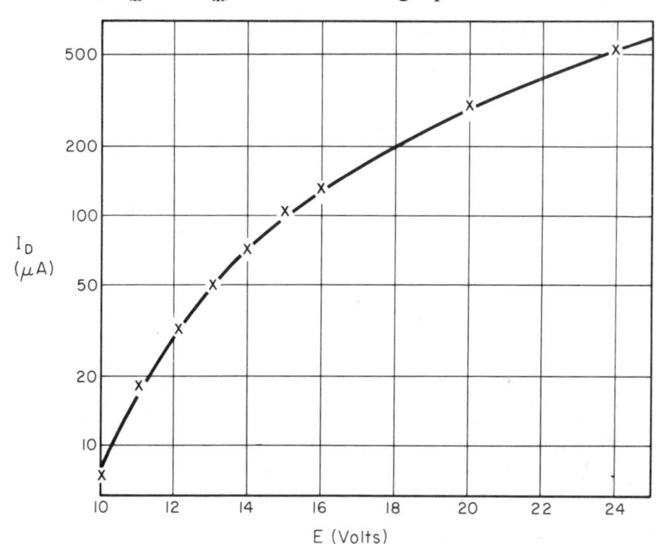

Fig. 14.16. Comparison of theoretical relationship between I_D *and* E *(curve) with experimental observations (points) for one stage of a Type 6022 ALMOST amplifier (Courtesy Marconi-Elliot)*

between E and I_D, predicted by equation (14.18), as a full line, whilst experimentally measured results, shown as crosses, fall precisely on the line.

Also from equation (14.17),

$$V_{DS,1} = V_{GS,1} = V_{th} + \frac{E - 2V_{th} - \Delta V_{th}}{1 - A_V} \tag{14.19}$$

Obviously, equation (14.19) indicates that the peak positive going swing at the output point, and hence the peak positive signal swing is

$$\hat{v}_+ = V_{DS,1} - V_S = (V_{th} - V_S) + \frac{E - 2V_{th} - \Delta V_{th}}{1 - A_V} \tag{14.20}$$

In the common special case where, to close approximation, $V_{th} = V_S$ equation (14.8) simplifies to

$$\hat{v}_+ = \frac{E - 2V_{th} - \Delta V_{th}}{1 - A_V} \qquad (14.21)$$

Both equations (14.20) and (14.21) indicate that the peak output signal gets smaller the larger the magnitude of A_V, bearing in mind that for a CS amplifier the gain is of negative sign.

The maximum positive signal excursion is

$$\hat{v}_+ = V_{DS,1} - V_S$$

whilst the corresponding negative excursion is

$$\hat{v} = (E - V_{DS,1}) - V_S$$

Thus the maximum peak-to-peak output signal is

$$v_{P\text{–}P} = E - 2V_S \qquad (14.22)$$

Thus a necessary condition on the value of E is

$$|E| > |2V_S| \qquad (14.23)$$

In general the values of $\hat{v}_-$ and $\hat{v}_+$ cannot be equated, thus the maximum undistorted P–P output can only be achieved for an assymetrical waveform, or, what amounts to the same thing, if the input terminal is allowed to be at a slightly different mean DC level from the output. If the requirement for the same input and output DC levels, already stated, is adhered to, the maximum peak-to-peak output signal is reduced to $2\hat{v}_+$, or

$$v_{P\text{–}P} = \frac{2E - 4V_{th} - 2\Delta V_{th}}{1 - A_V} \qquad (14.24)$$

Obviously, the larger the magnitude of A_V the smaller the permissible output voltage swing.

Amplifiers of this type, giving only a modest proportion of the total supply voltage as available output signal excursion, are obviously best suited to small signal applications, and in particular will give poor power efficiency used as output stages.

The input conductance of the amplifier is virtually zero. The output conductance, however, can be fairly large. Since the active MOST is operating virtually as a constant current device, the output conductance of the amplifier is substantially $g_{fs,2}$, but

$$\begin{aligned} g_{fs,2} &= \beta_2(V_{GS,2} - V_{th,2}) \\ &= (2\beta_2 I_D)^{\frac{1}{2}} \end{aligned} \qquad (14.25)$$

or, using equation (14.18),

$$g_{fs,2} = -\frac{A_V}{1-A_V}\beta_2(E - 2V_{th} - \Delta V_{th}) \qquad (14.26)$$

Since A_V is negative this corresponds to a positive output conductance, and as the magnitude of A_V will usually be greater than unity, the value of $g_{fs,2}$ will only be weakly dependent on A_V, taking values only a little less than $\beta_2(E - 2V_{th})$. The range of output resistance with currently available devices is upwards from under one hundred ohms, with a few kilohms or tens of kilohms as the commonest values. As would be expected, increasing I_D lowers the output resistance, however since the effect obeys a half power law the dependence is not too marked.

14.10.2 FREQUENCY RESPONSE OF CS ALMOST AMPLIFIERS

As with all FET circuits in common use, the frequency response of CS ALMOST amplifiers is determined by interelectrode capacitance and not by transit time effects in the device. Thus it is impossible to specify the performance of such an amplifier except in terms of a particular signal source impedance and a particular load.

The input impedance of such a device may be represented by a capacitor C_{in}, across which is a shunt conductance, which may, however, in almost all cases be neglected. The value of C_{in} is given by

$$C_{in} = C_{GS,1} + (1 - A_V)\,C_{GD,1} \qquad (14.27)$$

where the second term is due to Miller effect, and the '1' subscript denotes the active device. Usually the MOST is symmetrical and thus

$$C_{GS,1} = C_{GD,1} = C_1$$

then

$$C_{in} = (2 - A_V)\,C_1 \qquad (14.28)$$

The output capacitance, which is shunted by the output conductance equal to $g_{fs,2}$, is similarly given by

$$C_{out} = C_{GS,2} + \left(1 - \frac{1}{A_V}\right)C_{GD,1} \qquad (14.29)$$

However, with normal MOST structures

$$\frac{C_{GS,2}}{C_{GS,1}} = \frac{\beta_2}{\beta_1} = \frac{1}{A_V^2}$$

and, once again assuming a symmetrical device

$$C_{out} = \frac{1}{A_V^2} + 1 - \frac{1}{A_V} C_1 \tag{14.30}$$

Assuming specific source and load impedances, equations (14.25) and (14.27) can be used to calculate the frequency response of an amplifier. The value of A_V used in these equations can be that given by equation (14.1), namely

$$A_V = -\left\{\frac{\beta_1}{\beta_2}\right\}^{\frac{1}{2}}$$

provided that the admittance of C_{out} is small compared with $g_{fs,2}$. This implies that the frequency shall not be too high: specifically that

$$\omega \ll \frac{\beta_1(E - 2V_{th} - \Delta V_{th})}{C_1\left(A_V^2 - 2A_V + 2 - \frac{1}{A_V}\right)} \tag{14.31}$$

A particular case of special interest is when a number of identical ALMOST CS amplifiers are cascaded. The interstage node capacitance is then

$$C_n = C_{out} + C_{in}$$

or,

$$C_n = \left\{3 - A_V - \frac{1}{A_V} + \frac{1}{A_V^2}\right\}. C_1$$

For $|A_V| > 2$ the error in neglecting the last two terms in the bracket is less than 15%, hence to this accuracy

$$C_n = (3 - A_V)\, C_1$$

$$\text{for } |A_V| > 2.$$

To the same degree of approximation, the cut-off frequency for the individual stages is then

$$\begin{aligned}\omega_1 &= g_{fs,2}/C_n \\ &= \frac{-\beta_1(E - 2V_{th} - \Delta V_{th})}{A_V(1 - A_V)} \cdot \frac{1}{(3 - A_V)\, C_1} \\ &= \frac{-\beta_1(E - V_{th} - V_{th})}{C_1} \frac{1}{3A_V - 4A_V^2 + A_V^3}\end{aligned} \tag{14.32}$$

Clearly, the cut-off frequency of the amplifier falls sharply as A_V is increased in magnitude. The quantity $\beta_1(E - 2V_{th} - \Delta V_{th})\, C_1$ which has the dimensions of frequency is a bandwidth 'figure of merit' for amplifiers of this kind.

As an example, the E6022 (Marconi) uses three cascaded ALMOST CS amplifiers. Taking $E = -15$ V, $V_{th} = -4$ V, $\beta_1 = 72\ \mu A/V^2$, and $C_1 = 0{\cdot}5$ pFd, the value of $\beta_1(E - 2V_{th} - \Delta V_{th})/C_1$ is 10^9 radians per second, or about 161 MHz. Since A_V is $-3{\cdot}3$, this suggests a -3 dB frequency for the interstage coupling, (corresponding to the -6 dB frequency for the amplifier as a whole, since there are two such couplings) of 1·8 MHz. This figure is confirmed by experimental measurements. At higher values of E the frequency response would be extended: for example the -6 dB frequency for the amplifier is doubled if E is raised to 22 V. By contrast, designing for higher gain will reduce the cut-off frequency, thus at $A_V = -10$ and $E = -15$ V, the -6 dB frequency for the amplifier is 112 kH.

The upper frequency limit for an ALMOST amplifier can also be determined quite independently if the output terminal is loaded by a capacitor, and where this is the case the output stage MOST widths must be increased to lower the output resistance, or some other precaution, such as the use of a suitable buffer amplifier stage, must be adopted. Positive current feedback will also reduce the amplifier output resistance and due to the excellent gain stability, may be used safely.

14.10.3 OTHER AMPLIFIER CONFIGURATIONS

So far only the common source configuration has been considered, but common drain circuits are widely used in discrete component technology, so possible applications in ALMOST form ought also to be considered. However, the CD configuration has one serious disadvantage when considered as an integrated circuit: its input and output DC levels differ by an amount slightly greater than the threshold voltage of the device used, which will be several volts for an enhancement device. Also the low output impedance characteristic for which the CD circuit is widely used in discrete components technology is less important in the ALMOST case, since the output impedance of CS amplifiers can also be made low by choosing the β of the load MOST large enough. For a given output impedance, however, the CD stage occupies little less chip area than a CS equivalent, and the main point of difference is that the CD amplifier has lower voltage gain than is possible with the CS, and does not phase invert.

Consider Fig. 14.17, which is a typical ALMOST CS stage. A variant of the circuit is also possible in which the gate of T_1 is connected to the output terminal instead of to $-E$: the effects of this will be considered later.

The analysis of the amplifier is identical with that for a CD FET amplifier given in chapter 6, but with an infinite load resistor, since T_1 is assumed to operate as a true constant current device. The voltage gain is very close to $+1$, and the output impedance is almost exactly $g_{fs,2}$. The input capacitance does not contain a term in $C_{GS,2}$ due to the bootstrapping effect of the source voltage. Instead the input impedance is just due to $C_{GD,2}$ shunted by a negligibly small conductance. For similar reasons, the output capacitance of

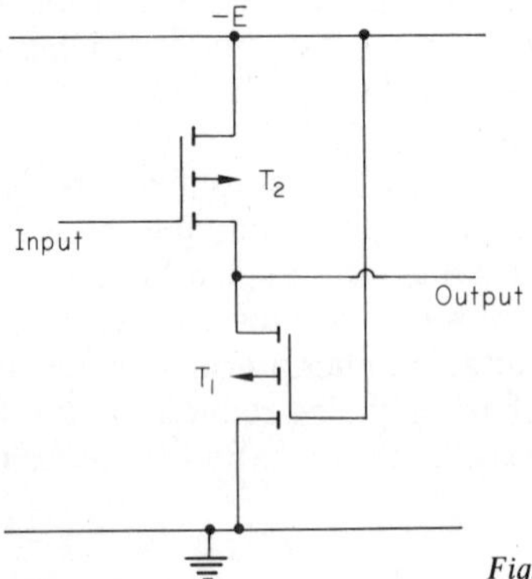

Fig. 14.17. A CD ALMOST amplifier

the amplifier is $C_{GD,1}$. Thus since T_2 is a much larger device than T_1, the output, but also both are much smaller than in the CS case.

Most important of all, however, it is impossible to design a CD amplifier of this type to have the same DC level at input and output terminals: for a *p*-channel device the latter must be more positive than the former by an amount at least as great as the threshold voltage, otherwise the MOST does not conduct. Thus, in addition to its other properties the circuit is a 'downward' DC level shifter, and is often used in circuits for just this purpose. For example, in an amplifier comprised of several cascaded CS stages, the signal handling capacity can be improved if, in the *p*-channel case, the output DC level of the CS stages is slightly more negative than the input level. A chain of amplifiers of this kind can be arranged to shift the mean signal level in the negative direction, but by terminating it with a CD stage a positive shift can be introduced, restoring the output DC level of the amplifier to that at the input, which is convenient, if the amplifier working point is to be stabilised by overall DC negative feedback.

An alternative version of the CD amplifier to that considered also exists, in which the gate of the load MOST is connected to its drain, instead of to the negative supply rail (in the p-channel case). In this case the effective value of the load falls, and in consequence the voltage gain falls below unity and the input capacitance of the amplifier contains a term in the gate-source capacitance of the active MOST, which is not entirely bootstrapped out. Generally, this form of the DC amplifier has no particularly advantageous

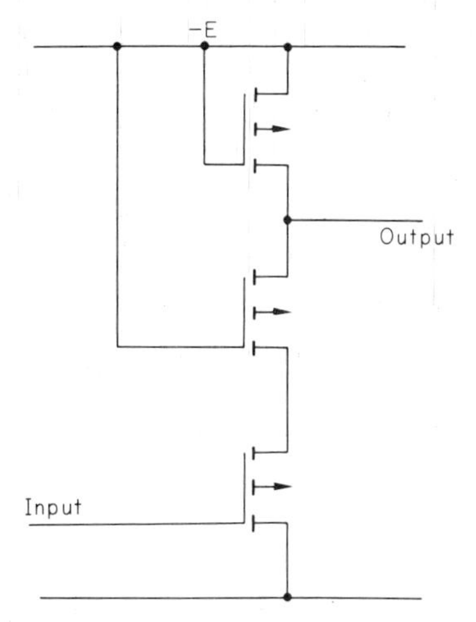

Fig. 14.18. A cascade ALMOST amplifier

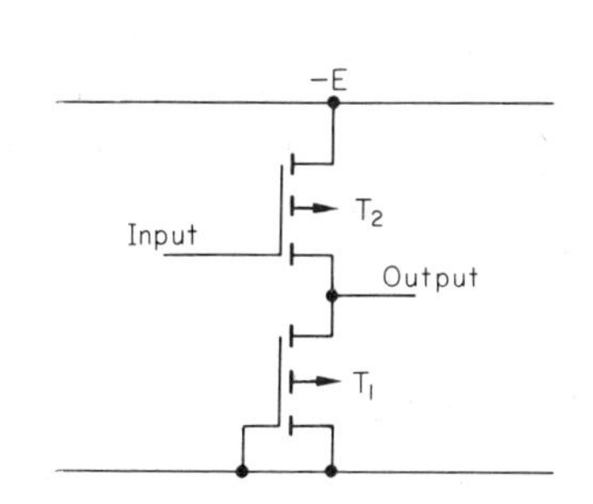

Fig. 14.19. A non level-shifting CD amplifier using depletion MOSTs. T_1 *and* T_2 *should be identical*

properties and will be used only where it helps to simplify a difficult circuit layout.

So far all the circuits considered use only two MOSTs connected in series between the supply rails. If three are used the supply voltage must, of course, be increased accordingly, but the range of possible circuit configurations is greatly extended. One of the most useful is the cascode circuit; familiar from discrete circuit technology. The circuit is as in Fig. 14.18, and its principal properties are that the Miller effect, and hence the input capacitance, are much reduced, whilst at the same time the DC level is shifted 'upward' (i.e. negative for p-channel devices) by an amount slightly greater than V_P. Due to the level shift, amplifying stages of this kind cannot be directly cascaded, but an extremely wide band amplifier can be built by cascading CD and cascode amplifiers. Since these introduce DC level shifts in opposite sense, a pair can be designed to give zero total shift.

Only circuits using enhancement MOSTs have been described in this chapter: however, depletion-mode devices are also very useful in linear integrated circuits. Common drain amplifiers without DC level shift are one use (Fig. 14.19) but CS amplifiers with much higher gain than for the normal ALMOST enhancement configuration are

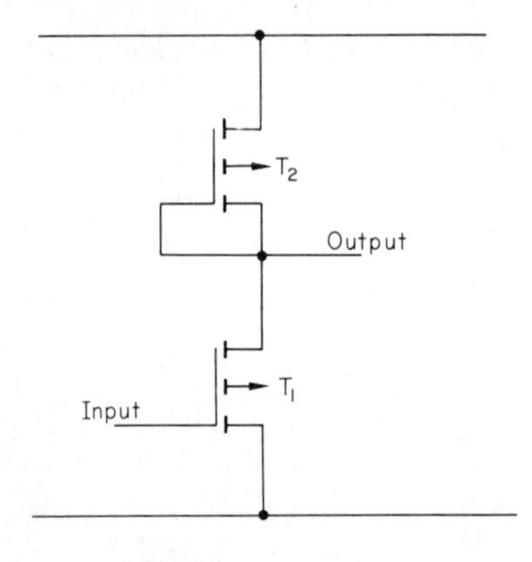

Fig. 14.20. A CS amplifier using a depletion MOST(T_2) *which gives large voltage gain*

also possible (Fig. 14.20) although in this case the gain is much less well defined than for the enhancement MOST circuits.

14.11. Conclusions—the ALMOST linear integrated circuits

Although ALMOST linear integrated circuits are, as yet, in their infancy, it is hardly possible to discern the limits of their commercial significance. They are simple and can have highly desirable properties. Although at first, because of the feature of well-defined gain, they will be likely to find application in military and professional equipment (especially instruments) in the longer run intrinsically very low cost should result in deep penetration into the entertainment markets, with consequent very high volume production.

Further advances depend principally on the introduction of MOSTs with still lower threshold voltage and, particularly, lower noise. Since depletion mode devices are fully acceptable for linear applications it may well be that Schottky barrier devices have an important role here. Another important factor in linear ALMOST exploitation is advances in circuit design. The potentialities of this technology have by no means been exhausted as yet: circuit ingenuity has a real contribution to make here.

REFERENCES

1. Warner, R. M. and Fordemwalt, J. N. (editors). *Integrated Circuits, Design Principles and Fabrication*, McGraw-Hill (1965).

2. Lynn, D. K., Meyer, C. S. and Hamilton, D. J. (editors). *Analysis and Design of Integrated Circuits*, McGraw-Hill (1967).
3. Evans, J. D. 'Integrated MOST circuits', *Microelectronics and Reliability,* **7**, No. 1, 11–36 (February 1968).
4. Watkins, B. G. 'A low-power multiphase circuit technique', *Journ. of IEEE Solid State Circuits,* **SC2**, 213–220 (December 1967).

15

THE PHOTO-FET

15.1. Principles of operation

When a semiconductor is exposed to light, hole–electron pairs are generated if the quantum energy of the light is sufficient to effect the transference of electrons across the relevant energy gap. For an extrinsic, or doped, semiconductor, this gap will be small (of the order of 0·01 eV), which means that radiant energy well into the infra-red region will suffice.

If a junction is present, the charge carriers will be separated, and a current will flow if an external circuit is completed. This is the principle of the photo-diode, whose characteristics are sketched in Fig. 15.1. Here, it will be seen that under dark conditions, the diode exhibits normal behaviour; that is, it presents a low forward impedance and a high reverse impedance. The leakage (or dark) current, being due to thermal agitation, is an exponential function of temperature.

When irradiated, but not polarised by an external voltage, a photo-current and a photo-voltage appear in the directions shown; that is, some of the incident radiation is converted to electrical energy. Under short-circuit conditions, the photo-current is found to be directly proportional to the irradiation intensity; whereas the open-circuit photo-voltage is approximately logarithmic.

When the photo-diode is reverse-biased—that is, the third quadrant of Fig. 15.1 is relevant—the two components of the observed current can be represented as in Fig. 15.2. Here, I_D is the reverse leakage, or dark current, which is essentially an exponential function of temperature; whilst I_P is the photo-current, which although primarily a direct function of the irradiation intensity, also exhibits a small temperature coefficient. Both currents are functions of the junction area, as is the capacitance C_P. This means that whereas a photo-diode having a large junction area will exhibit high

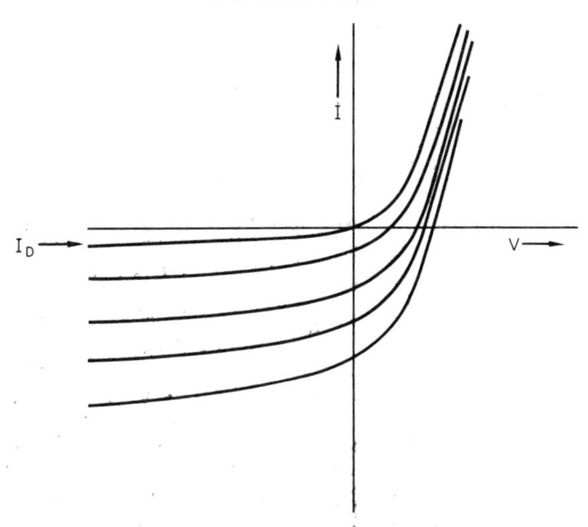

Fig. 15.1. Photodiode characteristics

sensitivity, it will also have a comparatively large dark current and an impaired frequency response.

The junction capacitance C_P will also be an inverse function of applied voltage, for a high potential difference will increase the thickness of the depletion layer.

The remaining, ohmic component of the leakage current is represented by the very high resistance R_P. The bulk resistance of the inactive part of the semiconductor material is R_T, which for a good geometry device, is low. The time-constant $R_T C_P$ largely determines the frequency response.

Because a photon has an energy which is an inverse function of its wavelength, the quantum efficiency (that is, the probability of the creation of a hole–electron pair by the photon) of a photo-diode is very dependent upon the wavelength of the incident radiation.

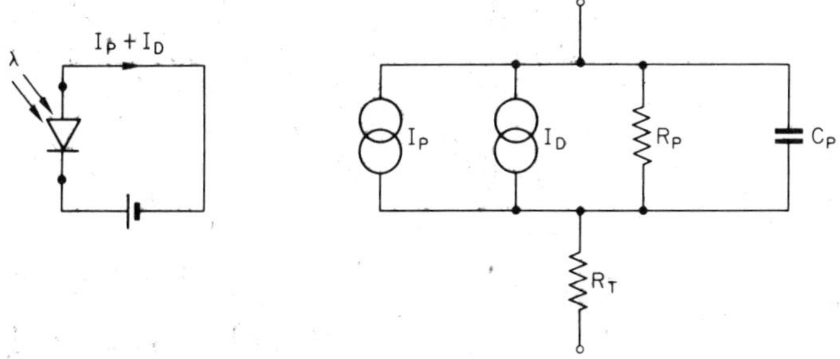

Fig. 15.2. Equivalent circuit for the reverse-biased photodiode

Consequently, the value of the photo-current will vary with wavelength as shown in the spectral response curve of Fig. 15.3. However, for an invariant spectral distribution, I_P must be directly proportional to the intensity of this incoming radiation.

Murphy and Kabell[1] have reported that the temperature coefficient of I_P is also a function of the wavelength of the incident radiation,

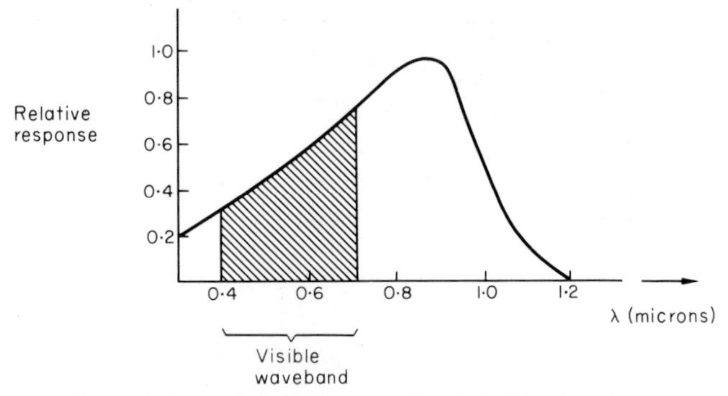

Fig. 15.3. Spectral response curve of typical silicon junction

and give a curve similar to that of Fig. 15.4. Here, the temperature coefficient (of about 0·1% per °C) is seen to rise rapidly at both extremes of the spectral response characteristic.

The rise at the UV end is attributed to temperature effects at the surface due to absorption of the short wavelengths near this surface; and that at the IR end to changes in bulk carrier lifetime engendered

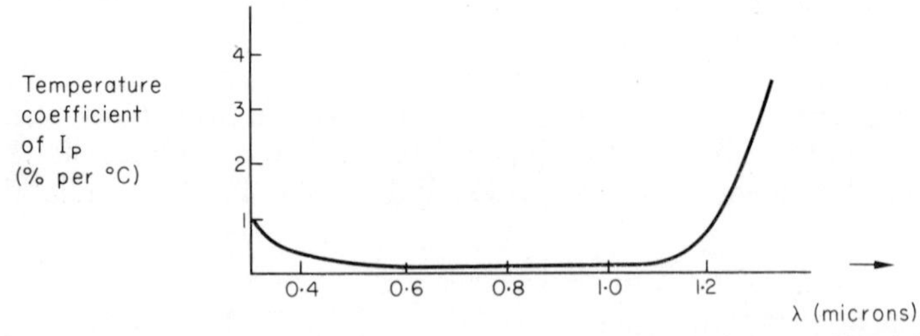

Fig. 15.4. Temperature coefficient of photocurrent. (After Murphy and Kabell[1])

by the absorption of the longer wavelengths deeper within the material.

The foregoing discussion is relevant to the photo-FET because it consists essentially of a junction formed on a semiconductor channel. In fact, it differs from a normal FET only in that the encapsulation is designed to allow irradiation of the junction area, so producing a photo-current in addition to I_G. Often, a lens is incorporated to collect and focus as much light as possible near the junction.

The effect of I_P in a simple circuit may be deduced from the self-biased source-follower configuration of Fig. 15.5. Here, the sum of the (large) photo-current I_P and the (small) leakage current I_G flows through R_G and produces a voltage drop which tends to increase the channel current.

Thus, the circuit constitutes a source-follower having radiation as its input rather than voltage, and providing that the spectral distribution of this radiation remains invariant, the output voltage at the source will vary linearly with the intensity.

If the dark, or leakage current is much smaller than I_P, its variation with temperature will not contribute significantly to the overall

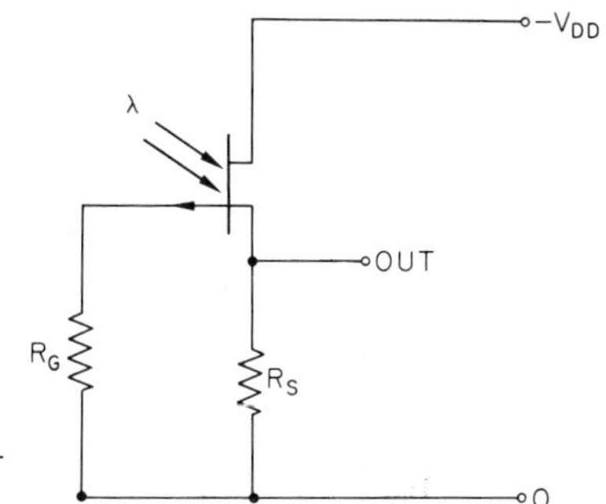

Fig. 15.5 The photo-FET in CD configuration

temperature coefficient. If, therefore, the spectral bandwidth of the incident light is limited either inherently, or by means of a filter, then I_P should exhibit the low temperature coefficient observed by Murphy and Kabell for a simple photo-diode. However, this will be augmented by the usual drifts due to temperature in an FET, and in order that the overall temperature coefficient should be minimal, the working point should lie at the optimum position as explained in **chapter 8, page 168.**

It is, of course, possible to chop the incident light so that the AC component of the output may be selected, thus avoiding inaccuracies due to drift. The gain of the source-follower will remain fairly constant and will approach unity if the chopping frequency is not excessively high (see equations (6.8) through (6.10)).

15.2. Photo-FET characterization

15.2.1 METHODS OF SPECIFYING SPECTRAL SENSITIVITY

There are, unfortunately, numerous methods of specifying the spectral sensitivity of photo-detectors[2]. The meaningful production of a calibrated spectral sensitivity curve of the type sketched in Fig. 15.3 is quite difficult, and manufacturers usually compromise by quoting the sensitivity under specified narrow-band conditions. Typical of these is the 'colour temperature' convention, wherein the radiation is derived from a tungsten filament lamp working at a specified colour temperature, this being defined as the temperature of a black-body whose spectral distribution is closest to that of the lamp in question. Such standardised lamps are commercially available.

Another method is to quote the sensitivity at some narrow band defined by an interference filter.

In both cases, the units of sensitivity are normally given in terms of photo-current per unit irradiation power per unit area. For example, the Siliconix P-102 photo-FET has a gate-current sensitivity S_{IN} which is typically 1·2 μA/mW/cm^2 for radiation at 0·9 microns (9000 Å).

Sometimes, units of illumination rather than irradiation are quoted, which implies that a photometer having a spectral sensitivity similar to that of the 'average' human eye is used to measure the illumination at the photo-FET. Such a unit is the foot-candle (or lumen per square foot), and it should be noted that its use does not imply that the photo-FET receives only irradiation within the sensitivity band of the human eye. For example, the typical gate current sensitivity of the Dickson PFN 3069 is 8 nA/foot-candle when irradiated by an incandescent lamp at a colour temperature of 2800°K. This mixture of conventions means that when the irradiation by the lamp in question produces a gate current of 8 nA in the photo-FET, it also gives rise to an illumination of 1 foot-candle at the same distance.

An attempt to clarify the confused situation regarding the specification of optical sensitivity appears in Reference 2, and this will not be repeated here except insofar as recommendations are concerned. Primary among these is that for detectors having a spectral sensitivity bandwidth significantly greater than that of the human eye (including photo-FETs), the units of irradiation should be used rather than those of illumination. This means that sensitivity should be quoted in terms of response per milliwatt per unit area; the spectral distribution of the light source being defined separately. Here, the colour temperature of a filament lamp is entirely adequate to define such a distribution, but otherwise, the nature of the light source should be quoted. For example, monochromatic laser light could be used, as could narrow-band light obtained via a (specified) interference filter.

15.2.2 NEP AND DETECTIVITY

The light received by a photo-FET is often passed through a chopper so that the AC component of the output may be extracted. Under these circumstances, the minimum detectable signal becomes of importance, and this is related to the noise voltage appearing at the gate. It will be recalled that the photo-FET must be associated with a gate resistor R_G as a necessary condition for operation, and it is therefore valid to quote the total input noise voltage for a real condition which includes the noise generated by R_G. A family of curves for the Siliconix P-102 is reproduced in Fig. 15.6, from which

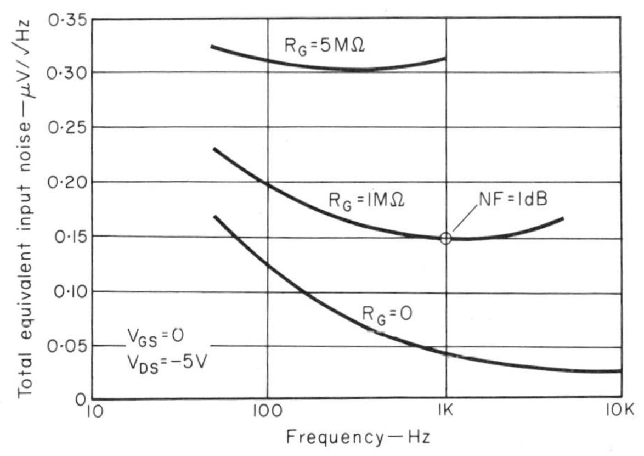

Fig. 15.6. Noise data for the P-102 photo-FET (Courtesy Siliconix Inc.)

it will be seen that the noise voltage is given in units of $\mu V/Hz^{\frac{1}{2}}$, which are relevant to unity bandwidth. This is because both thermal and shot noise calculations show that the mean square noise voltage is proportional to the bandwidth, so that for r.m.s. noise, and unity bandwidth, the units must be as indicated.

For a given operating condition, it is now possible to calculate the noise equivalent power (*NEP*), which is defined as the irradiation level which would produce a signal voltage equal to the noise voltage under the same external conditions. In other words, it is the irradiation level for a signal-to-noise ratio of unity.

For the photo-FET, it is reasonable to quote signal and noise voltages referred to the input, or gate, because both will be subject to whatever amplification is dictated by the mode of operation.

As an example, consider the P-102 irradiated at a 100 Hz chopping frequency, and using a 1 MΩ gate resistor. The effective area of the device, A, is quoted as $7{\cdot}9 \times 10^{-2}$ cm^2; the 'typical' sensitivity is

$1{\cdot}2\ \mu A/mW/cm^2$; and Fig. 15.6 gives the total input noise voltage as $0{\cdot}2\ \mu V/Hz^{\frac{1}{2}}$. The *NEP* will therefore be:

$$NEP = \frac{V_N \cdot A}{S_{IN} \cdot R_g} \tag{15.1}$$

which, in this case is,

$$NEP = \frac{0{\cdot}2 \times 7{\cdot}9 \times 10^{-2}}{1{\cdot}2 \times 10^6} = 1{\cdot}32 \times 10^{-8}\,\mathrm{mW/Hz^{\frac{1}{2}}}$$

$$\text{or } 1{\cdot}32 \times 10^{-11}\ \mathrm{W/Hz^{\frac{1}{2}}}$$

(Note that when the noise bandwidth is *not* unity, the term $\sqrt{(\Delta f)}$ must appear in the denominator of equation (15.1).)

A figure-of-merit which has come into general usage, particularly within the context of IR detectors, is the detectivity, D^*. This is defined[3] as the unity bandwidth signal-to-noise ratio for a cell of unit effective area irradiated with unit power. That is, using the foregoing symbols, and recalling that V_N is already defined for unit bandwidth,

$$D^* = \frac{S_{IN} \cdot R_g}{V_N}$$

But from equation (15.1), the *NEP* of a cell of unit area is $V_N/S_{IN} \cdot R_g$ so that

$$D^* = 1/NEP$$

Knowing that the signal-to-noise ratio for a photo-cell is inversely proportional to the square root of the effective area[4], then for any photo-cell,

$$D^* = \frac{\sqrt{A}}{NEP} \tag{15.2}$$

For the P-102, using the foregoing parameters,

$$D^* = \sqrt{(7{\cdot}9 \times 10^{-2})}/1{\cdot}32 \times 10^{-11} = 2{\cdot}13 \times 10^{10}\,\mathrm{cm\text{-}Hz^{\frac{1}{2}}/W}$$

When detectivity is quoted, the conditions under which it is defined must be stated, so that valid comparisons with other cells may be made. In the case of measurements made using the colour-temperature system, these conditions are often given in brackets as follows:

$$D^*(T°\mathrm{K}, f, \Delta f)$$

where $T°\mathrm{K}$ = colour (or black-body) temperature
f = frequency of light chopper
Δf = electrical bandwidth of noise voltage.

15.2.3 EQUIVALENT CIRCUIT AND FREQUENCY RESPONSE

To the normal complete equivalent circuit for the junction FET must be added the components of the incremental photo-current signal i_P, and this has been done in Fig. 15.7. The way in which i_P divides between the drain and source circuits is dependent upon the configuration of the depletion layer within the channel, and hence is a function of V_{GS}. This in turn is a function of the incident irradiation if the voltage drop in R_g due to I_P is significant compared with any applied bias voltage.

Normally $V_{gs}g_{fs} \gg i_p$, otherwise there would be no reason to use a photo-FET rather than a photo-diode. In other words, R_g is

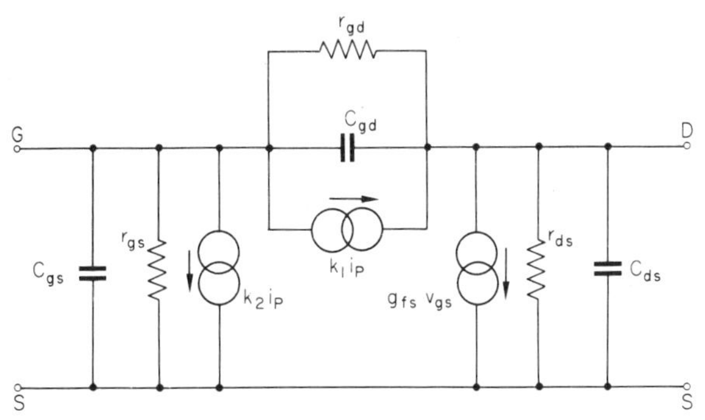

Fig. 15.7. Equivalent circuit for photo-FET

normally large enough to ensure that the change in channel current is much greater than the corresponding change in gate current. Consequently, both $k_1 i_P$ and $k_2 i_P$ may be ignored in most cases, and in particular for the source-follower case where $k_1 i_P$ flows only to the common line.

The bulk series resistance in the gate region, corresponding to R_T in the photo-diode equivalent circuit of Fig. 15.2 has been neglected because it will be very small compared with R_G. The high-frequency cut-off point is therefore:

$$f_h = \tfrac{1}{2}\pi C_{\text{in}} R_G$$

where C_{in} is a function of the mode of connection. For the CS case,

$$C_{\text{in}(CS)} \simeq C_{iss} - A_v C_{rss} \tag{15.3}$$

where A_v has a negative numerical value.

For the source-follower,

$$C_{\text{in}(CD)} \simeq C_{iss} + A_v(C_{oss} - C_{iss}) \tag{15.4}$$

where $A_v \to +1$

For the source-follower case, the light-chopping frequency can be comparatively high, as evidenced by the interelectrode capacitances for the Siliconix P236 range, which are $C_{iss} = 25$ pF(max), and $C_{oss} = 5$ pF(max). However, it must be recalled that for 'good geometry' light chopping, a close approach to a square wave can be made, which means that for low-distortion amplification of this square wave, approximately the tenth harmonic should have the same frequency as f_h.

In the above discussion, it has been tacitly assumed that the chopped light gives rise to a small, or incremental signal; that is, the photo-FET has been operated under properly biased conditions at a pre-determined quiescent point. If, however, the incident chopped light leads to a large-signal response, the rise and fall times, t_r and t_f, become the operative parameters.

Conventionally, t_r is defined as the time taken for the drain current to rise from 10% to 90% of its final value upon the application of a step input (of light). Similarly, t_f is the time taken for the drain current to fall from 90% to 10% for a step removal of light input. Because large signals are concerned here, the photo-FET should be biased to pinch-off under dark conditions. For the Siliconix P-102, t_r and t_f are given respectively as typically 0·1 and 1·5 μs for a 330 Ω load.

An injection laser, or a recombination-type light emitting diode are useful for rise and fall-time measurements, because both exhibit very fast responses. Otherwise, a fast Xenon flash-tube may prove adequate.

15.3. Photo-FET applications

15.3.1 SUMMARY

There are three basic situations in which a photo-sensitive transducer may be used:

(i) for the detection or measurement of a chopped-light input; that is, where an AC output is acceptable,
(ii) for the measurement of continuous-light intensity, where a DC output is necessary, and
(iii) for threshold detection, where a given light intensity will trigger regenerative action.

The fact that the photo-FET is applicable to all these conditions is an indication of its versatility: as far as semiconductor detectors are concerned, only the photo-diode can compete, the photo-resistor being both slow and highly non-linear. The photo-FET is in fact comparable to a photo-diode followed by an FET amplifier, but by virtue of the two functions being combined in one device, some improvement in overall noise figure is possible.

The sensitivity of the photo-FET must obviously be greater than that of the photo-diode, but it is also slower, for the limitations of the FET structure itself militate against fast response. At the time of writing, the best photo-diodes have PIN structures, the *p*-layer being very thin so that the incident photons release hole-electron pairs largely in the high-field intrinsic region. This makes for excellent detectivity, and in fact D^* may be almost two magnitudes better than that for contemporary photo-FETs. It is hoped that progress in photo-FET technology will improve this situation.

15.3.2 THE CHOPPED-LIGHT MODE

For the measurement of small changes in light intensity, or the detection of low-intensity cyclically varying light signals, it is possible to use the photo-FET in the simple CS or CD configurations. The CD configuration will not only be more linear, but will also accommodate higher frequency signals than will the CS mode. This latter capability is due to the lower Miller capacitance discussed in section 15.2.3. The CD connection is therefore often mandatory for 'good-geometry' chopping where fast rise and fall times of the incident square wave of light are produced.

Under these small-signal circumstances, the drift of the stage is not critical providing that the working point is reasonably well stabilised, because the output signal may be taken off via a capacitor. Consequently, the biasing procedure may be carried out as described in chapter 6, and often R_G may be connected from the gate to the common line, which makes $V_B = 0$. Under these circumstances, the actual value of R_G may be determined solely by noise, sensitivity and speed criteria.

Again assuming that only incremental signals are involved, the 'zero-drift' method of DC stabilisation may be used[5], this being described in section 9.2.1. By this means, DC coupling may be achieved, which can be useful if the incoming waveform has a very low chopping frequency. However, it must be established that the incoming light signals are too small to change the quiescent working point significantly. To ensure that this condition applies, it may be necessary to make the value of R_G quite low, and hence to tolerate the resultant low output signal.

15.3.3 THE DC MODE

If it is desired only to detect the presence of large DC light levels, such as is necessary in punched-card readers, for example, it is convenient to bias the photo-FET OFF, and allow the light to switch it hard ON. The basic circuit for achieving this is shown in Fig. 15.8, which is essentially a non-regenerative switching circuit. The bias voltage V_B should be preset at a value slightly higher than V_P.

The measurement of light intensity where the input-signal cannot be treated as incremental, is a more complex problem. For environ-

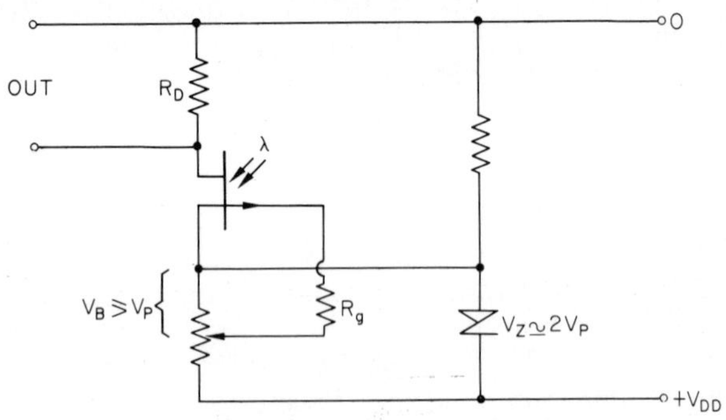

Fig. 15.8. Simple photo-FET switch

ments where temperature changes are small, and where adjustments prior to measurement may be carried out, a simple circuit such as that given by Shipley[6] may be used (see Fig. 15.9). Here, only simple bias stabilisation is employed, and the quiescent source voltage is backed off by a resistive potential divider.

To achieve better bias compensation, it would be necessary to use a matched pair of photo-FETs as shown in Fig. 15.10, where one unit

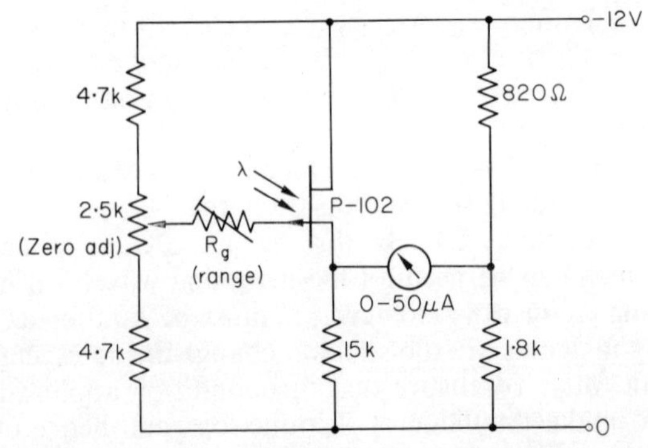

Fig. 15.9. Simple photometer circuit (After Shipley[6]) (Courtesy Siliconix Inc.)

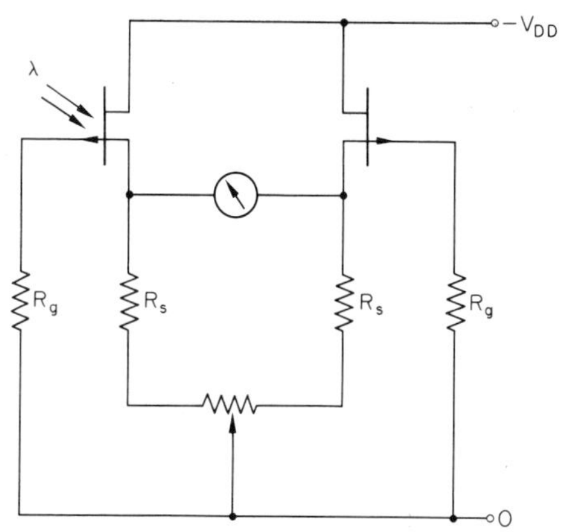

Fig. 15.10. Drift compensated photometer circuit

is occluded. As is usual with such circuits, both good initial matching and good tracking is essential, and at the time of writing, matched pairs of photo-FETs (or one photo-FET and one normal FET) are not easily obtainable.

An interesting circuit using a pair of (preferably matched) photo-FETs is given in Fig. 15.11. Here, the timing resistors in a normal

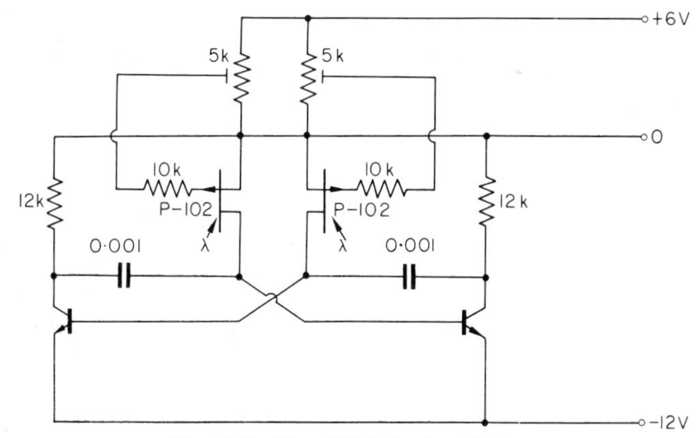

Fig. 15.11. Photo-FET-timed multivibrator

bipolar transistor multivibrator are replaced by the channels of the photo-FETs. If the irradiation at each photo-FET changes by an equal amount (that is, a common-mode signal exists), then the frequency of the multivibrator will alter; but if the irradiation at one photo-FET changes with respect to the other (a differential signal) then only the mark-space ratio will vary. Thus, by following the multivibrator with the proper measuring circuitry, both common-mode and differential signals may be detected.

15.3.4 THRESHOLD DETECTION

Where it is desired to obtain an output when a predetermined irradiation level is reached, a circuit of the form shown in Fig. 15.12 may be used. Here, the normal state is that both the photo-FET and the bipolar transistor are held OFF, because the gate is at a voltage

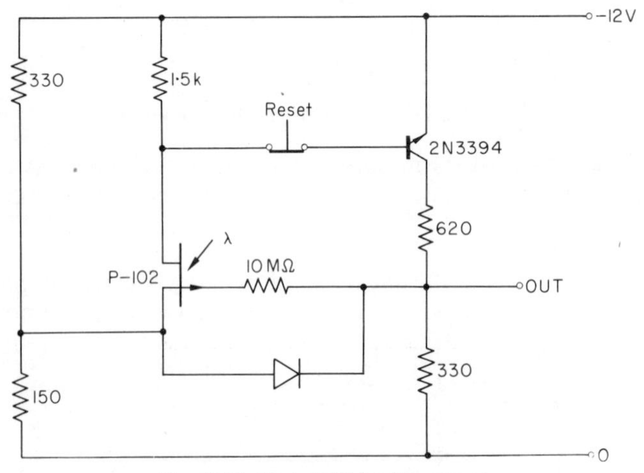

*Fig. 15.12. Photo-*FET *latching circuit*

greater than V_P when the bipolar transistor is not conducting. When the photo-FET is irradiated at a sufficiently high level, its drain current energises the base of the bipolar transistor, which then conducts, so bringing the gate voltage below V_P. Consequently, the photo-FET is turned harder ON and both transistors become saturated.

Notice that this is inherently a one-shot circuit which must be manually reset: this is because the positive feedback loop involves electrical signals alone, for the incident irradiation merely triggers the regenerative action, and can have no further effect upon the circuit when both transistors are ON. This is, in fact, entirely analogous to the operation of a photo-thyristor.

For applications requiring automatic resetting when the irradiation falls below a predetermined level, it is not feasible to include the photo-FET within the positive feedback loop, otherwise latching will occur as described above. Instead, the photo-FET should be maintained in an operative condition, and its output voltage applied to a level-sensitive device such as a voltage comparator monolithic amplifier (e.g. the Fairchild μA 710), or a Schmitt trigger. An example of the latter is given in Fig. 15.13 which illustrates a CD-connected photo-FET operating a simple bipolar transistor Schmitt trigger.

If a capacitor is substituted for the gate resistor of a photo-FET connected to a level-sensitive circuit, then this capacitor will be

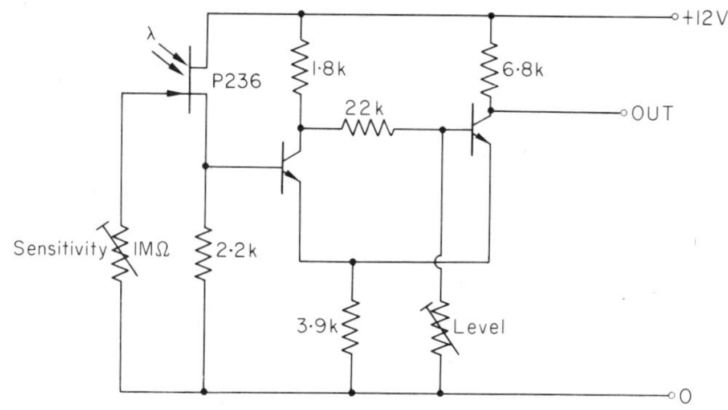

*Fig. 15.13. Photo-*FET *Schmitt trigger*

linearly charged by $(I_P + I_D)$. Eventually the voltage on the capacitor will become sufficient to trigger the regenerative circuit, and the time taken for this to occur will clearly be a measure of the magnitude of $(I_P + I_D)$. If V_{th} is the threshold voltage, and T_{ch} is the charging time, then,

$$(I_P + I_D) = \frac{CV_{th}}{T_{ch}} \tag{15.5}$$

In this application, it is advisable to make $I_P \gg I_D$ so that the effect of the highly temperature-sensitive leakage current is minimised, and the charging time becomes essentially a function of I_P alone.

15.4. Integrating-mode photodetection

In addition to their work on planar photo-diodes as such, Murphy and Kabell[1] have shown how a wide range of irradiation levels can be measured using these photo-diodes in the integrating mode. The principle of this mode is that a capacitor is allowed to charge (or discharge) via a reverse-biased photo-junction. The rate-of-change of voltage across the capacitor will be linear, because the reverse current of a photo-diode is the sum of the photo and leakage currents, both of which can be represented by current generators as shown in Fig. 15.2. However, if the capacitor in question is that which is formed by the reverse-biased junction itself (C_D in Fig. 15.2), then some departure from linearity will occur due to the value of C_D being an inverse function of the reverse voltage. Over a limited voltage change, however, this effect will not be serious.

The method employed by Murphy and Kabell is shown in Fig. 15.14. Here, $2C_D$ is the capacitance of two matched photo-diodes connected as shown. One of these diodes, D2, is occluded so that it

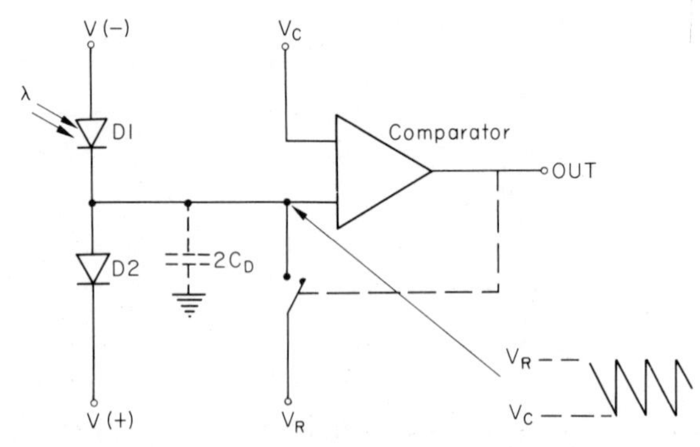

Fig. 15.14. Principle of irradiation-frequency converter (After Murphy and Kabell[1])

passes only leakage, or dark current, and this current is subtracted from the total current of D_1. Consequently, only the photo-current of D_1 is available to charge $2C_D$, so that the voltage across $2C_D$ goes negative along an essentially linear ramp as shown. When this voltage reaches a threshold value, a voltage comparator closes a switch which discharges C_D and then opens, so allowing the cycle to recommence.

Notice that at any instant on the ramp, the voltage is proportional to the total amount of light which has fallen on the photo-diode since the commencement of the charging cycle.

That is, the instantaneous voltage is proportional to the integral of the light received by the photo-diode up to that time in the cycle. This accounts for the term 'integrating mode'.

When the voltage across the capacitance reaches the same level as the comparator voltage V_C, however, the charging cycle is terminated, so that the complete charging time T is proportional to the *average* irradiation level $D_{r(av)}$ over that time. (This assumes, of course, that $2C_D$ can be considered to be constant.)

If the difference between the reference voltage V_R, and the comparator voltage V_C is ΔV, then,

$$\Delta V = \frac{I_{P(av)}T}{2C_D} = \frac{SD_{r(av)}T}{2C_D}$$

where S is the photo-diode sensitivity.

Hence, if the capacitance discharge at the end of each cycle is assumed to occur instantaneously, then the frequency of the output wave must be:

$$f = \frac{1}{T} = \frac{S_{Dr(av)}}{2C_D\Delta V} \qquad (15.6)$$

Murphy and Kabell go on to describe an irradiation-to-frequency converter using discrete components plus a voltage comparator, while Noble[7] discusses the somewhat modified monolithic version of the device shown in Fig. 15.15. This is the Plessey O.P.T.1, which

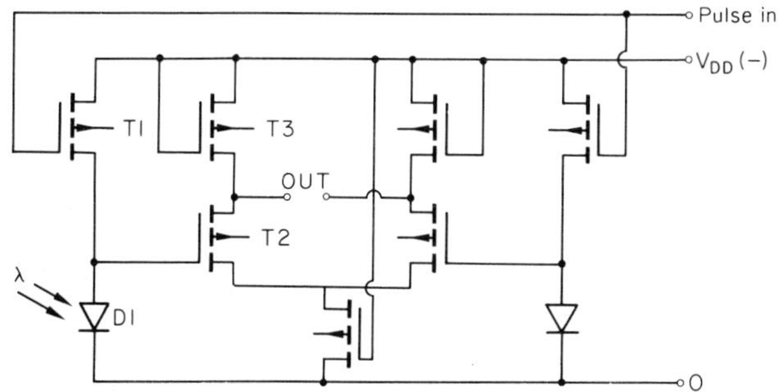

Fig. 15.15. Monolithic integrating light detector (Plessey O.P.T.I., described by Noble[7, 8])

is a monolithic microcircuit consisting of two planar photo-diodes (one occluded with an aluminium film) and seven *p*-channel enhancement IGFET structures.

When a negative pulse is applied to the gate of T_1, the internal capacitance of D_1 is charged to V_{DD} volts. This capacitance is discharged by the photo and leakage current in D_1, and the decaying voltage waveform is applied to the gate of T_2. This IGFET has an active load T_3, so that the amplified, inverted voltage waveform appears at its drain.

The complete circuit is duplicated on the chip, the alternate half involving the occluded diode so that the output on that side is a function of dark current only. The two output IGFETs are connected as a long-tailed pair with a constant-current IGFET in the source lead, so that the signals corresponding to the dark currents appear as common-mode signals and are much attenuated. The difference output signal is therefore essentially due only to the photo-current in D_1

The O.P.T.1 integrating photo-detector has an NEP which is slightly inferior to a photomultiplier at 25°C but is entirely comparable when cooled to −40°C (using a Peltier cooler for example). The output sensitivity is, however, better than that of most common photo-multipliers.

REFERENCES

1. Murphy, H. E. and Kabell, L. J. 'An integrating digital light meter', *I.E.E.E. Journ. of Solid-State Physics*, **SC-1**, No. 1 (September 1966).
2. Watson, J. *Semiconductor Circuit Design*, Chapter 9, Hilger & Watts. Adam Hilger (1970).
3. Smith, Jones and Chasmar, *Detection and Measurement of Infrared Radiation*, Oxford (1957).
4. Zworykin, V. K. *Photoelectricity and is Applications*, p. 256, Wiley (1949).
5. Evans, L. L. 'Biasing FETs for zero DC drift', *Electro-Technology* (August 1964).
6. Shipley, M. 'Photo-FET characteristics and applications', *Solid State Design* (April 1964).
7. Noble, P. J. W. 'Photodiodes in integrated circuits have integrating and logic applications', *Electronics and Communications* (February 1968).
8. Noble, P. J. W. 'Self-scanned silicon image detector arrays', *I.E.E.E. Trans. on Electron Devices*, **ED-15**, No. 4, 202–209 (April 1968).

INDEX